Shallow-Water Acoustics

Springer
London
Berlin
Heidelberg
New York
Barcelona
Hong Kong
Milan
Paris
Santa Clara
Singapore
Tokyo

Boris G. Katsnelson and Valery G. Petnikov

Shallow-Water Acoustics

Springer

Published in association with
Praxis Publishing
Chichester, UK

PRAXIS

Professor Boris G. Katsnelson
Universitetskaya sq 1,
Voronezh University,
394693 Voronezh, Russia

Professor Valery G. Petnikov
Vavilova str 38, Wave Research Centre
General Physics Institute,
Russian Academy of Science, 119991 Moscow, Russia

Translator: Dr Stephen S. Wilson, Technical Translation Services, 31 Harp Hill, Cheltenham, Gloucestershire, UK

SPRINGER–PRAXIS BOOKS IN GEOPHYSICAL SCIENCES
SUBJECT *ADVISORY EDITOR*: Dr Philippe Blondel, C.Geol., F.G.S., Ph.D., M.Sc., Senior Scientist, Department of Physics, University of Bath, Bath, UK

ISBN 1-85233-184-4 Springer-Verlag Berlin Heidelberg New York

British Library Cataloguing in Publication Data
Katsnelson, Boris G.
Shallow water acoustics. – (Springer–Praxis books in geophysical sciences)
1. Underwater acoustics
I. Title II. Petnikov, Valery G.
620.2′5
ISBN 1-85233-184-4

Library of Congress Cataloging-in-Publication Data
Katsnelson, B. G. (Boris Grigorevich)
[Akustika melkogo moria. English]
Shallow water acoustics / Boris G. Katsnelson and Valery G. Petnikov ; [translator, Stephen S. Wilson].
p. cm. – (Springer–Praxis books in geophysical sciences)
Includes index.
ISBN 1-85233-184-4 (acid-free paper)
1. Underwater acoustics. 2. Sea water – Acoustic properties.
I. Petnikov, V. G. (Valeriĭ Georgievich) II. Title. III. Series.
QC242.2.K3713 2001
620.2′5 – dc21 2001032058

Printed by MPG Books Ltd, Bodmin, Cornwall, UK

Cover design: Jim Wilkie

Printed on acid-free paper supplied by Precision Publishing Papers Ltd, UK

Contents

Preface to the English edition

This book is a translation into the English language of the monograph *Akustika Melkogo Morya* published in Russia by Nauka in 1997. The first edition was produced with financial support from the Russian Foundation for Basic Research (RFBR) (project 96-02-30051). In preparing the English edition we have extended the content of the book somewhat, adding to it two sections which are important for shallow-water acoustics. These relate to acoustic noise and the diffraction of acoustic waves on localised anomalies in shallow-water sound channel. It is also important to note that since the publication of the first Russian edition of the book interest in shallow-water acoustics has increased even more. This is evidenced by the succession of acoustic experiments carried out on the sea shelf in recent years, for example, the SWARM (Shallow Water Acoustics Random Media) experiment (Headrick *et al.* (2000)). To our great regret, we have not been able to give these experimental results the coverage they deserve, but we hope that this book will prove useful for their interpretation and in the design and execution of future studies on the sea shelf.

The authors are grateful to their colleagues Dr J.F. Lynch (Woods Hole Oceanographic Institution, USA) and Dr M. Porter (Naval Research Laboratory, USA) for making their material available to us.

In the preparation of the English edition of the book, certain results obtained in studies supported by the RFBR (grants 99-02-17671 and 00-05-64752) were also used.

Preface to the Russian edition

Sound waves are the only waves of a known physical nature capable of propagating under-water over a distance of many hundreds of kilometres. This unique capability of acoustic waves is the reason for the practical interest in the properties of the propagation of sound in the ocean exhibited by academics and engineers in various disciplines: geophysicists, specialists in underwater location and communication, ecologists and oceanologists.

This interest is mainly associated with the possibilities for acoustic monitoring of large-scale hydrodynamic disturbances in the ocean: eddying, meandering currents, tidal and internal waves, frontal zones. This monitoring, in principle, offers the possibility of determining the state of the ocean from acoustic sensing data and, ultimately, of predicting local and global climatic changes on our planet.

The propagation of sound waves underwater is an extremely complex process, which is difficult to describe, because of the great diversity of the phenomena and properties, inherent to different regions of the World Ocean. In the first instance, this relates to the shelf zone of the ocean, where the properties of the propagation of sound waves are associated with their interaction with the sea bed, and also with the specific nature of the hydrodynamic disturbances. In addition, the ocean shelf is very important for man's vital activity and, consequently, above all, of interest to researchers. The possibility of practical work on the acoustics of the ocean shelf has arisen as the result of the emergence over the last ten years of corresponding technologies such as acoustic wave transmitters and receivers, processing devices, and the corresponding computer technology.

A large number of experimental and theoretical results have now been obtained as a result of the efforts of various groups of researchers, making it possible to identify shallow-water acoustics as an independent branch of ocean acoustics. It suffices only to recall the results of unique large-scale experiments in sensing of the hydrodynamics of disturbances on the shelf of the Sea of Japan (Bondar *et al.* (1994), Caille *et al.* (1997)), on the shelf of the Barents Sea (Andreev *et al.* (1996), Belov *et al.* (1997), Lynch *et al.* (1996)), on the North American shelf of the Atlantic Ocean (Badiey and Jaya (1994), Beardsley *et al.* (1997)) and on the northwest shelf of the

British Isles (Field *et al.* (1997)). Overall, coarsely speaking, one can identify the following directions in shallow-water acoustics:

- propagation of sound in a stationary sea medium, including in a randomly inhomogeneous medium (deterministic and statistical approaches);
- effect of dynamic processes (tides, internal waves, eddies, frontal zones etc.) on the propagation of acoustic waves;
- possibility of diagnostics of the state of shallow water based on acoustic sensing data.

This monograph considers each of these directions both from the point of view of a theoretical description of the phenomena, and through an analysis of the results of experimental studies. The authors hope that this book will be useful in the development of the foundations for acoustic tomography of the shelf zone of the World Ocean, which has very recently come to the fore.

The authors are sincerely grateful to F.B. Bunkin, Yu.A. Kravtsov and L.G. Kulapin, with whom a large number of results reflected in this book were jointly obtained, and also to colleagues in the Wave Phenomena Division of the General Physics Institute of the Russian Academy of Sciences (GPI RAS), whose help in the organisation, execution and interpretation of full-scale experiments on the shelf of the Barents Sea has been invaluable.

A considerable proportion of the results presented in the monograph was obtained as part of research supported by the Russian Foundation for Basic Research (RFBR, grants 96-02-17194, 97-05-64878) and the International Science Foundation (ISF, grant MJ 5000).

List of figures and colour plates

Colour plates (between pages 128/129)

1

Shallow water as a medium for the propagation of sound waves

1.1 THE NOTION OF 'SHALLOW WATER'

The notion of 'shallow water' is recent in underwater acoustics. The reason for this lies in the properties of the propagation of sound in a medium of this type, mainly associated with the interaction with the seabed. Properly speaking, the first model of a hydroacoustic waveguide in such a medium was formulated (Pekeris (1948)) precisely for shallow water and is now more widespread for the carrying out of theoretical estimations. We are concerned with low-frequency (less than 500 Hz) sound waves, propagating to a distance of several tens of kilometres.

The values of the characteristic parameters of the waveguide, which is the 'shallow water', may vary within very broad ranges. This depends not only on the geometric dimensions of the waveguide, but also on the sound frequency. Naturally, it is difficult to determine the boundary of depths distinguishing 'deep' water from 'shallow' water with accuracy. A certain quantitative criterion for this distinction can be provisionally introduced based on simple physical considerations. It is well known (see, for example, Brekhovskikh (1980)) that the field of a point source in the ocean waveguide can be described using a mode or ray approach. For a waveguide with a constant sound speed and an absolutely rigid bottom, the maximum number of energy-carrying modes (normal waves) is determined by the simple relation (Brekhovskikh and Lysanov (1991)) $M \approx 2H/\lambda$, where H is the depth of the waveguide and λ is the length of the sound wave. On the other hand, an analogous estimate for the maximum number of energy-carrying rays can be written in the form (Kravtsov (1988)) $M' \approx 2r/H$, where r is the distance between the sound source and receiver. (Here, by energy-carrying rays, we mean rays passing along a waveguide path of less than $\sqrt{2}r$). Comparison of these two estimates shows that, when the following condition is satisfied

$$r^2 \gg \frac{2H^2}{\lambda} \tag{1.1}$$

the number of modes is greater than the number of rays. Accordingly, the energy in an individual normal wave is significantly greater than in each ray. This condition usually holds in shallow-water regions of the World Ocean. In acoustic monitoring of these regions, spatial (using long arrays) or temporal (by measuring the signal propagation time) selection of individual rays is impossible. Acoustic signals propagating on adjacent rays, will overlap. In this case, acoustic monitoring is based on the identification of normal waves. In the converse case, which usually occurs in the deep ocean, the situation changes correspondingly to the opposite. Thus, using equation (1.1) one can provisionally separate shallow water from deep ocean from the acoustic point of view. Another fundamental distinguishing trait of shallow water is the considerable effect of the bottom on the formation of the sound field.

1.2 OCEAN SHELF ZONE: CHARACTERISTIC HYDROLOGICAL CONDITIONS

The ocean shelf is the zone around a continent, stretching from the low-water line to depths at which there is a sharp increase in the slope of the bottom in the direction of great depths. Then, there is the continental slope, characterised by a comparatively sharp change in depth.

The shelf is a comparatively uniform plane of width around 70 km, with a slope of approximately 1.9% towards the sea. At a depth of approximately 130 m a sharp increase in the slope is usually observed. This area is called the shelf-break. The shelf-break is the boundary of the shelf and the upper continental slope, and the boundary of the open sea. Some shelves extend to depths of 180–550 m. Currently, the shelf boundary is taken through a depth of 200 m. Shelves are covered by a thin layer of characteristic deposits of sands, aleurites, and mud, with a mottled distribution. From the geological point of view, the shelf is the extension of the continent and comprises a single whole with the coastal plain. The shelf owes its origin to the activity of waves in shallow water, the deposit of alluvium in the coastal zone and fluctuations of sea level. Shelves have a global distribution and occupy around 5% of the total area of the planet (an area of approximately 27.5 million km^2) with an average depth of 50 m. Shelf widths vary from several tens (western coast of the USA) to several hundreds of kilometres (in the region of northern Australia). The table gives a number of examples of shelves taken from an encyclopedia (Parker, (1980)):

Location	*Depth (m)*	*Width (km)*
Gulf of Mexico	70	250
Barents Sea	200	1,200
North Sea	180	500
Hudson Bay	100	1,000

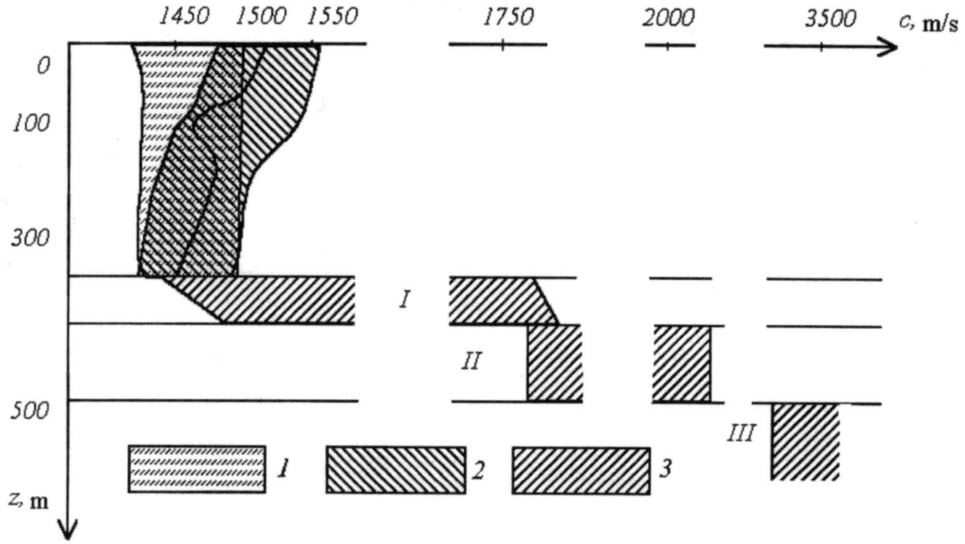

Figure 1.1. Model of a shallow-water sound channel. 1 – region of winter profiles of the sound speed; 2 – region of summer profiles of the sound speed; 3 – region of sound speed in bottom sediments. The non-monotonic curve corresponds to the profile of the sound speed with a possible internal waveguide.

Continental shelves are primary subjects for the organisation of geological exploration work for oil and gas (Buryakovskii *et al.* (1991)).

From the point of view of the theory and practice of acoustic sensing, the ocean shelf is a waveguide, limited by a pressure-release surface and an absorbing (bottom) boundary. A variety of factors affect the propagation of sound in such waveguides. The most important of these are the shape of the sound-speed profile and the geoacoustic properties of the bottom. Other parameters which alter the sound field include roughnesses of the bottom, surface disturbance, random inhomogeneities in the water layer, sea currents, etc. In comparison with the first two factors, the role of the latter is on the whole less marked. It is important that, in a real situation, the two main factors are subject to very large variations as a function of geographical location, time of year, meteorological conditions, etc. Nevertheless, for the majority of real tracks, a generalised geoacoustic model, representing a water layer with a depth of up to several hundred metres, lying on a multilayer bottom (Figure 1.1) is applicable. The layers which make up the bottom differ in a qualitative respect: the upper layer is the so-called layer of unconsolidated sediments (layer I), 1–10 m thick; below that is the layer of semi-consolidated sediments (layer II) with a thickness of 10–100 m; finally, the lower half-space is the foundation or the basement (layer III). This four-layer model can be simplified if the lower-lying layers II and III have a weak effect on the propagation of sound, or made more complex if details of the internal structure of the bottom are determined which are important for the propagation of sound.

Figure 1.1 shows the characteristic values of shallow-water parameters. Let us consider the properties of each layer in more detail.

1.2.1 Water layer

The sound speed in seawater is measured using special devices or computed from special empirical formulae, using measured values of the temperature, salinity and hydrostatic pressure (see, for example, Chen and Millero (1977)). From the literature, the vertical profile of the sound speed $c(z)$ is known to lie approximately within the limits shown in Figure 1.1. The regions and the times of year corresponding to the profile boundaries can be tentatively indicated. A continuous line denotes profiles of $c(z)$ predominant in the summer, while a hatched line represents winter, where the right-hand boundary of the profile corresponds to observations made in warm (southern) seas and the left-hand boundary corresponds to cold (northern) seas. One can see that the summer profiles are characterised by small negative gradients of the sound speed: $a \equiv dc/dz \approx -10^{-2}c^{-1}$; larger values of $|a|$ up to 0.5–1c^{-1} may be realised in a narrow surface layer. The winter profiles have a zero or small positive gradient of the sound speed: $a \approx +10^{-2}c^{-1}$. These changes in the gradient of the sound speed affect the nature of the propagation in the channel. The sound channel in summer is mainly near the bottom, so that the propagation of sound takes place (in the language of rays) by sequential reflections of refracting rays from the bottom. In this case one must *a fortiori* expect high propagation losses. In winter, the propagation of sound takes place either in a channel with a constant sound speed, when it is described by bottom–surface rays, or in a near-surface channel. We also note that sometimes there may be (Beebe (1980)) an internal sound channel (see Figure 1.1) in which sound is retained only by virtue of refraction. In addition to the 'deterministic' (average) part of the refractive index of sound in water, the propagation of low-frequency sound is most affected by random inhomogeneities with characteristic sizes ≥ 1 m. Random inhomogeneities with vertical scales of 1–10 m and horizontal scales of 100–1,000 m are mainly due to the fine thermohaline structure (Babii (1983), Fedorov (1976), Fedorov (1978)) and to internal waves (Konyaev (1992), Fedorov (1978)). The latter constantly perturb the sound-speed field, mixing and deforming the isodensity surfaces, as a result of which a propagating acoustic signal acquires fluctuations of phase and amplitude (Zhou (1991), Zhou (1996)), of a stochastic nature.

Without analysing in detail in this work questions relating to the formation and propagation of the field of internal waves on the shelf, we note that in shallow water, this has a number of very specific characteristics:

- a considerable inhomogeneity and non-stationarity, due to the characteristic trains of intense soliton-like internal waves against a relatively weak background;
- clearly expressed anisotropy, determined by the bottom relief, when waves propagate mainly in the direction towards the coast, perpendicularly to the outside edge of the shelf;

- synchronicity of the vertical fluctuations of all fluid layers, attesting to a predominance of the first gravitational mode.

The model due to Garrett and Munk (1975) used to describe the field of internal waves in a deep ocean is unsuitable in this case. Evidently, because of the large diversity of the characteristics of shelf zones it is completely impossible to create an analogous universal model for shallow water. A number of models based on experimental data for specific shallow-water regions have been considered in the works of Daugherty and Lynch (1990), and Katsnelson and Pereselkov (1996). According to these, the radius of correlation of the sound speed in the longitudinal direction amounts to 10^2–10^3 m, and in the transverse direction to < 10 m.

As far as the fine thermohaline structure is concerned, according to the results presented in Babii (1983), Fedorov (1976) and Fedorov (1978), the ocean is a finely stratified medium, in which there exist layers with thickness from tens of centimetres to tens of metres, with comparatively homogeneous properties, separated from each other by thin boundary layers with sharp changes in the thermodynamic characteristics (the vertical gradients of the physical properties in these layers may be 10–100 times greater than their average values). If we move from the random inhomogeneities of different characteristics of the water layer (density, salinity, temperature), using various known quasi-empirical formulae (Flatte (1979)) to the sound-speed field, as a function of coordinates and time, then from the point of view of statistical analysis the sound-speed field in shallow water is the sum of the average and fluctuating components, $c = \bar{c} + \delta c$, where δc describes the fine structure and is obtained by subtracting some average part from the observed structure of the sound-speed field. We note that this subdivision requires the prior selection of characteristic spatial scales (longitudinal and transverse) over which to perform the averaging, in other words, the selection of maximal spatial scales for the inhomogeneities, which we have decided to assign to the fine structure, which is in some sense inexact. The properties of the sound-speed field as a random function of spatial and temporal variables have been analysed (Babii (1983)) based on direct measurements. Using special sound-speed meters with a large resolution, the existence of sharp fine-scale changes in the vertical dependence of the sound speed was established. It is noted that the sound-speed field can be three-dimensionally isotropic only in the region of very small scales (of the order of the internal scale of the turbulence) where external forces and forces of inertia dominate over forces of buoyancy. As the spatial scale increases this relationship is violated and, starting from some value, the so-called critical scale $I_{\mathrm{cr}} \approx 1$ cm, the inhomogeneities become anisotropic. The characteristic vertical radius of correlation ζ_c at the 0.5 level is 4–6 m, which is also confirmed in Fedorov (1976), Fedorov (1978) and Nozdrin (1983). An almost complete absence of correlation is observed for $\Delta z \geq 10$ m. The value of ζ_c is very stable for the upper layer of the ocean. The values of the fluctuations of the sound speed and its variance σ_c^2 may change markedly, even within the upper layer. Overall σ_c^2 grows with depth, varying in the range 10^{-3}–10^{-1} $(\mathrm{m/s})^2$. At the same time for individual instances the value of the fluctuations δc may be 2 m/s. Data on the horizontal scales of inhomogeneities, given in Fedorov (1976), Fedorov

(1978) and Nozdrin (1983), show that they are significantly greater than the vertical scales. The ratio of these scales is $\zeta_c/\rho_c \approx 10^{-3}$–$10^{-2}$.

The aforementioned properties of the randomly inhomogeneous sound-speed field in the water layer make the creation of a universal model, describing this field, very problematic. In this situation in the modelling of fluctuations of acoustic waves propagating in shallow water, specific (taken from experiment) random realisations of δc, measured using special sound-speed meters or calculated from random realisations of the temperature field in shallow water, are often used.

Roughnesses of the water layer boundaries (roughnesses of the bottom and a disturbed surface) may also have a marked effect on the propagation of sound in shallow water (Ashley (1987)). Like random inhomogeneities in the water layer, they lead to scattering of sound (mode transformations), which may be taken into account in different methods. The characteristic scales of the mesorelief (variance of the discrepancies) amount to 1–10 m in the vertical, while in the horizontal the characteristic scales of roughnesses are also 100–1,000 m.

Losses on the propagation of low-frequency (up to a few kHz) sound are due to various mechanisms (absorption, scattering, geometric divergence). Experimenters have isolated the mechanisms responsible for the losses, to the extent of determining the physical mechanisms. The absorption of sound in clear water, propagating at low frequencies, is physically mainly due to the conversion of the sound energy to heat and is the result of the chemical composition of seawater, which is a complex electrolyte. A change in sound pressure leads to a periodic change in its ionic composition, which affects the volume viscosity. The absorption relaxation mechanism in this case is well described by a formula (Thorp (1973)), which yields the absorption coefficient β

$$\beta = \frac{0.11f^2}{1+f^2} + \frac{44f^2}{4100+f^2} \tag{1.2}$$

where the frequency is measured in kHz, and the absorption coefficient is given in dB/km. In particular, in the region of interest to us, namely approximately from 100 to 1,000 Hz, the coefficient increases monotonically from 10^{-3} to 0.06 dB/km.

1.2.2 Layer of unconsolidated sediments (I)

This layer consists mainly of bottom deposits of the mud type. Its parameters, like the parameters of the other layers, depend on the geographical region. The shear modulus is practically equal to zero, which explains the name of the layer (unconsolidated, fluid) and the absence of shear waves in it. One characteristic property of the layer is the existence of abrupt random inhomogeneities. These include layered (intermittent and tapered) structures of length up to tens of kilometres, and vertical channels, associated with the venting of gases and of diapirs (dome-shaped folds, in which rocks with a high plasticity are extruded from below). The properties of this layer have been studied in various works (see Hampton (1973), Kuperman and Jensen (*ed.*) (1980) and references therein) using profile shooting and samples. The density of earth in the layer usually amounts to on average $\rho \approx 1.5$–$2\,\text{g/cm}^3$. The sound speed (longitudinal waves) in the layer fluctuates from 1,470 to 1,880 m/s, and, according to certain data (Hampton (1973), Bunkin (1984)), may have a very marked positive gradient: $a = dc/dz \approx +1 - 2c^{-1}$. The layer is absorptive, where at frequencies up to 1 kHz, the coefficient of absorption of longitudinal waves may be coarsely assumed to be proportional to the frequency: $\beta \approx \alpha f$, where the values α lie in the range 0.03–0.05 dB/(km.Hz); however, at low frequencies, the frequency dependence is sharper.

1.2.3 Layer of semi-consolidated sediments (II)

This layer consists of denser sedimentary rocks, boulder clay, sandstones, etc. (Hampton (1973), Kuperman and Jensen (*ed.*) (1980)). In this layer the speed of longitudinal waves is $(2-3) \times 10^3$ m/s, where a small positive gradient with respect to depth is possible. The shear modulus is non-zero, so that in layer II transverse waves, whose velocity $c_s \approx 0.2c$ may reach values of several hundred metres per second, may propagate. This layer is also absorptive. The absorption coefficients for longitudinal and transverse waves differ and are distinguished by a large spread. Judging from the literature, the values of α and α_s lie in the range from 0.01 to 0.1 dB/(km.Hz) (Hampton (1973)).

1.2.4 Layer of consolidated sediments (III) basement

This layer, which in the given model extends to $z = \infty$, consists of basic rocks (basalt, granite, etc.). It is characterised by a high speed of both longitudinal ($c \approx (4-6) \times 10^3$ m/s) and transverse ($c_s \simeq (1-3 \times 10^3$ m/s) waves where the attenuation coefficients are estimated as $\alpha \approx 0.1$ dB/(km.Hz), $\alpha_s \approx 0.01-0.1$ dB/(km.Hz).

The thicknesses of the layers and the depths at which they lie vary over very wide ranges. In a number of regions, layers of sediments I and II may not be present. On the other hand, sometimes, geological data point to a more complex bottom structure. In this case, for the calculations, one might use models with more than three layers (Kuperman (1980), Lebedev (1996)). The degree of complexity of the model used for the calculations is determined not only by the available data on the geological structure of the bottom, but also by a number of other factors. First, one must bear in mind the extent to which deep layers affect the formation of the sound field in the channel in the given experimental conditions. For example, in the Barents Sea, there exist regions where the thickness of the first highly absorptive layer is greater than 300 m. In such a situation, for a frequency of the acoustic signals greater than 100 Hz, the effect of the subsequent layers can be neglected. Second, the choice of model depends on the degree of accuracy and the nature of the experiment to determine the propagation loss. For example, if one is interested in an average law for the drop in intensity of sound from a low-frequency source at great (> 100 km) distances, then it suffices to consider the bottom in the form of an homogeneous fluid half-space, since no specific layered structure of the bottom can exist over the whole track. In other words, the absorptance of the bottom is modelled by just one parameter, averaged, as it were, over the track. When one considers the properties of the frequency dependence of sound propagating in the channel, then the layered nature of the bottom should be taken into account (Kuperman and Jensen (*ed.*) (1980)).

In addition to the possible complication of the model with the introduction of extra layers, a more detailed analysis of the acousto-mechanical properties of the sediments themselves is also possible. Currently, there are a number of theories of the propagation of sound in sediments (Hampton (1973)), to which the theory of Biot (1956) is fundamental. In this theory, sediments are viewed as a two-component medium, consisting of a solid skeleton and a fluid component. These theories use a large number of parameters (porosity, average grain size, mean-square deviation from the average size, etc.), which determine the acoustic properties of a porous medium. This model allows one to consider the speed and attenuation coefficients of different types of waves. One of the most important characteristics of sediments, which can be calculated is the frequency dependence of the attenuation coefficient of a longitudinal or transverse wave. For longitudinal waves, in particular, this dependence is given by the model equation

$$\beta_f = \beta_q f^q \tag{1.3}$$

where the exponent depends on the average grain size M_Φ (usually measured in units of Φ), the mean-square deviation of the size σ_Φ, the degree of porosity (%), the temperature T, the pressure, etc. The coefficient of proportionality β_q also depends on the properties of the sediments. In addition, the coefficient of absorption, like the sound speed c, may depend on depth. The nature of this dependence was considered in the monograph by Kuperman and Jensen (*ed.*) (1980).

1.3 DESCRIPTION OF THE SOUND FIELD

The main quantity which is measured and studied in this work is the real field of the sound pressure $p(\vec{R}, t)$, formed in space by a source (located at the point $\vec{R}_0$), about which we shall make the following assumptions:

1 the dimensions of the source are much less than the radiation wavelength, and the source is a point source;
2 the source is not directional (directional diagram is constant in all directions);
3 the spectrum is narrow.

In Isakovich (1973), the source is represented in the form of a pulsating sphere, characterised by a volume velocity $V(t)$. In an homogeneous medium with constant sound speed c and density ρ, without dispersion, it generates a spherical spreading wave

$$p(\vec{R},t)=\frac{f\left(t-\dfrac{|\vec{R}-\vec{R}_0|}{c}\right)}{4\pi|\vec{R}-\vec{R}_0|} \tag{1.4}$$

where $f(t)=\rho\partial V/\partial t$. The function $f(t)$ will be called the real radiated signal, and the spectrum of the radiated signal will be denoted by $S(\omega)$:

$$S(\omega)=\frac{1}{2\pi}\int_{-\infty}^{\infty}f(t)e^{i\omega t}dt \qquad f(t)=\int_{-\infty}^{\infty}S(\omega)e^{-i\omega t}d\omega \tag{1.5}$$

For a quasi-monochromatic signal $f(t)=a(t)\cos(\omega_0 t-\varphi_0)$ with envelope $a(t)$, and carrier frequency and initial phase, ω_0 and φ_0, respectively, the spectral function $S(\omega)$ is a spike function, with maximum at the point ω_0 and width $\Delta\omega\ll\omega_0$.

In what follows, we shall also consider distributed (not point-like) non-monochromatic sources which, however, will be formed based on point-like monochromatic ones.

The main energy-related characteristic of a source will be assumed to be the radiated power, which is the flux density of the power (intensity) generated by the field $p(\vec{R},t)$, integrated over the whole (spherical) surface, surrounding the source and, in addition, in our case, a quasi-monochromatic emitted signal, averaged over a time greater than the period of the sound fluctuations:

$$W_0=\frac{\langle p^2\rangle}{\rho c}4\pi|\vec{R}-\vec{R}_0|^2=\frac{\langle f^2(t)\rangle}{4\pi\rho c} \tag{1.6}$$

where the angular brackets denote averaging over time[1]. For quasi-monochomatic signals this approximately corresponds to averaging over the whole pulse length $T=2\pi/\Delta\omega$, and then

$$\langle f^2(t)\rangle\approx\frac{1}{T}\int_{-\infty}^{\infty}f^2(t)dt=\Delta\omega\int_{-\infty}^{\infty}|S(\omega)|^2d\omega=2\Delta\omega\int_0^{\infty}|S(\omega)|^2d\omega \tag{1.7}$$

Using the spectrum (1.5) we can define the complex radiated signal

$$F(t)=2\int_0^{\infty}S(\omega)e^{-i\omega t}d\omega \tag{1.8}$$

which is an analytic function of t. The real part of the complex signal is equal to the function $f(t)=\operatorname{Re}F(t)$; it is related to the imaginary part by a Hilbert transform.

In an inhomogeneous medium with $c=c(\vec{R})$, the field of this source will be determined by the solution of the wave equation[2]

[1] The power of sound transmitters normally used for low-frequency acoustic monitoring in shallow water is usually 10^2–10^3 W.

[2] The medium will be assumed to be non-dispersive; that is, c does not depend on frequency.

$$\left[\Delta - \frac{1}{c^2}\frac{\partial^2}{\partial t^2}\right] p(\vec{R}, t) = -f(t)\delta(\vec{R} - \vec{R}_0) \tag{1.9}$$

with corresponding boundary and initial conditions.

Analogously to (1.5) and (1.8), the spectral function $S(\vec{R}, \omega)$ and the complex field $P(\vec{R}, t)$ are introduced, which depend on the spatial coordinates

$$S(\vec{R}, \omega) = \tfrac{1}{2}\pi \int_{-\infty}^{\infty} p(\vec{R}, t)\mathrm{e}^{i\omega t}\,dt \tag{1.10}$$

$$P(\vec{R}, t) = 2\int_{0}^{\infty} S(\vec{R}, \omega)\mathrm{e}^{-i\omega t}\,d\omega \tag{1.11}$$

$$p(\vec{R}, t) = \mathrm{Re}\, P(\vec{R}, t) \tag{1.12}$$

The equation for the complex field $P(\vec{R}, t)$ has the form (1.9) with the complex emitted signal $F(t)$ on the right-hand side. For an individual spectral component (in other words, for a value of $S(\vec{R}, \omega)$), as a function of the spatial coordinates, by virtue of the linearity of (1.9), one obtains the equation

$$\left[\Delta + \frac{\omega^2}{c^2}\right] S(\vec{R}, \omega) = -S(\omega)\delta(\vec{R} - \vec{R}_0) \tag{1.13}$$

If we define the Green function (transfer function) of the medium $\Psi(\vec{R}, \vec{R}_0; \omega)$ as the solution of the equation

$$\left[\Delta + \frac{\omega^2}{c^2}\right] \Psi(\vec{R}, \vec{R}_0; \omega) = -\delta(\vec{R} - \vec{R}_0) \tag{1.14}$$

the spectral function and the complex field are derived from the equations:

$$S(\vec{R}, \omega) = \Psi(\vec{R}, \vec{R}_0; \omega)S(\omega) \tag{1.15}$$

$$P(\vec{R}, t) = 2\int_{0}^{\infty} \Psi(\vec{R}, \vec{R}_0; \omega)S(\omega)\mathrm{e}^{-i\omega t}\,d\omega \tag{1.16}$$

The field is, correspondingly, obtained to be equal to

$$p(\vec{R}, t) = \int_{-\infty}^{\infty} \Psi(\vec{R}, \vec{R}_0; \omega)S(\omega)\mathrm{e}^{-i\omega t}\,d\omega \tag{1.17}$$

As we know, in the case of quasi-monochromatic fields, in order to compute the energy-related quantities (and, in general, those quadratic in the field), one can use complex signals, without transferring to their real parts. In particular, the intensity of the sound field can be determined in the form:

$$I = \frac{1}{\rho c}\langle p^2 \rangle = \frac{1}{2\rho c}|P|^2 \tag{1.18}$$

Based on the main concepts, formulae and relations presented in this chapter, in what follows, we shall formulate the foundations of the theory of the propagation of sound in shallow water (Chapter 2) and subsequently consider experimental facts.

2

Foundations of the theory of the propagation of sound

2.1 FIELD OF A POINT SOURCE IN A LAYERED WAVEGUIDE WITH ABSORBING BOUNDARIES

The coordinate system in the problem corresponds to Figure 2.1, with the z-axis pointing downwards. We shall use the notation $\vec{R}$ for the radius vector of the viewing point with coordinates (x, y, z). When the problem has cylindrical symmetry, which is usual, we shall use the cylindrical coordinate system (see Figure 2.1):

$$r, \theta, z : \quad r = \sqrt{x^2 + y^2}, \quad \tan\theta = \frac{y}{x}$$

The main quantity with which we shall be dealing in what follows is the sound pressure, generated by some source, as a function of coordinates and time. The theory of this section presupposes a harmonic dependence on time for the source (we shall use the expression 'pure-tone source') and, consequently, for the resulting field. This corresponds to a factor $\exp(-i\omega t)$ on time-dependent quantities (ω is the angular frequency, we shall also use the frequency in hertz $f = \omega/2\pi$). Work with signals which are modulated or generated in a special way by unpitched sources (for example, explosions) presupposes explicit use of Fourier analysis based on the theory of this section (we are working with noise intensities corresponding to linear acoustics and satisfying the linear superposition principle).

We shall consider a model of shallow water in the following form. A water layer of depth H with density ρ[1] and dependence of the sound speed $c(z)$ lies on a fluid, homogeneous half-space with density ρ_1 and sound speed c_1.

[1] A dependence of density on the coordinates can be easily taken into account using a change of functions (see Section 3.1.). The possibility of a more complicated model of the bottom is also discussed.

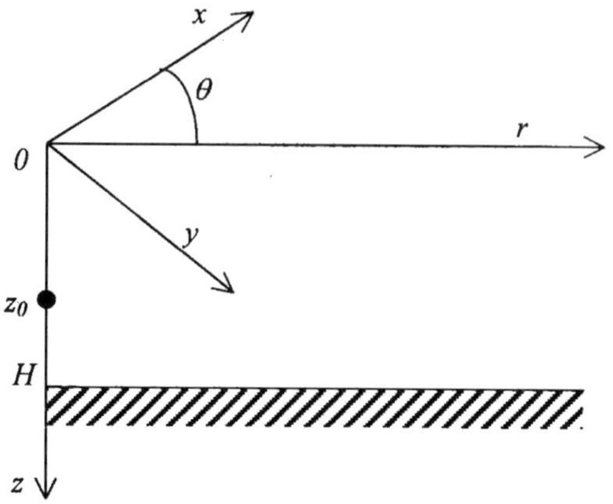

Figure 2.1. Model of a shallow-water waveguide.

Let us make some preliminary remarks concerning the description of the absorption of sound in the bottom. We shall characterise the absorptive properties of the bottom by a number $\alpha > 0$, determined as follows: assuming in what follows a sound speed in the bottom c_1 and a real-valued refractive index of the bottom, we shall assume a complex-valued wave vector in the bottom k_1:

$$k_1 = \frac{2\pi f}{c_1}\left(1 + i\frac{\alpha}{2}\right) = kn_1\left(1 + i\frac{\alpha}{2}\right) \tag{2.1}$$

Here, $n_1 = c(H)/c_1 < 1$ is the refractive index of sound in the bottom, $k = 2\pi f/c(H)$. The absorption coefficient of sound in the bottom, used in the literature, measured, for the most part, in decibels per kilometre, is related to the parameter α by the equation $\hat{\beta}[dB/km] = 27.3\alpha f/c_1$ (frequency in hertz, speed in km/s). In the case when α can be assumed to be frequency independent, this coefficient gives a linear frequency dependence of the absorption of sound in the bottom, and the coefficient of absorption of sound in the bottom, over frequency, $\hat{\beta}_f[dB/kmHz] = 27.3\alpha/c_1$ is also used; in addition, the coefficient of absorption of sound in the bottom, over wavelength, is also used, which quantity is given by the expression $\hat{\beta}_\lambda[dB/\lambda] = 27.3\alpha$.

The problem of determining the sound speed[2] of a point source of unit amplitude, located at the point with coordinates $x = x_0 = 0$, $y = y_0 = 0$, $z = z_0$, is formulated as the solution of the equation

$$[\nabla^2 + k^2(z)]\Psi(\vec{R}) = -\delta(\vec{R} - \vec{R}_0) \tag{2.2}$$

satisfying the radiation conditions at infinity. Here $\vec{R}_0(0, 0, z_0)$ is the radius vector of the source, $k(z) = 2\pi f/c(z)$. Another formulation of the problem is possible when

[2] This field is the Green function of the waveguide, determined by the general equation (1.14); in what follows, when not required in the problem, we shall omit the arguments $\vec{R}_0$, ω.

the right-hand side of equation (2.2) is equal to zero, and for the function $\Psi(\vec{R})$ as $\vec{R} \to \vec{R}_0$ we have the following limit:

$$\Psi(\vec{R}) \to \Psi_0(\vec{R}) = \frac{\exp(ik(z_0)|\vec{R} - \vec{R}_0|)}{4\pi|\vec{R} - \vec{R}_0|} \tag{2.3}$$

Both problems correspond to construction of the Green function for the wave equation, corresponding to a divergent wave. The problem additionally has the boundary conditions determined by the pressure-release surface

$$\Psi|_{z=0} = 0 \tag{2.4}$$

and the conditions of continuity of the pressure and of the vertical component of the vibration rate on transition from the water layer to the bottom

$$\Psi|_{z\to H-} = \Psi|_{z\to H+} \qquad \frac{1}{\rho}\frac{\partial\Psi}{\partial z}\bigg|_{z\to H-} = \frac{1}{\rho}\frac{\partial\Psi}{\partial z}\bigg|_{z\to H+} \tag{2.5.1}$$

where the notation $z \to H-$ ($z \to H+$) denotes the limit as z tends to the boundary from the water and from the bottom layer, respectively. Sometimes, this condition is written as the condition for continuity of the variable $(1/\rho\Psi)(\partial\Psi/\partial z)$ on transition through the boundary

$$\frac{1}{\rho\Psi}\frac{\partial\Psi}{\partial z}\bigg|_{z\to H-} = \frac{1}{\rho\Psi}\frac{\partial\Psi}{\partial z}\bigg|_{z\to H+} \tag{2.5.2}$$

Because of the cylindrical symmetry, equation (2.2) can be rewritten in cylindrical coordinates, without involving the angle θ. In this case, for $\Psi(r,z)$ we obtain

$$\frac{1}{r}\frac{\partial}{\partial r}\left(r\frac{\partial\Psi}{\partial r}\right) + \frac{\partial^2\Psi}{\partial z^2} + k^2(z)\Psi = -\delta(z - z_0)\frac{\delta(r)}{2\pi r} \tag{2.6}$$

The right-hand side of equation (2.6) can be easily obtained from the δ-function on the right-hand side of equation (2.2):

$$\begin{aligned}\delta(\vec{R} - \vec{R}_0) &= \delta(z - z_0)\frac{1}{(2\pi)^2}\int_{-\infty}^{\infty} dk_x \int_{-\infty}^{\infty} dk_y \exp(ik_x x + ik_y y) \\ &= \delta(z - z_0)\frac{1}{(2\pi)^2}\int_0^{\infty} \xi d\xi \int_0^{2\pi} d\vartheta' \exp[i\xi r\cos(\vartheta' - \vartheta)] \\ &= \delta(z - z_0)\frac{1}{2\pi}\int_0^{\infty} J_0(\xi r)\xi d\xi = \delta(z - z_0)\frac{\delta(r)}{2\pi r}\end{aligned} \tag{2.7}$$

where $\xi = \sqrt{k_x^2 + k_y^2}$ and $J_0(\xi r)$ is the Bessel function. The angle is not involved, since $\vec{r}_0 = 0$. Let us construct a solution of equation (2.6) using a Hankel transform, according to which the image $\overline{\Psi}(\xi, z)$ is related to the pre-image $\Psi(r,z)$ by the equation:

$$\overline{\Psi}(\xi, z) = \frac{1}{2\pi}\int_0^\infty J_0(\xi r)\Psi(r, z)rdr \tag{2.8.1}$$

$$\Psi(r, z) = 2\pi\int_0^\infty J_0(\xi r)\overline{\Psi}(\xi, z)\xi d\xi \tag{2.8.2}$$

Multiplying both sides of equation (2.6) by $(2\pi)^{-1}J_0(\xi r)r$ and integrating over r, we obtain

$$2\pi\int_0^\infty \frac{\partial}{\partial r}\left(r\frac{\partial\Psi(r, z)}{\partial r}\right)J_0(\xi r)dr + \frac{\partial^2\overline{\Psi}(\xi, z)}{\partial z^2} + k^2(z)\overline{\Psi}(\xi, z) = -\frac{1}{(2\pi)^2}\delta(z - z_0) \tag{2.9}$$

Computing the integral by parts, we obtain[3]

$$\frac{d^2\overline{\Psi}(\xi, z)}{dz^2} + [k^2(z) - \xi^2]\overline{\Psi}(\xi, z) = -\frac{1}{4\pi^2}\delta(z - z_0) \tag{2.10}$$

with boundary condition on the surface:

$$\overline{\Psi}(\xi, z)|_{z=0} = 0 \tag{2.11}$$

We shall make the boundary condition on the bottom (2.5.1) specific based on ideas about the bottom and about the fluid, homogeneous, absorptive half-space. The solution of equation (2.10) for $z > H$

$$\frac{d^2\overline{\Psi}(\xi, z)}{dz^2} + [k_1^2 - \xi^2]\overline{\Psi}(\xi, z) = 0 \tag{2.12}$$

satisfying the conditions at infinity, modulo a constant factor has the form

$$\overline{\Psi}(\xi, z) = \exp\left(i\sqrt{k_1^2 - \xi^2 z}\right) \tag{2.13}$$

Using equation (2.5.2), in order to obtain the boundary condition for the function $\overline{\Psi}(\xi, z)$, we rewrite it in the form

$$\left[\overline{\Psi} + g(\xi)\frac{d\overline{\Psi}}{dz}\right]_{z=H} = 0 \tag{2.14.1}$$

$$g(\xi) = \frac{m_1}{\sqrt{\xi^2 - k_1^2}} \tag{2.14.2}$$

where $m_1 = \rho_1/\rho$. The function $g(\xi)$ is determined by the properties of the bottom[4] and can be computed in an analogous fashion for more complex models of the

[3] Generally speaking, we should write $\overline{\Psi}(\xi, z, z_0)$; however, when it is not necessary we shall omit the second argument.
[4] The function $g(\xi)$ is related to the impedance of the bottom by the equation $g(\xi) = (i/\omega\rho)Z_1$, where the impedance is defined according to Brekhovskikh (1980); the change in sign from Brekhovskikh (1980) is due to the direction of the z-axis.

bottom, taking into account shear waves, the layered structure, etc. Thus, we need to construct the solution of the problem for $0 \le z \le H$:

$$\frac{d^2\overline{\Psi}(\xi, z)}{dz^2} + [k^2(z) - \xi^2]\overline{\Psi}(\xi, z) = -\frac{1}{4\pi^2}\delta(z - z_0) \tag{2.15.1}$$

$$\overline{\Psi}(\xi, z)|_{z=0} = 0 \tag{2.15.2}$$

$$\left[\overline{\Psi} + g(\xi)\frac{d\overline{\Psi}}{dz}\right]_{z=H} = 0 \tag{2.15.3}$$

One way of constructing this solution (one-dimensional Green function) involves a search for two linearly independent solutions of the homogeneous equation corresponding to (2.15), $V(\xi, z)$ and $U(\xi, z)$, satisfying, respectively, the first and the second boundary conditions:

$$\frac{d^2U(\xi, z)}{dz^2} + [k^2(z) - \xi^2]U(\xi, z) = 0 \tag{2.16.1}$$

$$\left[U + g(\xi)\frac{dU}{dz}\right]_{z=H} = 0 \tag{2.16.2}$$

and also

$$\frac{d^2V(\xi, z)}{dz^2} + [k^2(z) - \xi^2]V(\xi, z) = 0 \tag{2.17.1}$$

$$V(\xi, z)|_{z=0} \tag{2.17.2}$$

The solution of the problem (2.15) in this case has the form:

$$\overline{\Psi}(\xi, z) = \frac{U(\xi, z_>)V(\xi, z_<)}{4\pi^2 W(\xi)} \tag{2.18}$$

where $z_< = \min(z, z_0)$, $z_> = \max(z, z_0)$, $W(\xi)$ is the Wronskian of the system of solutions

$$W(\xi) = \begin{vmatrix} U(\xi, z) & V(\xi, z) \\ U'(\xi, z) & V'(\xi, z) \end{vmatrix} \tag{2.19}$$

which does not depend on z and therefore can be computed at any point. For the following, it will be convenient to compute it at the point $z = H$:

$$W(\xi) = -U'(\xi, H)[V(\xi, H) + g(\xi)V'(\xi, H)] \tag{2.20}$$

Here, the prime denotes the derivative with respect to z. An expression for the sound field at the viewing point can be obtained after substitution of equation (2.18) in (2.8.2) using a Hankel function instead of the Bessel function:

$$J_0(\xi r) = \frac{1}{2}\left[H_0^{(1)}(\xi r) + H_0^{(2)}(\xi r)\right]$$
$$H_0^{(2)}(\xi r) = -H_0^{(1)}(e^{i\pi}\xi r) \tag{2.21}$$

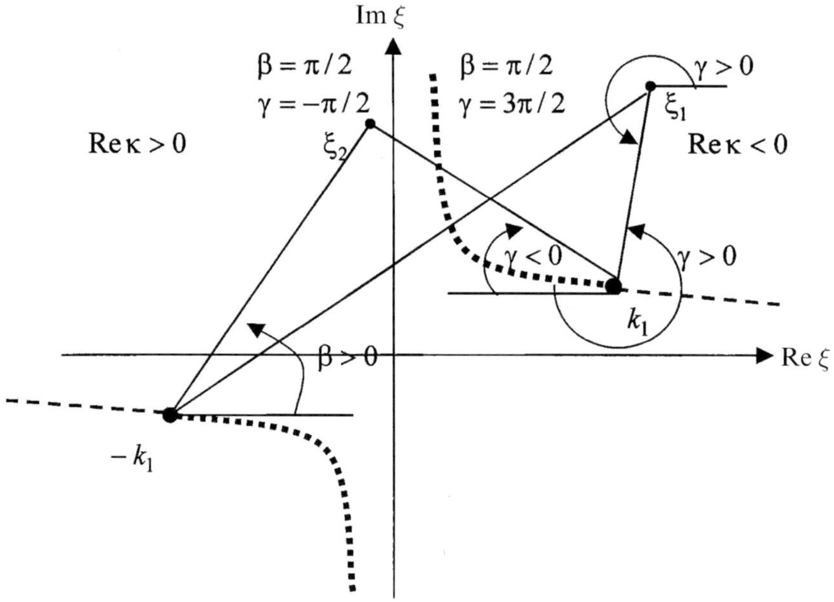

Figure 2.2. Analytic properties of the function $\sqrt{k_1^2 - \xi^2}$ in the complex plane.

Finally, we obtain

$$\Psi(r, z) = \frac{1}{4\pi} \int_{-\infty}^{\infty} \frac{U(\xi, z_>) V(\xi, z_<)}{W(\xi)} H_0^{(1)}(\xi r) \xi d\xi \tag{2.22}$$

The integral is computed as a path integral in the complex plane, with a proper circuit round the poles and branch points of the integrand.

The poles ξ_l are determined by the zeros of the Wronskian (2.20); in other words, by the expression:

$$V(\xi_l, H) + g(f_e) V'(\xi_l, H) = 0 \tag{2.23}$$

In addition to the contribution of poles to the integral (2.22), we must take into account branch points of the integrand, due to the presence of the root in the expression for $g(\xi)$ and, correspondingly, in the boundary conditions for solutions. These branch points are $\xi = \pm k_1$. Since the lower medium is assumed to be absorptive, the branch points lie in the complex plane, in the first and third quadrants, respectively, near the real axis. When the contour is closed in the upper half-plane, we must remain on the 'physical' sheet of the Riemann surface, determined by the decay of the waves in the lower half-space. Let us consider the properties of the function $\kappa(\xi) = \sqrt{k_1^2 - \xi^2}$ in the complex plane in more detail, on the assumption that the value of k_1 is complex (equation (2.1)). Let us define two angles (arguments) for the quantities $k_1 + \xi = |k_1 + \xi| e^{i\beta}$, $k_1 - \xi = |k_1 - \xi| e^{i\gamma}$ (see Figure 2.2). Positive values of the arguments β and γ are taken in the anticlockwise direction with respect to the horizontal axes, beginning at the points $-k_1$ and k_1, respectively. For the angle β, the direction of this horizontal axis coincides with the positive direction of the x-axis, while for the angle γ the corresponding direction coincides with the negative direction of the x-axis. In connection with the direction of the coordinate axes and the condition of attenuation of the field at infinity, the physical sheet will be $\text{Im}\sqrt{k_1^2 - \xi^2} > 0$; therefore we draw the section with the lines $\text{Im}\sqrt{k_1^2 - \xi^2} = 0$, which gives the equation $\text{Re}\xi \cdot \text{Im}\xi = \alpha k^2 n_1^2/2$, defining part of a hyperbola, beginning at the branch points $\xi = \pm k_1 = \pm k n_1(1 + i\alpha/2)$. These parts of the branches (shown in the figure by dots) are found to the left of the point $\xi = k_1$ and to the right of the point $\xi = -k_1$. In the remaining part of the hyperbola (shown by dashes) $\text{Re}\sqrt{k_1^2 - \xi^2} = 0$. In order to determine the value of the function in the upper half-plane on the right- and left-hand shores of the section, let us indicate the values of the angles β and γ in this region. Namely, on the right-hand shore of this region (where the point ξ_1 is shown) $0 < \beta < \pi/2$, $\pi < \gamma < 2\pi$, i.e. $\pi/2 < (\beta + \gamma)/2 < 5\pi/4$, which gives negative values for the real part of the function $\kappa(\xi)$: $\text{Re}\sqrt{k_1^2 - \xi^2} = |k_1^2 - \xi^2|^{1/2}\cos(\beta + \gamma)/2 < 0$. This means that for integration directly along the section line, where the function $\kappa(\xi)$ is real, on the right-hand shore of the section, we must take the root with the '$-$' sign. Correspondingly, on the left-hand shore of the section (where the point ξ_2 is shown), we have values $0 < \beta < \pi/2$, $-\pi/2 < \gamma < 0$ and $\text{Re}\sqrt{k_1^2 - \xi^2} = |k_1^2 - \xi^2|^{1/2}\cos(\beta + \gamma)/2 > 0$, which implies a '+' sign for the root. Thus, for the sound field, we obtain an expression in the form of a sum of contributions of residues at the poles and an integral along the section line, which we shall not yet make concrete

$$\Psi(r, z) = \frac{i}{2}\sum_l \frac{U(\xi_l, z_>)V(\xi_l, z_<)}{W'(\xi_l)} H_0^{(1)}(\xi_l r)\xi_l - \frac{1}{4\pi}\int_B \frac{U(\xi, z_>)V(\xi, z_<)}{W(\xi)} H_0^{(1)}(\xi r)\xi d\xi$$

The contribution of the integral over the shores of the section corresponds to the so-called continuous spectrum in the expression for the field of a point source. Its role at comparatively large distances is small and in our subsequent analysis we shall often neglect it. However, in some situations, the contribution of the continuous spectrum must be taken into account; there are many papers on this (Brekhovskikh and Godin (1992), Gazaryan (1958), Sticler *et al.* (1980)). We shall touch upon the question of the continuous spectrum once again in considering the Pekeris model and also in considering the field of a distributed source (noise).

Let us now consider the discrete spectrum of the field in more detail. Since we are neglecting the continuous spectrum, we retain the same notation Ψ for the component of the field, corresponding to the discrete spectrum. Calculating the residues at the poles, we obtain

$$\Psi(r, z) = \frac{i}{2}\sum_l \frac{U(\xi_e, z_>)V(\xi_l, z_<)}{W'(\xi_l)} H_0^{(1)}(\xi_l r)\xi_l \tag{2.24}$$

Expression (2.24) can be transformed somewhat, taking into account the fact that when $\xi = \xi_l$ the functions $U(\xi, z)$ and $V(\xi, z)$ become linearly dependent, since the Wronskian becomes zero. We shall suppose that $U(\xi_l, z) = AV(\xi_l, z)$. The equation for the functions U, V becomes an equation in the eigenvalues

$$\frac{d^2\psi_l(z)}{dz^2} + [k^2(z) - \xi_l^2]\psi_l(z) = 0 \tag{2.25.1}$$

$$\psi_l(z)|_{z=0} = 0 \tag{2.25.2}$$

$$\left[\psi_l(z) + g(\xi_l)\frac{d\psi_l(z)}{dz}\right]_{z=H} = 0 \tag{2.25.3}$$

where we have denoted $\psi_l(z) = V(\xi_l, z)$. Equation (2.25.1) with boundary conditions (2.25.2) and (2.25.3) determines the Sturm–Liouville equation or Sturm–Liouville problem, where $\psi_l(z)$ and ξ_l are eigenfunctions and eigenvalues, respectively. Using equations (2.16) and (2.25) we can obtain a simple equation for an arbitrary Wronskian $W'(\xi_l)$. For this, we multiply equations (2.16.1) and (2.25.1), respectively, by $\psi_l(z)$ and $U(\xi, z)$; then, subtracting the one equation from the other, we integrate both sides of the equation from zero to H, taking into account the gradient conditions, and obtain

$$U(\xi, H)\psi_l(H)\left[\frac{1}{g(\xi_e)} - \frac{1}{g(\xi)}\right] + U(\xi, 0)\psi_l'(0) = (\xi^2 - \xi_l^2)\int_0^H U(\xi, z)\psi_l(z)dz \tag{2.26}$$

Using the expression for the Wronskian at the point $z = 0$:

$$W(\xi) = U(\xi, 0)V'(\xi, 0) \tag{2.27}$$

as a result of which

$$\begin{aligned}\frac{W(\xi)}{\xi - \xi_l} = \frac{V'(\xi, 0)}{\psi_l'(0)}\Bigg[(\xi + \xi_l)\int_0^H U(\xi, z)\psi_l(z)dz \\ + U(\xi, H)\psi_l(H)\frac{g^{-1}(\xi) - g^{-1}(\xi_l)}{\xi - \xi_l}\Bigg]\end{aligned} \tag{2.28}$$

In the limit $\xi \to \xi_l$ we obtain derivatives on the left-hand side of the equation and in the last term on the right-hand side. Computing the derivative of the variable $g^{-1}(\xi)$ at the point $\xi = \xi_l$, we obtain

$$\begin{aligned}W'(\xi_l) &= 2A\xi_l\left[\int_0^H \psi_l^2(z)dz + \frac{\psi_l^2(H)}{2\xi_l}\left(\frac{dg^{-1}(\xi)}{d\xi}\right)_{\xi=\xi_l}\right] \\ &= 2A\xi_l\left[\int_0^H \psi_l^2(z)dz + \frac{\psi_l^2(H)}{2m_1\sqrt{\xi_l^2 - k_1^2}}\right]\end{aligned} \tag{2.29}$$

Substituting this expression in the equation for the sound field (2.24), we finally obtain

$$\Psi(r,z) = \frac{i}{4}\sum_l \frac{\psi_l(z_0)\psi_l(z)}{\int_0^H \psi_l^2(z)dz + \frac{\psi_l^2(H)}{2m_1\sqrt{\xi_l^2-k_1^2}}} H_0^{(1)}(\xi_l r) \tag{2.30}$$

If we take the normalisation of the eigenfunctions of the Sturm–Liouville problem (2.25)

$$\int_0^H \psi_l^2(z)dz + \frac{\psi_l^2(H)}{2m_1\sqrt{\xi_l^2 - k_1^2}} = 1 \tag{2.31}$$

then for the field of a point source we have the decomposition:

$$\Psi(r,z) = \frac{i}{4}\sum_l \psi_l(z_0)\psi_l(z)H_0^{(1)}(\xi_l r) \tag{2.32}$$

The meaning of the second term on the left-hand side of equation (2.31) is easy to understand, if one includes the half-space $z > H$ in the search for eigenfunctions. In this case, the solution in this region, satisfying the boundary conditions, will have the form $\psi_l(z) = \psi_l(H)\exp[-\sqrt{\xi_l^2 - k_1^2}(z - H)]$, and the normalisation condition (2.31) coincides with the condition:

$$\int_0^H \psi_l^2(z)dz + \frac{\rho}{\rho_1}\int_H^\infty \psi_l^2(z)dz = 1 \tag{2.33}$$

Similarly, we obtain the orthogonality condition for the functions $\psi_l(z)$:

$$\int_0^H \psi_m(z)\psi_l(z)dz = -\frac{\psi_m(H)\psi_l(H)}{m_1\left(\sqrt{\xi_m^2 - k_1^2} + \sqrt{\xi_l^2 - k_1^2}\right)} \tag{2.34}$$

Sometimes, instead of (2.32) the normalisation condition

$$\int_0^\infty \frac{\psi_l^2(z)}{\rho(z)}dz = 1 \tag{2.35}$$

is used, with the corresponding orthogonality condition[5]

$$\int_0^\infty \frac{\psi_m(z)\psi_l(z)}{\rho(z)}dz = 0 \quad \text{for } m \neq l \tag{2.36}$$

Hence, equation (2.32) gives the field of a point source in a waveguide. The summation in the formula extends over all possible values of ξ_l. Values $\mathrm{Re}(\xi_l) > \mathrm{Re}(k_1)$ correspond to wave vectors of so-called propagating (trapped) normal waves (modes); these modes decay weakly in the bottom. Eigenvalues $\mathrm{Re}(\xi_l) < \mathrm{Re}(k_1)$ correspond to so-called leaking modes (modes of the continuous spectrum). The number of normal waves propagating is determined by the

[5] It is not difficult to see that functions normalised according to (2.33) have dimension $\mathrm{m}^{-1/2}$, while those normalised according to (2.35) have dimension $\mathrm{kg}^{1/2}\,\mathrm{m}^{-2}$.

parameters of the problem: the sound frequency, the waveguide parameters (depth, profile of the sound speed), etc.

Of course, because of the presence of absorption a complex impedance appears in the boundary conditions and the problem becomes non-self-adjoint. The eigenvalues and eigenfunctions of the Sturm–Liouville problem (2.25) are complex:

$$\psi_l(z) = \mathrm{Re}\psi_l + i\mathrm{Im}\psi_l \quad \xi_l = q_l + i\frac{\gamma_l}{2} \tag{2.37}$$

A small absorption (small imaginary part of the refractive index at the bottom in comparison with the real part) leads to a small imaginary part of the eigenfunctions and eigenvalues in comparison with their real parts; however, while in practical calculations one can neglect the imaginary part of the functions in comparison with the real part, one cannot do this for the eigenvalues, since the value of the absorbed sound energy increases exponentially with distance.

The value of q_l determines the wave vector of the individual normal mode and the coefficient γ_l is called the mode attenuation coefficient. The phase and group velocities of the mode are associated with these quantities:

$$\nu_l^{\mathrm{ph}} = \left(\frac{q_l}{\omega}\right)^{-1} \quad \nu_l^{\mathrm{gr}} = \left(\frac{dq_l}{d\omega}\right)^{-1} \tag{2.38}$$

In problems of shallow-water acoustics, as a rule, the sound field is considered to be at very large distances from the source $kr \gg 1$. In this case, in equation (2.32), for the Hankel functions, we can use the asymptotic expressions

$$H_0^{(1)}(\xi_l r) \approx \sqrt{\frac{2}{\pi\xi_l r}}\exp[i(\xi_l r - \pi/4)] \tag{2.39}$$

Finally, we find the field in the waveguide $p(r, z)$, generated by an acoustic source with power W_0. This corresponds to some non-unit amplitude A, the factor in front of the δ-function on the right-hand side of (2.2) or in front of $\Psi_0(\vec{R})$ in (2.3). Let us determine the power of the source (or the connection between W_0 and A) as follows. Consider a neighbourhood of the source, sufficiently small ($\vec{R} \to \vec{R}_0$) that the medium in this region may be assumed to be homogeneous ($c = c(z_0) = c_0$, $\rho = \rho(z_0) = \rho_0$), and neglect the effect of the waveguide boundaries. The time-dependent field generated by the source in this neighbourhood is $A\Psi_0(\vec{R})\exp(-i\omega t)$. Calculating the power according to equation (1.4), we obtain

$$A = \sqrt{8\pi\rho_0 c_0 W_0} \tag{2.40}$$

Taking this into account and substituting (2.39) in (2.32) we obtain for the field of a source of power W_0 in the waveguide, a time-dependent expression for the pressure $P(r, z, t)$:

$$P(r, z, t) = i\sqrt{\rho_0 c_0 W_0}\sum_l \frac{\psi_l(z_0)\psi_l(z)}{\sqrt{\xi_l r}}\exp[i(\xi_l r - \pi/4 - \omega t) - \beta r] \tag{2.41}$$

which expression also takes into account the absorption in the water (1.2). We shall also refer to equation (2.41) in considering the complex envelope $P(r, z)$ of the field of an acoustic source, when we shall omit the factor $e^{-i\omega t}$ in this equation, and also, if the absorption is not significant, the quantity β.

2.2 THE PEKERIS MODEL

The simplest model of a shallow-water waveguide having all the main characteristic properties, is the model proposed by Pekeris (1948). In this model, the water layer is assumed to have a constant sound speed and density, $c(z) = c = \text{constant}$, $\rho(z) = \rho = \text{constant}$, respectively, and the bottom parameters are also assumed to be constants c_1, ρ_1, α. The Pekeris model is often used for putting forward qualitative arguments and giving quantitative estimates in the description of various phenomena in shallow water. A detailed analysis of the behaviour of the sound field in this model (in other words, in a two-layer waveguide) is given in the literature (Gazaryan (1958), Brekhovskikh (1980), De Santo, (1979)). We shall consider only those properties and relationships which we shall use later for estimates and analysis.

Equations (2.16), (2.17) for the construction of the field of a point source in this case have analytic solutions:

$$U(\xi, z) = \sin\sigma(z - H) - g\sigma\cos\sigma(z - H) \quad V(\xi, z) = \sin\sigma z \tag{2.42}$$

Here, we have denoted $\sigma = \sqrt{k^2 - \xi^2}$. The Wronskian of this system of solutions

$$W(\xi) = -\sigma(\sin\sigma H + g\sigma\cos\sigma H) \tag{2.43}$$

and its zeros (eigenvalues of the normal modes) are defined by the equation

$$\tan\sqrt{k^2 - \xi_l^2}H = -m_1\frac{\sqrt{k^2 - \xi_l^2}}{\sqrt{\xi_l^2 - k_1^2}} \tag{2.44}$$

Equation (2.44) has an infinite number of roots in the complex plane ξ (the real and imaginary axes are respectively $\mathrm{Re}\xi$ and $\mathrm{Im}\xi$), whose position depends on the relationships of the parameters of this equation, or, more precisely, on the quantity $\nu = (H/\pi)\sqrt{k^2 - k_1^2}$. It is convenient to carry out a preliminary analysis of equation (2.44) in the limiting case $k_1 = k$. For the poles of the integrand in (2.22), we obtain the equation

$$\tan\sqrt{k^2 - \xi_l^2}H = -im_1 \tag{2.45}$$

since for $z > H$, there can only exist outgoing waves, then, in accordance with the choice of the signs of the roots

$$\sqrt{\xi_l^2 - k_1^2} = -i\sqrt{k_1^2 - \xi_l^2} \tag{2.46}$$

We then seek solutions of (2.42) setting $\sqrt{k^2 - \xi_l^2} = \sigma_l' + i\sigma_l''$. Equation (2.45) gives

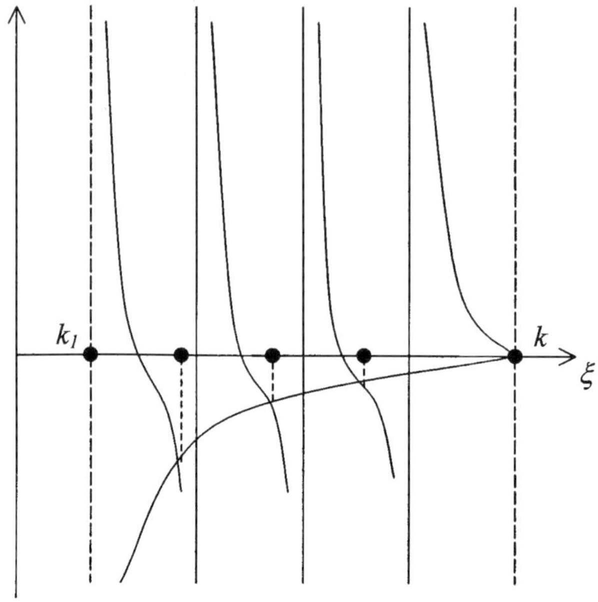

Figure 2.3. Graphical solution of the transcendental equation (2.44).

two families of points:

$$\tan(\sigma_l{}'H) = 0 \quad \tanh(\sigma_l{}''H) = -m_1 \tag{2.47}$$

and also

$$\cot(\sigma_l{}'H) = 0 \quad \tanh(\sigma_l{}''H) = -\frac{1}{m_1} \tag{2.48}$$

Since we are assuming that $m_1 > 1$, it follows that only the second remains. That is $\sigma_l' = (l + (1/2))(\pi/H)$. Moreover, from equation (2.48) it is clear that $\sigma_l'' = \operatorname{Im}\sqrt{k^2 - \xi_l^2} = -(1/H)\arctan(1/m_1) < 0$, which corresponds to a unphysical sheet of the complex plane. Moving to the values of ξ_l, we obtain

$$\xi_l^2 = k^2 + (\sigma_l{}'')^2 - \left[\left(l + \frac{1}{2}\right)\frac{\pi}{H}\right]^2 - i(2l+1)\frac{\pi\sigma_l{}''}{H}$$

It is clear that, since ξ_l^2 are located in the upper half-plane, the poles are found in the first and third quadrants of an unphysical sheet of the complex plane and for most values of the indices are located almost equidistantly along the imaginary axis. When $k > \operatorname{Re}(k_1)$ there may exist roots of equation (2.44) lying on a physical sheet of the complex plane.

The solution of the equation for real k_1 (i.e. when there is no absorption) can be illustrated graphically (Figure 2.3). It is clear that the poles (all of which are located between the points k_1 and k) are determined by the points of intersection of branches of the tangent (the function $\tan \sqrt{k^2 - \xi^2} H$) and the function $f(\xi) = -m_1 \sqrt{k^2 - \xi^2} / \sqrt{\xi^2 - k_1^2}$. As the difference $k - k_1$ increases, and, correspondingly, the point $\xi = k_1$ moves to the left relative to the point $\xi = k$, new poles appear, or more precisely, poles of the unphysical sheet move to the physical one. The first of these appears for $\sqrt{k^2 - k_1^2} > \pi/2H$, and the total number of poles M is determined by the number of branches of the function $\tan \sqrt{k^2 - \xi^2} H$, lying between the points $\xi = k_1$ and $\xi = k$, and is equal to

$$M = E\left\{ (H/\pi) \sqrt{k^2 - k_1^2} \right\} \tag{2.49}$$

where $E\{\,\}$ is the integral part of the function. We note further that from Figure 2.3, it is clear that for arbitrary variations of the parameters of the waveguide, the value of the transverse wavenumber belongs to the interval $(l - (1/2))(\pi/H) < \sigma_l < l(\pi/H)$ or can be written in the form

$$\sigma_l = \frac{l^* \pi}{H} \tag{2.50}$$

where l^* is a number such that $l - (1/2) \leq l^* \leq l$. The extreme values correspond to the limiting cases of a rigid and a pressure-release bottom, respectively. Another variant corresponds to a change in depth, i.e. $\sigma_l = l\pi/H^*$, where H^* is the effective depth.

In the case of fixed values of the depth and also of the sound speed in the bottom and in the water layer, the parameter determining the number of normal modes is the sound frequency. We shall call the quantity $f = f_l$, determined by the equation $2 f_l H \sqrt{c^{-2} - c_1^{-2}} = l - (1/2)$, the cut-off frequency (or critical frequency) for modes with index l; this means that if the frequency $f < f_l$, then there exist $l - 1$ propagating modes, and when the frequency passes through the value $f = f_l$, the mode with index l appears.

In the general case of a complex value of k_1, the poles are located in the upper half-plane. Assuming that the absorption is low ($\alpha \ll 1$) the poles are located near the real axis and our earlier considerations about their number and behaviour remain valid.

Let us briefly consider separately the two component fields of a point source for the Pekeris model, determined by the contributions of the discrete and continuous spectra, respectively.

The contribution of the discrete spectrum of the point source Ψ_d is thus determined by the expression[6]

$$\Psi_d(r, z) = \frac{i}{4} \sum_{l=1}^{M} \psi_l(z_0) \psi_l(z) H_0^{(1)}(\xi_l r) \tag{2.51}$$

[6] In other sections we shall neglect the contribution of the continuous spectrum, and the notation Ψ will be used for the field generated by the discrete spectrum.

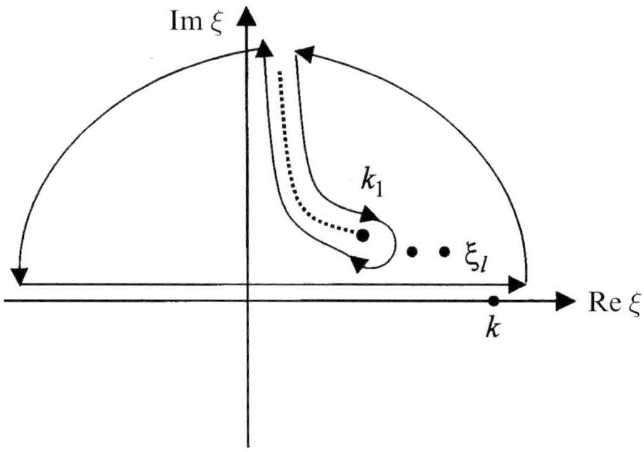

Figure 2.4. Contour of integration in the complex plane ξ for computation of the Green function. The branch point in the complex plane is the value of the wave vector in the bottom. The complex eigenvalues shown are zeros of the Wronskian.

where $\psi_l(z)$ are the eigenfunctions

$$\psi_l(z) = N_l \begin{cases} \sin(\sigma_l z) & 0 \le z \le H \\ \sin(\sigma_l H) \mathrm{e}^{-\kappa_l (z-H)} & z > H \end{cases} \tag{2.52}$$

where $\kappa_l^2 = \xi_l^2 - k_1^2$.

The normalisation coefficients for the eigenfunctions of the discrete spectrum normalised according to the formula (2.31) are determined similarly to those in the preceding section:

$$N_l = \left(\frac{H}{2} + \frac{m_1}{2\kappa_l} \frac{k^2 - k_1^2}{\kappa_l^2 + m_1^2 \sigma_l^2} \right)^{-1/2} \tag{2.53}$$

Let us now consider the contribution of the continuous spectrum (we denote it by Ψ_c). According to what has been said earlier, it is determined by the integral over the shores of the section, beginning at the branch point $\xi = \kappa_1$. For ease of computation the section line is deformed. There are a number of different ways of transforming the section line. For an asymptotic estimate of the integral, the section line is drawn in the direction of steepest descent (thus a lateral wave is defined); in Pekeris (1948) the section line was taken to be a straight line running parallel to the imaginary axis through the point $\xi = k_1$. Let us consider as the section, the upper part of the hyperbola $\mathrm{Re}\xi \cdot \mathrm{Im}\xi = (\alpha/2) k^2 n_1^2$, passing through a branch point (see Figure 2.4). A section of this type corresponds to the so-called EJP method (Ewing, Jardetzky and Press (1957)). Neglecting absorption in the lower half-space, the section line represents the region from the point $\xi = k_1$ to zero and from zero to infinity along the imaginary axis. In this case, the quantity $\sqrt{\xi^2 - k_1^2}$ becomes purely imaginary and we shall use the expression

$$g(\xi) = \frac{im_1}{\sqrt{k_1^2 - \xi^2}} \tag{2.54}$$

The sum of integrals over the shores of the section is thus given by the expression:

$$\Psi_c(r,z) = \frac{1}{4\pi}\int_{k_1}^{i\infty} \frac{U^{(+)}(\xi, z_>)V^{(+)}(\xi, z_<)}{W^{(+)}(\xi)} H_0^{(1)}(\xi r)\xi d\xi$$
$$+ \frac{1}{4\pi}\int_{i\infty}^{k_1} \frac{U^{(-)}(\xi, z_>)V^{(-)}(\xi, z_<)}{W^{(-)}(\xi)} H_0^{(1)}(\xi r)\xi d\xi \tag{2.55}$$

where the signs $(+)$ and $(-)$ denote the values of the functions on the left- and right-hand shores of the section, respectively. They are distinguished by the sign of the root $\sqrt{k_1^2 - \xi^2}$ in the impedance. Substituting the values of the functions into this expression, we obtain, for the contribution of the continuous spectrum

$$\Psi_c(r,z) = \frac{i}{4}\int_{k_1}^{i\infty} \frac{m_1}{\pi\kappa} \frac{\sin(\sigma z)\sin(\sigma z_0)}{\sin^2(\sigma H) + m_1\sigma^2\kappa^{-2}\cos^2(\sigma H)} H_0^{(1)}(\xi r)\xi d\xi \tag{2.56}$$

The meaning of the integrand can be understood similarly to how this was done in the previous section. For values $\xi^2 < k_1^2$ we define the modes of the continuous spectrum $\psi_\xi(z)$, sewing together the solution on the boundary $z = H$:

$$\psi_\xi = N_\xi \begin{cases} \sin \sigma z & 0 < z < H \\ \sin \sigma H \cos \kappa(z - H) + i\frac{m_1\sigma}{\kappa}\cos\sigma H \sin\kappa(z - H) & z > H \end{cases} \tag{2.57}$$

Here, ξ, σ and K takes a continuous series of values. Let us calculate the integral of the overlap of the two different functions of the continuous spectrum:

$$\int_0^H \psi_\xi(z)\psi_{\xi'}(z)dz + \frac{\rho}{\rho_1}\int_0^\infty \psi_\xi(z)\psi_{\xi'}(z)dz$$
$$= \pi N_\xi^2 \frac{\delta}{m_1}\left(\sin^2\sigma H + \frac{m_1^2\sigma^2}{\delta^2}\cos^2\sigma H\right)\delta(\xi^2 - (\xi')^2) \tag{2.58}$$

We have used the equations

$$\int_0^\infty \sin x\tau d\tau = P\frac{1}{x} \quad \int_0^\infty \cos x\tau d\tau = \pi\delta(x) \tag{2.59}$$

where P denotes Cauchys Principal Value integral.

The functions $\psi_s(z)$ will be normalised by the δ-function, if we take

$$N_\xi = \left[\frac{\pi\kappa}{m_1}\left(\sin^2\sigma H + \frac{m_1^2\sigma^2}{\kappa^2}\cos^2\sigma H\right)\right]^{-1/2} \tag{2.60}$$

Thus, the field of a point source, taking into account the continuous spectrum, can be written in the form

$$\Psi(r,z) = \frac{i}{4}\sum_{l=1}^{M} \psi_l(z_0)\psi_l(z)H_0^{(1)}(\xi_l r) + \frac{i}{4}\int_{k_1}^{i\infty} \psi_\xi(z)\psi_\xi(z)H_0^{(1)}(\xi r)\xi d\xi \tag{2.61}$$

As mentioned earlier, the continuous spectrum, as a rule, gives a small contribution to the field, and at sufficiently large distances it is neglected; however, there exist situations when it has to be taken into account. This question is studied in great detail in the literature (Brekhovskikh (1980), Brekhovskikh and Godin (1989), Gazaryan (1958), Sticler (1980), Kuperman and Ferla (1985)).

When absorption in the bottom is taken into account, the value of k_1^2 becomes complex, with a small imaginary part $\alpha k^2 n_1^2$. In this case, the imaginary part has a pole ξ_l, moving in the upper half-plane: $\xi_l = q_l + i\gamma_l/2$ where the coefficient γ_l is called the mode attenuation coefficient. The numerical value of α is usually two or three orders less than unity, so that $\gamma_l \ll q_l$. Assuming a low absorption in the bottom, one can define the absorption coefficient of a mode of the discrete spectrum γ_l. Here, the real values σ_l and κ_l have to be replaced by complex expressions. Retaining the notation σ_l and κ_l for the real parts of these quantities, we have, respectively, complex quantities $\sigma_l - i(q_l\gamma_l/2\sigma_l)$ and $\kappa_l + i(\gamma_l q_l - \alpha k^2 n_1^2)/2\kappa_l$. Substituting these expressions in the transcendental equation (2.44), we obtain for the mode attenuation coefficient in the Pekeris model

$$\gamma_l = \alpha \frac{N_l^2 \sigma_l^2 k^2 n_1^2 m_1}{2q_l\kappa_l(\kappa_l^2 + m_1^2\sigma_l^2)} \tag{2.62}$$

It is also possible to calculate the group velocities ν_l^{gr} of propagating modes

$$\nu_l^{\text{gr}} = \frac{d\omega}{dq_l} = c^2\frac{q_l}{\omega}\frac{\dfrac{m_1\sigma_l}{\kappa_l^2} + \dfrac{\kappa_l H}{\sigma_l}\left(1 + \dfrac{m_1^2\sigma_l^2}{\kappa_l^2}\right) + \dfrac{m_1}{\sigma_l}}{\dfrac{m_1\sigma_l}{\kappa_l^2}n_1^2 + \dfrac{\kappa_l H}{\sigma_l}\left(1 + \dfrac{m_1^2\sigma_l^2}{\kappa_l^2}\right) + \dfrac{m_1}{\sigma_l}} \tag{2.63}$$

If we consider in more detail the dependence of the spectrum of the problem on the index of the mode, the distance between the modes and the absorption coefficient in the Pekeris model grow with the index of the mode. This can be seen in the figure.

Let us consider the behaviour of the mode attenuation coefficients as a function of frequency. This can be done based on equation (2.62), where the normalisation coefficient is determined by equation (2.53). Suppose that for the given mode index l, the frequency approximates from above to the critical frequency for the given mode index, remaining somewhat larger. In this case $q_l \to kn_1$ and, accordingly, for the attenuation coefficient, $\gamma_l \to \alpha kn_1$. The meaning of the latter limit is that when the frequency approximates to the critical frequency, the mode becomes a mode of the continuous spectrum, i.e. it is unconstrained and propagates in the bottom to a greater extent. Accordingly, the attenuation coefficient approximates to the

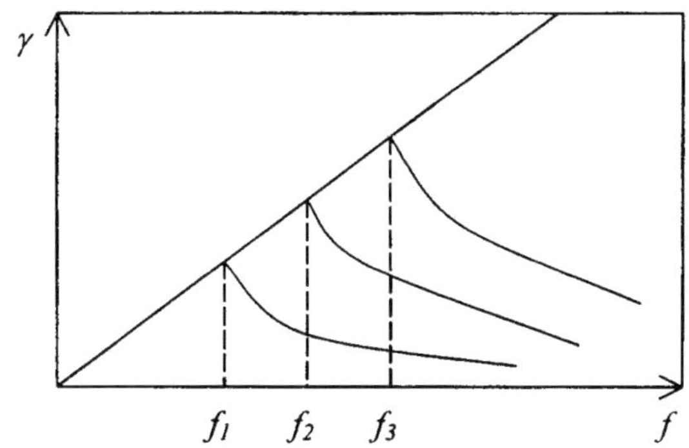

Figure 2.5. Frequency dependence of the mode attenuation coefficients. The straight line corresponds to linear dependence of the sound absorption coefficient in the bottom, determined by our condition for an absorptive bottom.

coefficient of sound absorption in the bottom at the given frequency. If for this same mode the frequency grows, then as can be easily seen from (2.62), the value of γ_l tends monotonically to zero, for large values of the frequency, in proportion to $1/f^2$. This is also easy to understand, since as the mode frequency increases it penetrates less and less in the bottom and the proportion of the function propagating in the bottom tends to zero. Figure 2.5 shows the behaviour of the mode attenuation coefficients as a function of frequency.

Analogously, we can consider the behaviour of the mode attenuation coefficients as a function of the depth H. If we fix some mode with index l and decrease the depth H, then as H nears some critical depth H_l, it is clear from equations (2.62) and (2.46) that the coefficient γ grows, giving, in the limit, a value $\alpha k n_1$, which is the coefficient of sound absorption in the bottom. The situation of decreasing depth can be interpreted as the propagation of the mode in a waveguide with a smoothly decreasing depth, for example, in a coastal wedge in the direction towards the shore. In this situation, the absorption of the mode increases, and when a section with the critical depth is reached the mode transfers to the continuous spectrum and is quickly absorbed by the bottom. The direction of energy flow in the bottom in the region of the critical section is determined by the vector $(k\sqrt{1 - n_1^2}, \sigma_l)$ in the water layer and, in the language of rays, corresponds to incidence of the ray at the angle of total internal reflection. The greater the mode index, the further from the coast its exit to the bottom occurs. Numerical analysis of such propagation based on a parabolic equation method has been carried out, for example, in Jensen (1981). As the depth increases, the attenuation coefficient of the mode decreases as H^{-3}.

To conclude this section, we note that, despite its simplicity, the Pekeris model is not only used quite often for estimates and qualitative arguments, but is also successfully applied for the analysis of concrete problems; for example, for studies of noise (Buckingham (1980)).

2.3 SIMPLIFIED THEORY AND PROPERTIES OF MODE ATTENUATION COEFFICIENTS

As we have already pointed out on a number of occasions, when considering the propagation of sound in a waveguide, it is of primary importance to take into account the interaction of sound with the bottom, or more precisely, absorption in the bottom, which leads to a comparatively rapid drop in the signal intensity with distance. Formally, this decay is due to the fact that complex eigenvalues occur in the Sturm–Liouville problem (2.25). This is equivalent to the fact that the constant of propagation ξ_l has an imaginary part (2.37), which determines the decay of the normal mode. This imaginary part, $\gamma_l = 2\mathrm{Im}\xi_l$ (i.e. the attenuation coefficient of the normal mode) depends on the mode index and on other parameters of the problem. Despite the fact that the values of γ_l are very easily obtained in numerical solution of the problem, a more detailed study of their behaviour as a function of the parameters of the problem, in the first place, as a function of the mode index l, is advisable, since this dependence is most significant for an explanation of the structure of the field formed at a large distance from the source, including its modal composition and total intensity. Moreover, in approximate calculations of the averaged intensity, simplified but none the less analytic formulae for γ_l are required. These simplified formulae can be obtained and interpreted in the framework of ray theory.

In ray theory the absorption of sound is, as it were, of a discrete nature, as energy is lost at the instant a ray is reflected from the bottom; this is described by losses on reflection. These losses can be associated with the mode attenuation coefficient, if one uses a spatial scale of interference $\Lambda_l \approx 2\pi|q_l - q_{l+1}|^{-1}$. This decay of the mode at wavelength Λ_l is precisely equal to the losses on reflection. In turn, Λ_l corresponds to the cycle distance D_l of the equivalent ray with a correction (if it is significant) by the displacement of the ray on reflection. In this approach, one can obtain the dependence of the quantity γ on the mode index l or, in ray theory, on the grazing angle of the ray χ corresponding to the given mode index.

These and related questions have been considered many times in the literature (Krupin (1972), Tindle (1980), Katsnelson (1986), Weston (1994)). In Tindle (1980, 1985, 1987) the connection between γ, D and the losses on reflection was considered based on the Pekeris model. In Koch (1983) this problem was solved for an arbitrary profile $c(z)$ based on numerical analysis. In Kuperman and Jensen (1980) the results of an approximate computation of γ_l were compared with those obtained based on solution of the Sturm–Liouville problem for a concrete hydro-acoustic waveguide. Based on this calculation a comparison with experiment was made. The importance of taking into account the displacement of the beam on reflection in a number of cases was first pointed out in Tindle (1980). In Krupin (1972) and Weston (1994) the connection between ray and mode representations was obtained based on the WKB method; in Macferson (1967) and Tindle (1985) stratification and inhomogeneity of the bottom were also taken into account. All the works cited, except for Katsnelson (1986), considered regular waveguides.

Let us give the derivation of the relation for γ_l, based mainly on the results of Krupin (1972), Tindle (1980) in the framework of the waveguide model formulated.

We multiply equation (2.25.1) by $\psi_l(z)$ and integrate both sides of the equation from 0 to H. Taking into account the boundary conditions we have

$$-\frac{\psi_l^2(H)}{g(\xi_l)} - \int_0^H (\psi_l'(z))^2 dz + \int_0^H k^2(z)\psi_l^2(z)dz - \xi_l^2 \int_0^H \psi_l^2(z)dz = 0 \tag{2.64}$$

We shall further hypothesise that the imaginary corrections are small in comparison with the real parts of the quantities, so that the eigenfunctions of the problem are real. In this approximation

$$\frac{1}{g(\xi_l)} = \frac{\sqrt{\xi_l^2 - k^2 n_1^2(1+i\alpha)}}{m_1} = \frac{\sqrt{q_l^2 - k^2 n_1^2}}{m_1} + i\frac{q_l\gamma_l - k^2 n_1^2 \alpha}{2m_1\sqrt{q_l^2 - k^2 n_1^2}} \tag{2.65}$$

Substituting this expression in the previous one and separating the imaginary part, we obtain

$$\gamma_l q_l \left[\int_0^H \psi_l^2(z)dz + \frac{\psi_l^2(H)}{2m_1\sqrt{q_l^2 - k^2 n_1^2}}\right] = \frac{\alpha k^2 n_1^2}{2m_1\sqrt{q_l^2 - k^2 n_1^2}}\psi_l^2(H) \tag{2.66}$$

Taking into account the normalisation of the proposed eigenfunctions, we obtain for the attenuation coefficients:

$$\gamma_l = \frac{k^2 n_1^2 \alpha}{2m_1 q_l\sqrt{q_l^2 - k^2 n_1^2}}\psi_l^2(H) = \frac{k^2 n_1^2 \alpha}{2m_1^2 q_l} g_l \psi_l^2(H) \tag{2.67}$$

Here, $g_l = g(q_l) = m_1/\sqrt{q_l^2 - k^2 n_1^2}$.

Equation (2.67) can be obtained from perturbation theory, which will be studied later (Chapter 3, from (3.11)), if for the integrand in (3.11) one uses the expression for the eigenfunction employed in the derivation of equation (2.33).

Let us now write down the real part of equation (2.64)

$$-\frac{\psi_l^2(H)}{m_1}\sqrt{q_l^2 - \frac{\omega^2}{c_1^2}} - \int_0^H (\psi_l')^2 dz + \omega^2 \int_0^H c^{-2}(z)\psi_l^2(z)dz - q_l^2 \int_0^H \psi_l^2(z)dz = 0 \tag{2.68}$$

Here, for $k^2(z)$ we have substituted the expression for it in terms of the sound frequency and speed profile in the water layer.

If in expression (2.68), the squares of the eigenvalues q_l and the frequency ω are taken to be variables, then it has the form

$$F(q_l^2, \omega^2) = 0 \tag{2.69}$$

whence

$$\frac{\partial F}{\partial q_l^2} dq_l^2 + \frac{\partial F}{\partial \omega^2} d\omega^2 = 0 \tag{2.70}$$

Assuming that the eigenvalues have a weak frequency dependence and calculating the partial derivatives, we obtain for the normalised functions

$$\frac{q_l}{\omega}\frac{dq_l}{d\omega} = \int_0^H \frac{\psi_l^2(z)}{c^2(z)}dz + \frac{\psi_l^2(H)}{2m_1 c_1^2\sqrt{q_l^2 - k^2 n_1^2}} \tag{2.71}$$

Replacing the last term in this expression using (2.31) and taking account of the fact that the phase and group velocities of the modes are determined according to (2.38), we obtain

$$\nu_l^{\mathrm{gr}}\nu_l^{\mathrm{ph}} = \left[c_1^{-2} + \int_0^H (c^{-2}(z) - c_1^{-2})\psi_l^2(z)dz\right]^{-1} \tag{2.72}$$

For the attenuation coefficient of a mode a simple estimation formula can be obtained using the ray approach. To find the value of $\psi_l(H)$ we use the WKB method to represent the function in the water layer; here, we shall neglect the absorption, according to the discussion of Section 2.2, and the eigenvalues q_l and functions are assumed to be real

$$\psi_l(z) = \frac{A_l}{\sqrt{\sigma_l(z)}}\sin[\phi_l(z)] \quad \phi_l(z) = \int_{\underline{z}}^{z} \sigma_l(z)dz + \zeta_l \tag{2.73}$$

Here, $\sigma_l(z) = \sqrt{k^2(z) - q_l^2}$, $\underline{z}$ is a turning point, and the phase correction ζ_l depends on the nature of the turning point. We note that in the water layer, a ray may undergo reflection from the surface, in which case, the value $\underline{z} = 0$ should be taken for the initial lower limit of the integral. If a ray does not reach the bottom (there is a second turning point) then that ray does not undergo absorption in this approach. The value of A_l is determined by the normalisation equation and by the boundary conditions. From the boundary conditions we obtain:

$$\tan\phi_l(H) + \sigma_l(H)g(q_l) = 0 \tag{2.74}$$

In what follows, we shall write down the functions without arguments for $z = H$: $\sigma_l = \sigma_l(H)$, $k = k(H)$, etc. The normalisation condition gives us the following value for A_l:

$$\begin{aligned} A_l &= \left[\int_{\underline{z}}^{H} \frac{\sin^2\phi_l(z)}{\sigma_l(z)}dz + \frac{\sin^2\phi_l}{2m_1\sigma_l\sqrt{q_l^2 - k^2 n_1^2}}\right]^{-1/2} \\ &= \left\{\frac{1}{2}\left[\int_{\underline{z}}^{H}\frac{dz}{\sigma_l(z)} + \int_{\underline{z}}^{H}\frac{\cos 2\phi_l(z)}{\sigma_l(z)}dz + \frac{\sigma_l g_l^3}{m_1^2(1+\sigma_l^2 g_l^2)}\right]\right\}^{-1/2} \\ &= \left\{\frac{1}{2}\left[\int_{\underline{z}}^{H}\frac{dz}{\sigma_l(z)} + \frac{k^2(1-n_1^2)g_l^3}{\sigma_l m_1^2(1+\sigma_l^2 g_l^2)}\right]\right\}^{-1/2} \end{aligned} \tag{2.75}$$

To compute the second integral, we have assumed that the function $\sigma_l(z)$ is sufficiently smooth in comparison with the cosine and have taken it out from under the integral. Substituting (2.75) and (2.73) in (2.67) we obtain in the WKB method the following expression for γ_l:

$$\gamma_l = k^2 n_1^2 \alpha \frac{\dfrac{\sigma_l g_l^3}{m_1^2 q_l (1+\sigma_l^2 g_l^2)}}{\displaystyle\int_z^H \frac{dz}{\sigma_l(z)} + \frac{k^2(1-n_1^2) g_l^3}{\sigma_l m_1^2 (1+\sigma_l^2 g_l^2)}}$$

$$\approx 2k^2 n_1^2 \alpha \frac{\alpha_l g_l^2}{m_1^2 D_l} \tag{2.76}$$

We then obtain the ray interpretation of equation (2.76). For this, we need to determine the connection between the modes of the waveguide and rays with corresponding Brillouin angles. This question has been studied in great detail in the literature (Brekhovskikh (1980), Tindle (1980), Weston (1994)), but we shall not reproduce this analysis and shall use only the connection between the eigenvalues and the grazing angle χ

$$\cos \chi_l(H) = q_l / k(H) \tag{2.77}$$

the angle between the direction of the ray, corresponding to the mode with index l and the bottom (horizontal axis at $z = H$). Other 'ray' characteristics, such as the cycle distance of the ray in the waveguide D_l, can be expressed in terms of this angle and the parameters of the waveguide. In subsequent formulae in this section, we shall omit the mode index (l) on the corresponding quantities. We shall assume that the dependence on the mode index enters through the value of the corresponding angle χ, determined according to the formula (we shall also omit the index l on the angle χ).

In what follows, we shall need the reflection coefficient V of a plane wave of the lower medium with wave vector (q, σ). Its value is easily obtained using the boundary condition (2.5):

$$V = \frac{\sigma g - i}{\sigma g + i} \tag{2.78}$$

When there is no absorption the value of $g(q)$ is real, $|V| = 1$, and the argument is determined by the expression

$$\varphi_V = \arg V = -2 \arctan\left(\frac{1}{\sigma g}\right) \tag{2.79}$$

In the presence of weak absorption, to first order in α, we obtain for $|V|$

$$|V| \approx 1 - \alpha k^2 n_1^2 \frac{\sigma g^3}{m_1^2 (1+\sigma^2 g^2)} \tag{2.80}$$

Let us consider the denominator of equation (2.76) in more detail. It is easy to see that the integral $\int dz/\sigma(z)$, corresponding to the phase change in the water layer, determines the horizontal displacement of the ray D on passing through the water layer (in other words, the cycle distance of the ray in the channel):

$$\int_{\underline{z}}^{H} \frac{dz}{\sigma(z)} = \frac{D}{2q} \tag{2.81}$$

The sense of the second term in the denominator of expression (2.76) becomes clear if we evaluate the phase of the coefficient of reflection of (2.79), differentiate it with respect to q, and then set $q = q_l$

$$\frac{d\varphi_V}{dq} = -2\frac{qk^2(1-n_1^2)g^3}{\sigma m_1^2(1+\sigma^2 g^2)} = -\Delta \tag{2.82}$$

where Δ is the horizontal displacement of the beam on reflection.

Substituting everything in equation (2.76), we finally obtain for the mode attenuation coefficient, the known expression

$$\frac{\gamma}{2} = \frac{1-|V|}{D+\Delta} = -\frac{\ln|V|}{D+\Delta} \tag{2.83}$$

(when $1-|V|$ is small we have $-\ln|V| \approx 1-|V|$). In most cases the displacement of the beam is small compared with D (see the estimated value of Δ at the end of the section) and can be neglected. As a result, we obtain the frequently used formula (see, for example, Breklovskich and Lysanov (1991))

$$\gamma/2 = -\ln|V|/D \tag{2.84}$$

By way of example, let us consider how the attenuation coefficient γ behaves as a function of the grazing angle χ in the Pekeris model. $V(\chi)$ is given by the known Fresnel formula (Brekhovskikh (1980))

$$V(\chi) = \frac{m_1 \sin\chi - \sqrt{n_1^2(1+i\alpha) - \cos^2\chi}}{m_1 \sin\chi + \sqrt{n_1^2(1+i\alpha) - \cos^2\chi}} \tag{2.85}$$

where typical values of n_1 lie in the range 0.9–1.0, corresponding to the region $\chi \leq 0.5$ for total reflection angles. In many cases of interest only small angles χ play the role; therefore, we give a decomposition of V in a power series in χ up to the third term, inclusively, which we shall need in what follows:

$$V(\chi) = -1 + \frac{2m_1}{\sqrt{n_1^2(1+i\alpha)-1}}\chi - \frac{2m_1^2}{n_1^2(1+i\alpha)-1}\chi^2 - \frac{m_1(n_1^2(1+i\alpha)-6m_1^2+2)}{3(n_1^2(1+i\alpha)-1)^{3/2}}\chi^3 + \dots \tag{2.86}$$

The cycle distance of a ray for the Pekeris model is defined by the expression

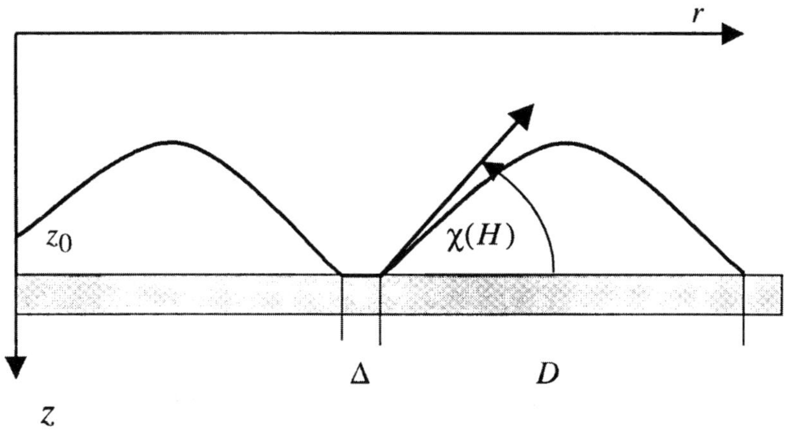

Figure 2.6. Ray path, corresponding to the lth normal mode, cycle distance D, ray displacement Δ amd grazing angle at the bottom $\chi(H)$ (or simply χ) are shown (index l is omitted).

$D = 2H\cot(\chi)$. For γ, we thus obtain the known expression

$$\gamma = \frac{s}{H}\chi^2 \quad s = \mathrm{Re}\frac{2m_1}{\sqrt{n_1^2(1+i\alpha)-1}} \approx \frac{m_1 n_1^2}{(1-n_1^2)^{3/2}}\alpha \tag{2.87}$$

From (2.87), it is clear that the values of the mode attenuation coefficient may grow sharply as the angle χ grows. Since for small angles χ_l, using (2.77) and (2.50), we can write $\chi_l = \pi l^*/kH$ and, correspondingly

$$\gamma_l = \frac{s\pi^2(l^*)^2}{k^2H^3} \tag{2.88}$$

it is clear that the coefficient γ grows as the square of the mode index. One practical consequence of these relations is the fact that in a waveguide corresponding to the Pekeris model, as the sound propagates the high modes die out very quickly and the far field is essentially determined by the first mode.

The dependence of the ray cycle distance D on the grazing angle is determined by the sound speed profile. The function $\gamma(\chi)$ is also modified according to the shape of the profile. For example, if the sound speed profile has a mainly negative gradient, then because of refraction, the ray is deflected towards the bottom and may, in principle, not touch the surface. A channel with such parameters will be called a near-bottom channel; as the grazing angle of a ray grows, its cycle distance decreases and the attenuation coefficient $\gamma(\chi)$ may decrease. The decrease (and in general, the non-monotonic nature) of the attenuation coefficient with increasing mode index in a waveguide with a negative sound speed gradient was first established and studied in Krupin (1972). There, it is shown that, depending on the value of the sound speed gradient, starting from some mode number, the corresponding attenuation coefficients have the same order of magnitude and initially even decrease with the mode index. Higher modes may be more strongly extinguished. Physically, this leads to the fact that the group of weakly decaying modes may contribute to the field at large distances from the source, which is very important for problems of ocean acoustics.

Simple analytic formulae, describing this phenomenon, can be obtained in a channel with linear dependence $c(z) = c_S(1 - \varepsilon z)$, with an absorptive, homogeneous bottom. In this case, the cycle distance of bottom rays D grows as a function of grazing angle, until the ray grazes the surface, while for rays which are reflected from the surface (bottom–surface rays) it decreases as χ grows. This is described by the formula:

$$D(\chi) = \mathrm{Re}\left(\frac{2c_H \tan\chi}{c_S\varepsilon} - \frac{c_H}{c_S\varepsilon\cos\chi}\sqrt{1 - \frac{c_S^2}{c_H^2}\cos^2\chi}\right) \tag{2.89}$$

where $c_H = c_S(1 - \varepsilon H)$ is the sound speed at $z = H$. The critical angle corresponding to a ray grazing the surface, has value $\chi_C \sim \sqrt{2c_S\varepsilon H/c_H} \sim \sqrt{2\varepsilon H}$. For $\chi < \chi_c$ the expression within the square root in (2.89) is negative and D is given by the first term only. For $\chi > \chi_c$ the second term in (2.89) leads to a decrease in the cycle distance.

The number of such modes M (rays not touching the surface) in a channel with linear sound speed can be coarsely estimated, based on the condition that for the turning point of this ray (modes with index greater than M) $\underline{z} = 0$.

A semiclassical quantisation condition for this mode is described by the expression $\int_0^H \sqrt{k_S^2(1 + 2\varepsilon z) - q_M^2}\,dz \approx M\pi$, where $k_S = 2\pi f/c_S$. Calculating the elementary integral and given that at the turning point $\underline{z} = 0$ and $k_S^2 - q_M^2 = 0$, we obtain

$$M \approx \frac{kH}{\pi}\sqrt{\frac{\Delta c}{c_S}} \sim \frac{kH}{\pi}\sqrt{\varepsilon H} \tag{2.90}$$

where $\Delta c = (c_S - c_H) = c_S\varepsilon H$ is the drop in sound speed between the bottom and the surface. This estimate can be used for an arbitrary sound speed profile with a negative gradient ($c(z)$ decreases in the direction towards the bottom).

Using expression (2.89) it is easy to show that the attenuation coefficients of bottom rays have value

$$\gamma = \frac{sc_S\varepsilon}{c_H} \sim s\varepsilon \tag{2.91}$$

($\chi < \sqrt{2\varepsilon H}$) and are weakly dependent on grazing angle. For bottom–surface rays γ grows linearly as the gradient εc_S grows. Figure 2.7 shows the behaviour of γ as a function of mode index for the Pekeris model. For a channel with a linear dependence $c(z)$ the behaviour of the attenuation coefficient is shown in Figure 2.8. The values of the parameters are given in the figures. The dependence as a function of grazing angle $\gamma(\chi)$ for the second case is shown in Figure 2.9, and discrete values γ_l corresponding to individual modes can be obtained based on the quantisation rules.

One can also consider the dependence of the quantity γ on the value of the sound speed gradient. This dependence is shown in Figure 2.10, from which it is clear that the attenuation coefficients of the individual modes at first decrease comparatively sharply down to some minimum value and then grow proportionally with the gradient of the sound speed according to formula (2.87). The decrease in the attenuation coefficient γ takes place at different times and non-uniformly for the

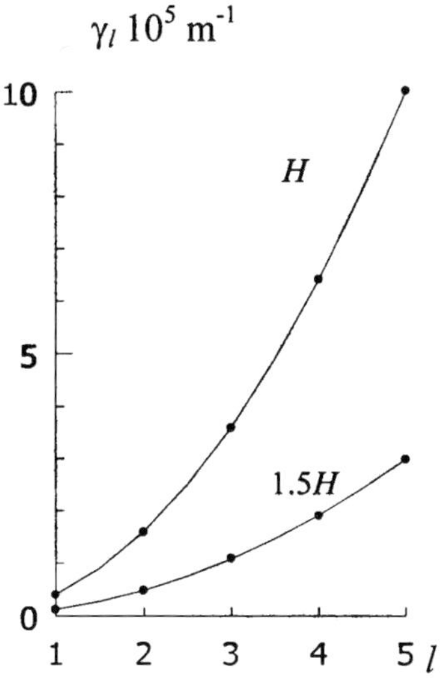

Figure 2.7. Dependence of γ_l on the mode index and the grazing angle for the Pekeris model. $s = 0.1$, $H = 100\,\mathrm{m}$, $f = 100\,\mathrm{Hz}$, $c = 1500\,\mathrm{m/s}$; depth values are shown on the curves.

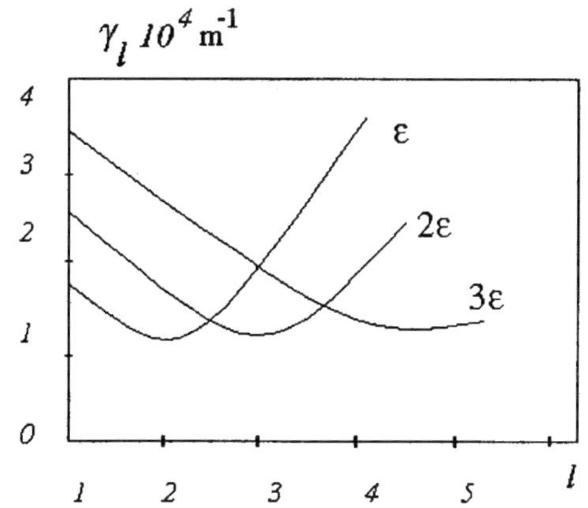

Figure 2.8. Dependence of γ_l on the mode index for a waveguide with a negative (linear) sound speed profile for various values of the gradient (shown on the curves) $f = 300\,\mathrm{Hz}$, $H = 100\,\mathrm{m}$, $C_H = 1500\,\mathrm{m/s}$, $\varepsilon = 3 \times 10^{-5}\,\mathrm{m}^{-1}$.

different modes. For each fixed index l (or the grazing angle χ of the equivalent ray) there is a value of the gradient ε_l where the decay of the given mode is minimal. Estimation of the gradient ε_l gives a value $\varepsilon_l \sim \chi_l^2/2H$, i.e. at which the given ray grazes the surface. Accordingly, if we fix the value of the gradient, then for a given ε there are one or more modes, whose decay is less than others, where the modes may have very high indices. Rays corresponding to these modes graze the surface or approach very close to it.

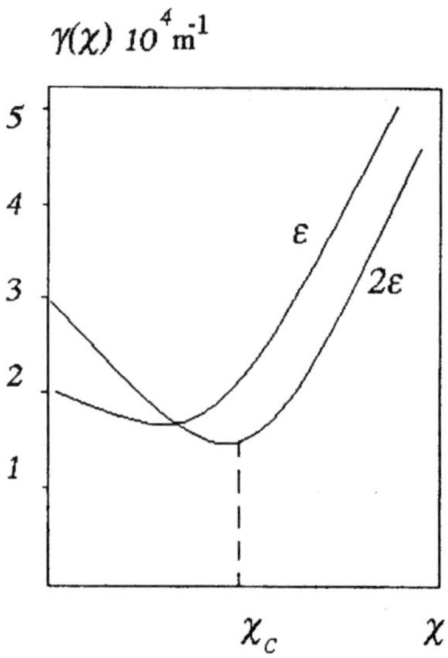

Figure 2.9. Dependence of γ_l on the grazing angle, calculated according to equation (2.8) for a waveguide with a negative (linear) sound speed profile for various values of the gradient (shown on the curves).

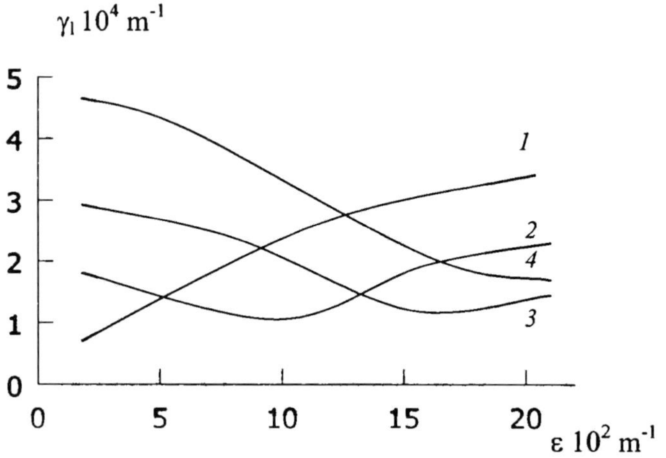

Figure 2.10. Dependence of γ_l on the gradient of the sound speed for the first four modes.

For a layer with linear dependence $c(z)$ the index of modes corresponding to given angles can be estimated from the quantisation rules. Assuming that the grazing angles are small, the estimate has the form

$$\chi_l \approx (3c_S\varepsilon l/2f)^{1/3} \tag{2.92}$$

Hence the index of the mode with the least decay can be estimated:

$$\bar{l} \approx \frac{2f}{3c_S\varepsilon}\left(\frac{2c_S\varepsilon H}{c_H}\right)^{3/2} \sim \frac{2fH\sqrt{\varepsilon H}}{c_H} \tag{2.93}$$

This expression agrees with (2.90), which means that the least decay in this waveguide occurs for the near-bottom mode with the smallest index, which is the mode which grazes the sea surface.

For example, if $H = 300\,\text{m}$, $c_H = 1500\,\text{m/s}$, $\varepsilon = 3 \times 10^{-5}\,\text{m}^{-1}$, then at frequency $f = 300\,\text{Hz}$, we have $l \sim 15$, i.e. there are approximately fifteen modes which give an approximately identical contribution to the far field.

Figure 2.10 also shows the results of calculation of the mode attenuation coefficients (imaginary parts of the eigenvalues), obtained based on precise solution of the Sturm–Liouville problem (2.25) with a linear sound speed profile in the water layer: $c(z) = c_S(1 - \varepsilon z)$. The bottom is assumed to be a fluid, homogeneous half-space. Results obtained from numerical calculation agree well with approximate estimates, except for a small range of values of the parameters ε (individually for each mode), where γ is minimal. This region again corresponds to values of ε for which the ray grazes the surface, and in this region ray theory does not apply.

In conclusion, let us consider what the value of the horizontal displacement of a ray Δ might be in real conditions. Assuming the grazing angles are small for the case of homogeneous layers of water and sediments, the following estimate for Δ is easily obtained:

$$\Delta \approx 2m \Big/ \left(\chi\sqrt{k^2 - k_1^2}\right) \tag{2.94}$$

If we take the values $c = 1500\,\text{m/s}$, $c_1 = 1600\,\text{m/s}$, $f \approx 250\,\text{Hz}$, $m_1 \approx 2$, $H = 150\,\text{m}$, then for an angle $\chi = 0.01$ radian corresponding to the lowest angle for such a waveguide mode, this displacement of the ray on reflection is of the order of $\Delta \sim 40\,\text{m}$, whereas the cycle distance of this ray is equal to approximately 30 km. It is clear that in this model the displacement of the ray can be neglected.

If there is a small negative sound speed gradient in the water – for example, if $\varepsilon \sim 3 \times 10^{-5}\,\text{m}^{-1}$ – then the angle corresponding to the fundamental mode is $\chi \approx 0.04$, and then $\Delta \approx 10\,\text{m}$. The ray cycle distance under these conditions is $D \approx 3\,\text{km}$. Consequently, under these conditions the value of Δ is significantly less than D and can be neglected.

The beam displacement is more pronounced if there is a layer of sediments with a positive sound speed gradient $c_1\varepsilon_1$. The smaller this gradient, the larger the beam displacement $\Delta \sim 2\chi/\varepsilon_1$. In this case, the correction to the cycle distance D and, correspondingly, to the mode attenuation coefficient γ is somewhat greater than for an homogeneous bottom, but still remains small in absolute value, since gradients in the bottom are at least an order of magnitude greater than the value of the gradient in the water layer.

The theory studied in this section enables us to estimate the transmission loss of sound in shallow water, due to the interaction with the bottom. For example, for typical values of the bottom parameters $m_1 \approx 2$, $c/c_1 \approx 0.83$, $\alpha \approx 0.02$ for a depth $H \approx 100\,\text{m}$, at frequency $f \approx 100\,\text{Hz}$, for the value of the mode attenuation coefficient (2.88) for the fundamental mode ($l^* \approx 1$) we obtain $\gamma_1 \approx 6.4 \times 10^{-3}\,\text{km}^{-1}$. If we take into account the fact that at large distances from the source, because of differences in the attenuation coefficients, just one fundamental mode remains (in the Pekeris model), then the overall decay of the observed field corresponds to the decay of this mode. In the case in question $\beta = 10\lg\gamma_1 \approx 0.056\,\mathrm{dB/km}$. Similarly, one can estimate the law for the transmission loss in a waveguide with a negative sound speed profile (very common case). Here, at large distances a group of approximately identically decaying modes, with attenuation coefficient determined by equation (2.91), remains. For the earlier example, with gradient of the sound speed profile $\varepsilon \approx 2 \times 10^{-4}\,\mathrm{m}^{-1}$ we obtain $\gamma_1 \approx 10^{-2}\,\mathrm{km}^{-1}$ or $\beta \approx 0.1\,\mathrm{dB/km}$. This value is in complete agreement with the results of experiments in shallow-water regions. We mention also that the analogous value for a deep-water underwater sound channel is considerably less ($\beta \approx 0.003$–$0.007\,\mathrm{dB/km}$) and, since the absorption of sound by the oceanic water medium makes a marked contribution to this decay, which generally takes place in shallow water, the absorption of sound by water in shallow-water channels can be legitimately neglected.

2.4 INCLUSION OF A LAYERED BOTTOM STRUCTURE

In Sections 2.1–2.3 we considered the construction of the sound field of a point source, on the assumption that the bottom is modelled by a fluid, homogeneous absorptive half-space. This assumption is, of course, an idealisation of the real complex structure of the bottom; however, when sound passes along long tracks, where the properties of the bottom vary in a highly arbitrary (sometimes random) manner along the track, such a description (introducing an effective value of the density, the sound speed and the absorption in the bottom) is natural. If, however, we are dealing with comparatively small regions in which a more or less regular (along the track) layer structure of the bottom can be established, which may in principle affect the characteristics of the sound field, the model of the bottom can be made more complicated, based on the assumption that the bottom consists of a certain number of, generally speaking, fluid/elastic, homogeneous absorptive layers, lying one above the other, which corresponds to experimental observations of the bottom structure (see earlier).

Thus, let us consider as the next step, a model of the bottom consisting of homogeneous, fluid, absorptive layers. The spatial region for $z > H$ is divided up by planes $z_n = H_n = H + h_n$, where h_n is thickness of the layer with index n, and we shall suppose that there are N layers, each of which has density ρ_n, sound speed c_n and attenuation coefficient α_n. In this case, the boundary conditions for the normal mode with index l in the water layer have the same form as in (2.25):

$$\psi_l(H) + g_1(\xi)\frac{d\psi_l}{dz}\bigg|_{z=H} = 0 \tag{2.95}$$

where the value of $g_1(\xi)$ is determined from a recursive system with sequentially decreasing indices, which is derived after sewing together the values $\rho_n\psi_l/\psi_l'$ at the boundaries of layers $n-1$ and n:

$$g_n = \frac{\rho_n}{\rho_{n-1}\kappa^{(n)}}\frac{\kappa^{(n)}g_{n+1} - \tanh\kappa^{(n)}h_n}{1 - \kappa^{(n)}g_{n+1}\tanh\kappa^{(n)}h_n} \qquad g_{N+1} = \frac{\rho_{N+1}}{\rho_N\kappa_{N+1}} \tag{2.96}$$

where $\kappa^{(n)} = \sqrt{\xi^2 - k_n^2}$, which is complex valued because k_l is complex. Note that in what follows, we shall also use the quantity

$$\sigma^{(n)} = \sqrt{k_n^2 - \xi^2} = i\kappa^{(n)} \tag{2.97}$$

When $N = 0$, we obtain the case of an homogeneous, fluid bottom, considered earlier. An analogous but more cumbersome relation is obtained for the boundary conditions in the case of fluid/elastic layers.

Let us consider, for example, the case in which there is a single fluid absorptive bottom layer of thickness h_1 (sound speed c_1, density ρ_1, absorption coefficient α_1), lying on a fluid, homogeneous, absorptive half-space (c_2, ρ_2, α_2). To be specific, we shall consider the non-exotic case $\max\{c(z)\} < c_1 < c_2$. In this situation it makes sense to distinguish between modes with phase velocities $\nu_l^{\text{ph}} < c_1$ and those with phase velocities $c_1 < \nu_l^{\text{ph}} < c_2$. The former correspond to modes, propagating primarily in the water layer ('water' modes), where these modes and the behaviour of the eigenvalues are qualitatively little different from those obtained, on the assumption of an homogeneous fluid bottom (except for the situation when $\nu_l^{\text{ph}} \approx c_1$); the latter also penetrate into the bottom layer ('bottom' modes), as a result of which they have, generally speaking, a large absorption coefficient. Thus, at a fixed frequency, there are three groups of modes: water modes (their number can be coarsely estimated using formula (2.49) for the Pekeris model, $M = H/\pi\sqrt{k^2 - k_1^2}$); bottom modes (their number is analogously determined by the sound speed in the underlying layer); and modes of the continuous spectrum. Of course, the presence of the layers is manifest to the greatest extent in the behaviour of the bottom modes, the dependences of whose absorption coefficients on the layer characteristics have certain properties and analysable consequences.

Bottom modes have the form

$$\psi_l(z) = \begin{cases} \psi_l(z) & 0 < z < H \\ \psi_l(H)\cos\sigma_l^{(1)}(z-H) + \dfrac{m_l\psi_l'(H)}{\sigma_l^{(1)}}\sin\sigma_l^{(1)}(z-H) & H < z < H + h_1 \\ \left[\psi_l(H)\cos\sigma_l^{(1)}h_1 + \dfrac{m_1\psi_l'(H)}{\sigma_l^{(1)}}\sin\sigma_l^{(1)}h_1\right] & \\ \times\exp[-\kappa_l^{(2)}(z-H-h_1)] & z > H + h_1 \end{cases} \tag{2.98}$$

where for modes in the water layer, solutions of the equation of type (2.25), the same notation $\psi_l(z)$ is retained, and the equation for determining the eigenvalues is

$$\frac{\psi_l(H)}{\psi_l'(H)} = \frac{m_1}{\sigma_l^{(1)}} \cdot \frac{m_2 \dfrac{\sigma_l^{(1)}}{\kappa_l^{(2)}} + \tan \sigma_l^{(1)} h_1}{m_2 \dfrac{\sigma_l^{(1)}}{\kappa_l^{(2)}} \tan \sigma_l^{(1)} h_1 - 1} \tag{2.99}$$

where $m_n = \rho_n / \rho_{n-1}$. The presence of the trigonometric functions in the formulae is the result of oscillating properties of the characteristics of the bottom modes.

Let us consider in more detail the dependence which is of more interest to us: namely, the behaviour of the mode absorption coefficients as a function of the parameters of the bottom layer. Since, as mentioned earlier, the absorption coefficients are well described by perturbation theory, we shall use an expression for the coefficients γ_l, taking the solution obtained on the assumption that there is no absorption in the bottom layer to be 'unperturbed'. The 'perturbation' in our case will be a correction to the wave vector $k^2 n^2 \alpha_1$. Thus, in first-order perturbation theory

$$\begin{aligned}
\gamma_l &= \frac{k^2 n_1^2 \alpha_1}{m_1 q_1} \int_H^{H+h_1} \psi_l^2(z) dz \\
&= \frac{k^2 n_1^2 \alpha_1}{m_1 q_1} \int_H^{H+h_1} \left[\psi_l(H) \cos \sigma_l^{(1)}(z - H) + \frac{m_1 \psi_l'(H)}{\sigma_l^{(1)}} \sin \sigma_l^{(1)}(z - H) \right]^2 dz
\end{aligned} \tag{2.100}$$

Since the unperturbed solutions are constructed on the assumption that there is no absorption, the eigenfunctions and eigenvalues in this formula are real.

As a simple example, let us consider a waveguide with ideal boundaries (pressure-release surface at $z = 0$ and rigid at $z = H + h_1$). The non-normalised eigenfunctions of the problem have the form:

$$\psi_l(z) = \begin{cases} \cos \sigma_l^{(1)} h \cdot \sin \sigma_l z \\ \sin \sigma_l H \cdot \cos \sigma_l^{(1)}(z - H - h_1) \end{cases} \tag{2.101}$$

and the eigenvalues are obtained from the transcendental equation

$$\tan \sigma_l^{(1)} h \cdot \tan \sigma_l H = m_1 \frac{\sigma_l}{\sigma_l^{(1)}} \tag{2.102}$$

To first order in perturbation theory, according to (3.11) the absorption coefficient is determined by the formula

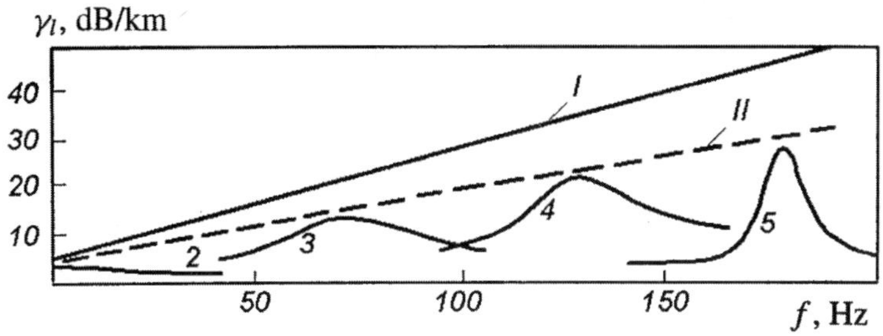

Figure 2.11. Non-monotonic dependence of the mode attenuation coefficients on frequency. I – coefficient of absorption of sound in the bottom layer; II – maximum value of the absorption coefficient in a two-layer waveguide; parameter values for the calculation: $H = 70\,\text{m}$, $h = 30\,\text{m}$, $c = 1500\,\text{m/s}$, $c_1 = 1550\,\text{m/s}$, $\alpha_1 = 0.015$.

$$\gamma_l = \alpha_1 \frac{k^2 n_1^2}{q_l} \times \frac{\sin^2 \sigma_l H \left[1 + \dfrac{\sin 2\sigma_l^{(1)} h_1}{2\sigma_l^{(1)} h_1} \right]}{m_1 \dfrac{H}{h_1} \cos^2 \sigma_l^{(1)} h_1 \left[1 - \dfrac{\sin 2\sigma_l H}{2\sigma_l H} \right] + \sin^2 \sigma_l H \left[1 + \dfrac{\sin 2\sigma_l^{(1)} h_1}{2\sigma_l^{(1)} h_1} \right]} \tag{2.103}$$

As an example, let us calculate the frequency dependence for the case $H = 70\,\text{m}$, $h = 30\,\text{m}$, $m_1 = 2$, $c = 1500\,\text{m/s}$, $c_1 = 1550\,\text{m/s}$. Figure 2.11 shows the behaviour of the attenuation coefficients of modes two to five in the region of the first maxima (referenced from the lowest frequency). The presence of maxima on all these curves is evident, where, as the mode index increases the maxima become increasingly well expressed and shift towards high frequencies, at which the next ones appear. At the same time, their numbers grow with the mode index. For each mode the first maximum is located somewhat to the right of the cut-off frequency, and the value of the absorption coefficient at the maximum approximates to the value of the absorption coefficient of sound in an absorbing medium $kn\alpha_1/2$. This value corresponds to a linear dependence of the absorption coefficient on frequency and is also shown on the figure. The envelope of the family of curves is determined by a linear dependence on frequency with coefficient of proportionality determined by the ratio of the absorbing layer to the total thickness of the waveguide. As the frequency increases further, the bottom mode becomes a water mode and its absorption coefficient approximates to the value in the Pekeris model and tends to zero.

The presence of resonance maxima is associated with the transmitting properties of the absorbing layer, and their number is determined by the relationship between the parameters of the problem (in other words between the layer thickness, and the differences in the sound speed and frequency) (Brekhovskikh (1980)). The dependence on the layer thickness can be constructed in an analogous manner.

Sometimes a non-monotonic dependence of a mode absorption coefficient can be seen from equation (2.103). The absorption coefficient is determined by the proportion of the intensity of the individual mode, confined to the absorbing layer. Of course, the distribution of the eigenfunction with depth varies in the dependence on frequency. If we fix the mode index and change the frequency, then the nature of the oscillations with depth changes, and the absorbing layer may contain a greater or lesser proportion of the function. The absorption coefficient will vary non-monotonically according to the oscillations. Here, we must note that as the frequency grows the phase velocity of the mode decreases and a time is reached when it becomes less than c_1. This means that the mode does not penetrate further into the absorbing layer (there is total internal reflection from the boundary of the absorbing layer) and the nature of the behaviour with increasing frequency approximates more and more to the case of a water layer on a fluid, infinite half-space. The greater the mode index, the greater the frequency corresponding to the boundary for it. The distance between maxima on the curve of the dependence of the mode attenuation coefficient can be coarsely estimated using equation (2.101):

$$\Delta f \approx \frac{\sigma_l c^3}{64 h_1 f \Delta c} \tag{2.104}$$

where $\Delta c = c_1 - c$ is the drop in the sound velocity in the water and the bottom layers. In the case in question, for the fifth mode we obtain a value of 17 Hz for the width of the peak.

The properties of the behaviour of the characteristics of the eigenmodes as functions of the parameters of the bottom layer and also the frequency dependence are determined by the integral (2.103), which is the proportion of the intensity of the mode, in the bottom layer. Since this proportion has a complex, non-monotonic behaviour as the parameters of the layer change, the characteristics of the mode (phase and group velocities, absorption coefficients) will also behave non-monotonically.

A detailed numerical analysis of the behaviour of the characteristics of the waveguide modes in a layered waveguide was given in Ageeva and Krupin (1980, 1981), in which the effect of various factors was considered. We further note that, generally speaking, the bottom layer (layers) may be inhomogeneous with depth, and a large proportion may have a positive gradient of the sound speed, but the model considered here may be used in this case, by splitting the inhomogeneous layer into a certain number of homogeneous layers, as has been done in a number of works (Ageeva *et al.* (1994), Zhuravlev *et al.* (1986), Porter and Reiss (1984, 1985), Porter (1982)).

Sometimes, an inhomogeneous layer can be considered by approximating the dependence of the sound speed on z, for example, by a linear function, when, by constructing the solution in this layer via the Airy function, one can find the solution $g(\xi)$ for the boundary conditions on the bottom, using special functions. Finally, the function $g(\xi)$ can be constructed numerically for various more complex models of the bottom.

Summarising Chapter 2, we can say that in it we have formulated the main comprehensible language in we shall, in what follows, describe the propagation of sound in a shallow-water waveguide. Here, we have obtained the main equations for calculating the characteristics of the field, which will be used to interpret experimental data. This description and the formulae are based on a 'modal' approach, in other words, on decomposition of the sound pressure field into the eigenfunctions (waveguide modes) of the corresponding Sturm–Liouville equation. The solution of this equation gives us the main characteristics of the waveguide modes (propagation constants, phase and group velocities, attenuation coefficients, etc.) which for the conditions of a real waveguide, can only be found numerically. However, using the WKB method (or other approximate methods), we have obtained (albeit approximate) analytic formulae for the waveguide parameters and have given an interpretation of these in the language of rays. Equations for the mode attenuation coefficients are particular important for subsequent understanding of certain effects and of experimental data.

3

Approximate methods in shallow-water acoustics

3.1 PERTURBATION THEORY

The purpose of theoretical calculations is to describe and explain experimentally observed phenomena. Of course, a real ocean environment (sound speed profile, bathymetry of the bottom, etc.) cannot be described by functions admitting an analytic solution, with at least sufficient accuracy for comparison with real experiments. Thus, various approximate methods for solving the sound propagation problem are very commonly used. We can coarsely divide the approximate methods into two classes. In the first class, we include solutions to problems such as corrections to some 'reference' solution, which is obtained within a soluble model, that is in some sense similar to our problem. We shall distinguish three possible situations, depending on the smallness parameters for computation of the corrections: 1) cases when the waveguide parameters (sound speed, bathymetry of the bottom, etc.) exhibit a small difference in absolute value from some 'reference' function which admits a solution; 2) cases when the waveguide parameters vary smoothly and the rate of change of the sound speed profile (depth) along the track is the smallness parameter; 3) another smallness parameter which can be used to construct an approximate method is the violation of cylindrical symmetry (i.e. smallness of the derivative of parameters with respect to the transverse coordinate).

The other class of approximate methods is taken to include cases in which we are not interested in the exact value of the amplitude of the sound field, but seek certain approximate (averaged) characteristics of the latter. In this case, in order to find these characteristics, it may not be necessary to solve exact equations, but to construct approximation algorithms, which may even admit analytic solution.

In case 1) in the calculation of corrections to the 'reference' solution – that is, essentially to the eigenvalues and eigenfunctions – perturbation theory may be used.

Perturbation theory is one of the most common means of calculating eigenfunctions and eigenvalues, in which the main equations for the corrections are represented in various forms (Burov (1991, 1992), Davidov (1973), Turbiner (1984)). In underwater acoustics, it is also used to calculate corrections to the spectrum due to changes in the waveguide characteristics (both in the water layer and in the bottom). Based on perturbation theory, a method for recovering waveguide characteristics (in particular, the sound speed profile and density in the bottom) has been proposed (Rajan *et al.* (1987)). The actual formulae of perturbation theory for shallow water exhibit properties resulting from the non-self-adjoint nature of the Sturm–Liouville problem, due to absorption. This contributes small corrections to the basic formulae in comparison with the known Rayleigh–Schrödinger formulae (Turbiner (1984)). Some works have considered different forms of representation for higher-order perturbative corrections for underwater acoustics (Burov (1982), Gindler (1990)).

To formulate the theory of perturbations, it is convenient to use a model of a shallow-water waveguide with no division into the water and bottom layer, representing the propagation in the medium as a dependence of the wave vector (sound speed) and the density on depth: $k(z), \rho(z)$. To construct a perturbation theory which enables us to take into account the change in density, which is important for shallow-water acoustics, we have to use a wave equation for a medium with variable (in this case depending on just one coordinate) density. The equation for the eigenfunctions in this case will have a somewhat different form to (2.25) (Isakovich (1973)):

$$\frac{d^2\psi_l(z)}{dz^2} + \rho(z)\frac{d}{dz}[\rho^{-1}(z)]\frac{d\psi_l(z)}{dz} + [k^2(z) - \xi_l^2]\psi_l(z) = 0 \tag{3.1}$$

where $\psi_l(0) = 0$ and $\psi_l(z) \to 0$ as $z \to \infty$. The eigenfunctions must satisfy the normalisation condition

$$\int_0^\infty \frac{\rho_S}{\rho(z)}\psi_m(z)\psi_l(z)dz = \delta_{ml} \tag{3.2}$$

where ρ_S is some fixed value of the density (in this case, near the water surface), and δ_{ml} is the Kronecker symbol. To construct the perturbation theory, we make the following substitution of functions in equation (3.1):

$$\tilde{\psi}_l(z) = \sqrt{\frac{\rho_S}{\rho(z)}}\psi_l(z) \tag{3.3}$$

as a result of which, for the function $\tilde{\psi}_l(z)$, we have the equation:

$$\frac{d^2\tilde{\psi}_l(z)}{dz^2} + [\tilde{k}^2(z) - \xi_l^2]\tilde{\psi}_l(z) = 0 \tag{3.4}$$

where

$$\tilde{k}^2(z) = k^2(z) - \frac{1}{2}\sqrt{\frac{\rho(z)}{\rho_S}}\frac{d^2}{dz^2}\sqrt{\frac{\rho_S}{\rho(z)}} \tag{3.5}$$

The functions $\tilde{\psi}_l(z)$ are normalised according to the condition (3.2)

$$\int_0^\infty \tilde{\psi}_m(z)\tilde{\psi}_l(z)dz = \delta_{ml} \tag{3.6}$$

Let $k^0(z)$ and $\rho^0(z)$, respectively, be the 'unperturbed' values of the wave vector and the density of the medium, for which the solution of the Sturm–Liouville problem (3.1) is known, being denoted here by $\psi_l^0(z)$ and ξ_l^0. According to our substitution, we have an unperturbed quantity $\tilde{k}^0(z)$ and spectra $\psi_l^0(z)$ and ξ_l^0. We shall assume that the quantities $k^0(z)$ and $k(z)$ are complex with a small imaginary part (in comparison with the real part). We shall further assume that the quantities $k(z)$ and $\rho(z)$ exhibit only small differences from their unperturbed values (we shall make the criteria for the smallness of these differences more precise in what follows):

$$k^2(z) - [k^0(z)]^2 = \delta k^2(z) \quad \rho(z) - \rho^0(z) = \delta\rho(z) \tag{3.7}$$

and, correspondingly, we have the perturbation $\tilde{k}^2(z) - [\tilde{k}^0(z)]^2 = \delta\tilde{k}^2(z)$, which can be expressed in terms of perturbations of the density $\delta\rho(z)$ and of the sound speed $\delta k^2(z)$, separately.

Using the standard procedure, the following expressions, to second order in the eigenvalues and to first order in the eigenfunctions, can be obtained:

$$\xi_l^2 = (\xi_l^0)^2 + V_{ll} + \sum_{m \neq l} \frac{V_{ml}^2}{(\xi_l^0)^2 - (\xi_m^0)^2} \tag{3.8.1}$$

$$\tilde{\psi}_l(z) = \tilde{\psi}_l^0(z) + \sum_{m \neq l} \frac{V_{ml}}{(\xi_l^0)^2 - (\xi_m^0)^2} \tilde{\psi}_m^0(z) \tag{3.8.2}$$

where we have denoted

$$V_{ml} = \int_0^\infty \tilde{\psi}_m^0(z)\delta\tilde{k}^2(z)\tilde{\psi}_l^0(z)dz = \int_0^\infty \frac{\rho_0}{\rho_0(z)} \psi_m^0(z)\delta\tilde{k}^2(z)\psi_l^0(z)dz \tag{3.9}$$

Let us consider in more detail the correction $\Delta\xi_l = \xi_l - \xi_l^0$ to the horizontal propagation constant of a mode with index l, which in first-order perturbation theory has the form:

$$\Delta\xi_l = \frac{1}{2\xi_l^0} \int_0^\infty \frac{\rho_S}{\rho^0(z)} [\psi_l^0(z)]^2 \delta\tilde{k}^2(z)dz \tag{3.10}$$

Because of absorption, the value of this correction is complex and its imaginary part gives a correction to the mode attenuation coefficient for modes with index l. In particular, we can consider the coefficient of absorption of sound in the bottom, which generally depends on depth, as a perturbation for the reference problem without absorption, that is, $\delta\tilde{k}^2 = ik^2 n_1^2(z)\alpha(z)$ for $z > H$, which gives us mode absorption coefficients:

$$\gamma_l = \frac{k^2}{q_l} \int_H^\infty \frac{\rho_S}{\rho^0(z)} [\psi_l^0(z)]^2 n_1^2(z)\alpha(z)dz \tag{3.11}$$

where q_l denote real eigenvalues of the unperturbed problem. For real parameters of the waveguide, calculations of the mode absorption coefficients according to equation (3.10) produce a completely acceptable accuracy, which makes perturbation theory a convenient means of calculating and studying the absorption of sound in the bottom.

For perturbation theory to be applicable, the corrections calculated from equation (3.10) must be much less than the characteristic distances between eigenvalues. This condition determines the order of magnitude of the quantity $\delta\tilde{k}^2$ which we shall take as a perturbation, or produces bounds for the other parameters of the problem. Of course, this estimate is different in each specific case. Let us consider, for example, the case when the Pekeris model is adopted for the unperturbed waveguide and the perturbation is addition of δc_1 to the sound speed in the bottom, taking place at a depth of from H to $H + \delta H$. If we suppose that the minimum distance between eigenvalues in the Pekeris model is estimated by $\min|\xi_{l+1} - \xi_l| \approx \pi^2/kH^2$, then the condition for applicability of perturbation theory has the form:

$$\frac{\pi^2}{kH^2} \gg \frac{\delta c_1}{c_1} \frac{1 - \exp(-2\delta Hk\sqrt{1 - n_1^2})}{m_1 H\sqrt{n_1^{-2} - 1}} \tag{3.12}$$

The possibility of using perturbation theory clearly depends on many parameters of the problem and, in particular, on the sound frequency. For example, in the case studied by Rajan *et al.* (1987), where $H = 40\,\text{m}$, $c = 1500\,\text{m/s}$, $c_1 = 1900\,\text{m/s}$, $\delta c_1 = 100\,\text{m/s}$, $\delta H = 40\,\text{m}$, $m_1 = 1.6$, the above estimate produces a condition for applicability of perturbation theory for frequencies $f \ll 1800\,\text{Hz}$. By fixing the frequency, one can obtain a condition for applicability for other parameters.

Equation (3.10), as in the work of Frisk (1988) and Rajan (1987), can be viewed from another point of view, if we suppose that the left-hand side of that equation – the spectrum of the problem – is known, while the perturbation of the sound speed profile – the function $\delta\tilde{k}^2(z)$ in the integrand – is unknown. In this case, equation (3.10) emerges as an integral equation for the unknown function $\delta\tilde{k}^2(z)$, or, more precisely, as a system of integral equations. Thus, given some initial model of the waveguide, considered to be unperturbed, and the measured spectrum of the real waveguide, by solving the system of integral equations, one can find the correction $\delta\tilde{k}^2(z)$ which produces the 'true' sound speed profile; that is, one can solve the inverse problem in that formulation. Of course, such a solution is possible only if the original parameters of the problem (and, of course, it is not just the sound speed which may vary) are sufficiently close to the final ones. In other words, for successful application of this algorithm the initial model of the waveguide must be a good 'guess'. Several examples of recovery of unknown functions $c(z)$, $\rho(z)$ of this type were analysed in Rajan *et al.* (1987).

3.2 SOUND FIELD IN A RANGE-DEPENDENT WAVEGUIDE: COUPLED MODES AND ADIABATIC APPROXIMATION

For real shallow-water waveguides, the supposition that the depth, the sound speed profile and the bottom are constants along the track is usually too coarse. In the majority of cases, the variations of the waveguide parameters are very significant. For example, the depth of a waveguide in real experiments (Bunkin *et al.* (1984, 1989), Katsnelson *et al.* (1986)) varies by a factor of two or more. Variations of depth in coastal regions are particularly marked. Variations of the sound speed profile and the bottom properties are less, but nevertheless affect the signal propagation properties. In addition, the variations of the waveguide parameters are often very smooth, which enables us to formulate approximate methods for calculating the sound field in a smoothly irregular waveguide, the main such method being that known in the Russian literature as the cross-sections method (Katsenelenbaum (1961), Pierce (1965), Brekhovskikh (1980), Maltsev (1970)).

Let us consider the case when the waveguide parameters vary with distance; that is, when they are functions of the distance to the source, r. For the moment, we shall assume that the cylindrical symmetry of the problem is retained, and we shall neglect 'cross' or horizontal variations of the waveguide parameters. The effect of these variations will be analysed below. We have

$$H = H(r) \quad c = c(r, z) \quad k = k(r, z) \quad g = g(r, \xi)$$

Other parameters may also depend on r.

The existence of such a dependence means that the field of a point source $\Psi(r, z)$

$$\frac{1}{r}\frac{\partial}{\partial r}\left(r\frac{\partial \Psi}{\partial r}\right) + \frac{\partial^2 \Psi}{\partial z^2} + k^2(r, z)\Psi = 0 \tag{3.13}$$

cannot be represented in the form of a sum of normal modes (2.32) whose amplitudes do not vary with distance. The so-called local modes $\psi_l(r; z)$, which are the solution of the Sturm–Liouville problem (2.25) and in which r enters as a parameter (this will be denoted by the separation of the parameters from the arguments within the brackets by a semicolon), are fundamental for the subsequent analysis:

$$\begin{gathered} \frac{d^2\psi_l(r; z)}{dz^2} + [k^2(r, z) - \xi_l^2(r)]\psi_l(r; z) = 0 \\ \psi_l(r; z)|_{z=0} = 0 \\ \left[\psi_l(r; z) + g(r, \xi_l)\frac{d\psi_l(r; z)}{dz}\right]_{z=H} = 0 \end{gathered} \tag{3.14}$$

The solution of the problem for $z > H(r)$ in the case of a model with a homogeneous fluid bottom can be constructed according to the conditions for continuity at the boundary:

$$\psi_l(r; z) = \psi_l(r; H)\exp\left\{-\sqrt{\xi_l^2(r) - k_1^2}(z - H(r))\right\} \tag{3.15}$$

If the model of the bottom is more complex – for example, a layered bottom – then this function can be constructed according to the theory of Section 2.4. Since the set of eigenfunctions $\psi_l(r;z)$ is a complete system for each fixed value of r, the field $\psi_l(r;z)$ can be decomposed in $\psi_l(r;z)$, where the amplitudes of the decomposition will, generally speaking, depend on r. This means that the field varies as it propagates along the track, as a result of the variation of the modes themselves, depending on r, as on a parameter, and also because of the variation of the specific weight of the modes (or the transfer of energy from one mode to another); that is, because of the mode coupling.

Thus, a solution of equation (13.13) with parameters depending on r at large distances from the source must be sought in the form:

$$\Psi(r,z) = \sum_l \frac{\psi_l(r;z)}{\sqrt{r\xi_l(r)}}\left\{C_l^+(r)\exp\left[i\int_0^r \xi_l(r')dr'\right] + C_l^-(r)\exp\left[-i\int_0^r \xi_l(r')dr'\right]\right\} \tag{3.16}$$

for a wavefunction $\Psi(r,z)$ depending on the horizontal range r and on the depth z. The term with a negative sign in the exponential exponent corresponds to waves propagating towards the source, i.e. reflected waves, which, strictly speaking must also be present in the decomposition. In principle, the summation in (3.16) presupposes integration over the continuous spectrum, which presupposes the taking into account of the possibility of additional attenuation due to the transition of energy to evanescent modes. In the theory considered below, we shall neglect these processes, and also reflection. This means that the summation in (3.16) is limited solely to propagating modes $l \leq M$.

The functions $\psi_l(r,z)$ are orthonormalised:

$$\int_0^\infty \frac{\rho_S}{\rho(r,z)}\psi_l(r;z)\psi_m(r;z)dz = \delta_{lm} \tag{3.17}$$

Substituting (3.16) in (3.13) and taking into account the smoothness of the variation of the waveguide parameters with distance, i.e. rejecting terms containing derivatives with respect to r of the waveguide parameters and also terms with reflected modes (i.e. we assume $C_l^- = 0$), we obtain for the unknown coefficients $C_l = C_l^+$ a system of simultaneous differential equations[1]:

$$\frac{dC_l(r)}{dr} = -\sum_{m=1}^{L}\sqrt{\frac{\xi_m(r)}{\xi_l(r)}}C_m(r)V_{ml}(r)\exp\left\{-i\int_0^r[\xi_l(r') - \xi_m(r')]dr'\right\} \tag{3.18}$$

One consequence of the irregularity of the waveguide (and a cause of the transfer of

[1] If second-order derivatives are taken into account, as well as possible reflection, corresponding to terms with C_l^{-1}, we would obtain a system of second-order differential equations (Brekhovskikh (1980)).

energy from one mode to another) is a non-zero mode coupling coefficient

$$V_{ml}(r) = \int_0^\infty \psi_l \frac{\partial \psi_m}{\partial r} \frac{\rho_S}{\rho(r,z)} dz \tag{3.19}$$

which defines the dependence $C_l(r)$. We note that this simple form for the coupling coefficients arises because in our analysis we are limiting ourselves to a forward-scattering approximation and terms of first-order smallness. This means that in some cases the coupling coefficient may turn out to be non-symmetric, i.e. $|V_{ml}| \neq |V_{lm}|$. A detailed analysis of questions relating to the coupling of modes in different situations can be found in Dacol (1997) and Godin (1998) (see also Boyles (1983) and McDaniel (1978, 1987)). In addition, the formula for V_{ml} enables us to estimate the degree of variation of the mode amplitudes $C_l(r)$. This can be done by integrating the system (3.18) from zero to r, assuming the functions involved are smooth. We obtain

$$C_l(r) = C_l(0) + i \sum_m \sqrt{\frac{\xi_m(r)}{\xi_l(r)}} C_m(r) \frac{V_{ml}(r)}{\xi_{ml}(r)} [\exp(-i\xi_{lm}(r)) - 1] \tag{3.20}$$

where we have denoted $\xi_{ml} = \xi_m - \xi_l$.

If ψ_l depends on r sufficiently smoothly that mode coupling can be neglected, then the sum in (3.20) can be discarded and the coefficients C_l are constants: $C_l = \text{constant} \equiv C_l^0$, and the desired solution is the sum

$$\Psi(r,z) = \sum_l \frac{C_l^0}{\sqrt{r\xi_l(r)}} \psi_l(r;z) \exp\left[i \int_0^r \xi_l(r')dr'\right] \tag{3.21}$$

The value of the coefficients C_l^0 is determined by the conditions at the source. In particular. for a monochromatic (acoustic) point source (2.3)

$$C_l^0 = \frac{\exp[i\pi/4]}{\sqrt{8\pi}} \psi_l(0;z_0) \tag{3.22}$$

We recall that the integrand in (3.21) is complex valued, where its imaginary part gives the mode attenuation coefficient which varies with distance.

For (3.21) to be applicable, as can be seen from (3.20), $|V_{ml}|$ must be small in comparison with $|\xi_{ml}| = |\xi_m - \xi_l|$:

$$\tilde{\kappa} \approx \frac{|V_{ml}|}{|\xi_m - \xi_l|} \ll 1 \tag{3.23}$$

The inequality (3.23) can be given in another form, if we take into account that $1/|\xi_{ml}|$ is of the order of magnitude of the cycle distance in the channel D, and $V_{ml} \approx \tilde{L}_k^{-1}$, where $\tilde{L}_k$ is the characteristic scale of the variation of the waveguide properties (for example, the scale of the variation of the wavenumber $k(r)$). Then, the inequality (3.23) takes the form:

$$\tilde{\kappa} \approx D|V_{ml}| \approx D/\tilde{L}_k \ll 1 \tag{3.24}$$

The parameter $\tilde{\kappa}$ can be called the constant of non-adiabaticity of the waveguide.

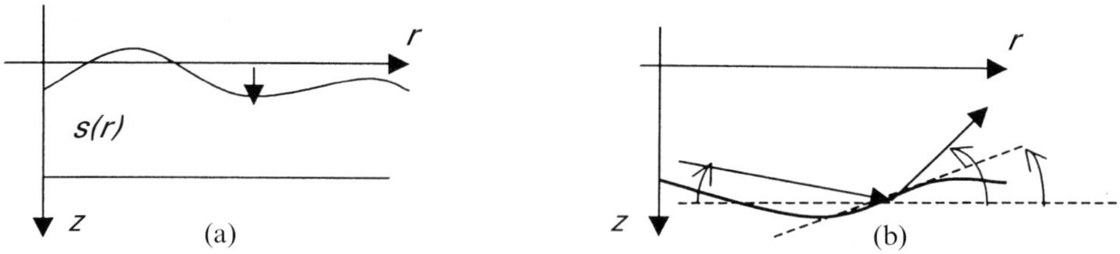

Figure 3.1. Irregular channel with variable depth: (*a*) due to surface waves, (*b*) due to variations in the bottom relief.

We shall estimate the coupling coefficient V_{ml}, resulting from the change in depth due to surface disturbance, within a very simple model in a channel with a constant sound speed (k = constant). This situation is described as follows. We shall assume that the boundaries of our waveguide are a bottom of constant depth and a surface, which depends on the coordinates and on time $s(r,t) \leq z \leq H$ (see Figure 3.1(*a*)), where the function $s(r,t)$, describing the elevation of waves, as a function of distance, is assumed to be smooth. (The dependence on time is assumed to be slowly changing, so that we shall consider the problem in a 'frozen time' approximation and the argument t will be omitted). In this case, the solution of equation (3.14) with the simplest boundary condition on the bottom $\partial\psi_l/\partial z|_H = 0$, corresponding to $g \to \infty$, and on the surface $\psi_l|_{z=s(r)} = 0$, has the form:

$$\psi_l(r;z) = \sqrt{\frac{2}{H - s(r)}} \sin\left[\frac{(l+\frac{1}{2})\pi}{H - s(r)}[z - s(r)]\right] \tag{3.25}$$

Substituting (3.25) in V_{ml}, we find

$$|V_{ml}| = 2\frac{s'}{H}\frac{(l+\frac{1}{2})(m+\frac{1}{2})}{|m-l|(m+l+1)} \tag{3.26}$$

Evidently, the coupling will be maximal for adjacent modes with indices $l = m \pm 1$:

$$|V_{m,m\pm1}| = 2\left|\frac{s'}{H}\right|\frac{(m\pm1+\frac{1}{2})^2}{(2m+1\pm1)} \approx m\left|\frac{s'}{H}\right| \tag{3.27}$$

The coupling of the same modes with index m with distant modes with $|m-l| \gg 1$ has value $|V_{m,\infty}| \approx 2|s'/H|$. For the scale of interferential beatings of rays in the channel $|\xi_{ml}|$ in the case of indices m, l which are not too large (this corresponds to a small angle of inclination of rays), we obtain the estimate:

$$|\xi_m - \xi_l| \approx \frac{\pi^2}{2H^2k}|m-l|(m+l+1) \tag{3.28}$$

The condition for applicability of the adiabatic approximation takes the form

$$\kappa = 2k|s'|H\frac{(l+\frac{1}{2})^2}{(m-l)^2(m+l+1)^2} \ll 1 \tag{3.29}$$

Clearly, this quantity has its greatest value for adjacent modes with indices $|m - l| = 1$, and in this subcase the following inequality must be satisfied:

$$\tilde{\kappa} \approx k|s'H/2| \ll 1 \tag{3.30}$$

We note that in the case when we have an irregular channel with a variable depth $H = H(r)$ (Figure 3.1(b)), the condition for applicability of the adiabatic approximation has the analogous form

$$\tilde{\kappa} \approx k|H'H/2| \ll 1 \tag{3.31}$$

Substituting typical values of $f = 300$ Hz, $H = 200$ m, $|s'| \approx 10^{-3}$, $c \approx 1500$ m/s in (3.31) we obtain the estimate $\tilde{\kappa} \approx 0.1$. If we take the frequency f as variable, then for the same parameters of the channel $\tilde{\kappa} \approx 4 \times 10^{-4} f$, in other words, for a given depth, mode transformations can be neglected up to frequencies $f \leq 10^3$ Hz. For higher frequencies, the coupling of modes must be taken into account.

Considering that H' characterises the angle of inclination of the bottom χ_b and that kH/π is the number of propagating modes M, the adiabaticity parameter κ for the given model of the channel can be written as $\pi M \chi_b$. Naturally, the greater the number of propagating modes, the closer their phase velocities will be and the stronger the couplings between them.

The same idea can be expressed in a different way. When there are M modes propagating in the channel, the average angle between adjacent Brillouin rays $\approx \pi/M$. When a ray inclined at angle χ_b is reflected from the surface, the direction of the ray (grazing angle) changes by $\approx \chi_b$. Clearly, the mode transformation will be insignificant if the change in the grazing angle by $\approx \chi_b$ is less than the angle between Brillouin rays. This condition can be written as $\chi_b \ll \pi/M$, which corresponds approximately to the previous adiabaticity condition.

In the case of a more complicated waveguide model, when not only depth changes with distance, one has to check whether it is possible to use the transverse sections method for a specific dependence of the waveguide parameters on the coordinates.

3.3 HORIZONTAL REFLECTION OF SOUND WAVES IN SHALLOW WATER

When calculating the sound field in a waveguide we previously (Sections 2.1–2.4 and 3.1–3.2) assumed that the problem had a cylindrical symmetry, which enabled us to reduce it to a two-dimensional problem, with coordinates r and z only. In other words, we assumed that in the horizontal plane, the propagation of sound energy from the point source takes place in a straight line, in a radial direction. If the parameters of the waveguide vary smoothly in the radial direction, then the sound field is the sum of independently propagating (or coupling) adiabatic normal modes (or local modes) with coefficients (weights) depending smoothly on the longitudinal coordinate r. As far as the three-dimensional picture is concerned, the wavefronts in this case represent cylindrical surfaces. Clearly, a real waveguide does not have cylindrical symmetry, the sound does not propagate in a straight line (in the horizontal plane) and the wavefronts have a different form. The problem now becomes three-dimensional, and the field at the point of receipt depends on one vertical coordinate (z) and two horizontal coordinates ($\vec{r} = (x, y)$). Assuming the waveguide parameters vary smoothly (depth, sound speed profile) as functions of the horizontal coordinates, the deviation of the propagation direction from a straight line can be taken into account in the so-called theory of vertical modes and horizontal rays (Barridge (1977), Brekhovskikh and Lysanov (1991)).

We shall obtain the main equations under this approach. We shall assume that we have the waveguide model described in Section 2.1 (water layer on a fluid, absorbing half-space) with parameters depending on the coordinates x, y and z. We shall write equations (2.2) and (2.3) in the cartesian coordinate system, assuming a smooth dependence on x and y

$$\left[\nabla_r^2 + \frac{\partial^2}{\partial z^2} + k^2(\vec{r}, z)\right]\Psi(\vec{r}, z) = 0 \tag{3.32}$$

$$\Psi(\vec{r}, z) \to \frac{\exp(ik_0|\vec{R} - \vec{R}_0|)}{|\vec{R} - \vec{R}_0|} \quad \text{as } \vec{R} \to \vec{R}_0$$

where $\nabla_r = (\partial/\partial x, \partial/\partial y)$, $\vec{R} = (x, y, z)$, $\vec{R}_0 = (0, 0, z_0)$.

We shall search for a solution of equation (3.32) in the form of a decomposition into local modes, $\psi_l(\vec{r}; z) = \psi_l(x, y; z)$, depending (unlike the functions in (3.13)–(3.15)) on two horizontal components via the dependence of the waveguide parameters on these components:

$$\Psi(\vec{r}, z) = \sum_l A_l(\vec{r})\psi_l(\vec{r}; z)\exp[i\theta_l(\vec{r})] \tag{3.33}$$

where

$$\begin{gathered}
\frac{\partial^2\psi_l(\vec{r}; z)}{\partial z^2} + [k^2(\vec{r}; z) - \xi_l^2(\vec{r})]\psi_l(\vec{r}; z) = 0 \\
\psi_l(\vec{r}; z)|_{z=0} = 0 \\
\left[\psi_l(\vec{r}; z) + g(\xi_l(\vec{r}))\frac{\partial\psi_l(\vec{r}; z)}{\partial z}\right]_{z=H(\vec{r})} = 0
\end{gathered} \tag{3.34}$$

Substituting (3.33) in (3.32) and neglecting the derivatives with respect to x and y of the functions $\psi_l(\vec{r}; z)$, and also the second derivatives with respect to x and y of $A_l(\vec{r})$, separating the real and imaginary parts, we obtain

$$\begin{aligned}
&\sum_n \Big\{A_l(\vec{r})\left[q_l^2(\vec{r}) - (\nabla_r\theta_l(\vec{r}))^2\right] + i\left[2\nabla_r A_l(\vec{r})\nabla_r\theta_l(\vec{r}) + A_l(\vec{r})\nabla_r^2\theta_l(\vec{r}) + q_l(\vec{r})\gamma_l(\vec{r})A_l(\vec{r})\right]\Big\} \\
&\quad \times \psi_l(\vec{r}; z)\exp\{i\theta_l(\vec{r})\} = 0
\end{aligned} \tag{3.35}$$

We shall assume that $\xi_l(\vec{r}) = q_l(\vec{r}) + i[\gamma_l(\vec{r})/2]$, where $\gamma_l \ll q_l$, whence we obtain the usual eikonal equations for rays in the horizontal plane and a transport equation taking into account absorption in the bottom:

$$(\nabla_r \theta_l)^2 = q_l^2 \tag{3.36.1}$$

$$2\nabla_r A_l \nabla_r \theta_l + A_l \nabla_r^2 \theta_l + q_l \gamma_l A_l = 0 \tag{3.36.2}$$

Equations (3.36) are equations of geometric optics, describing the propagation of a ray with amplitude $A_l(\vec{r})$ and eikonal $\theta_l(\vec{r})$ in the horizontal plane, where the eigenvalue depending on the horizontal coordinates plays the role of the refractive index; therefore, we shall speak of 'horizontal' rays.

Clearly, for each eigenvalue $q_l(x, y)$ we obtain a corresponding eikonal $\theta_l(x, y)$ and amplitude $A_l(x, y)$; in other words, for each eigenvalue and eigenfunction $q_l(x, y)$ and $\psi_l(x, y; z)$ we have a set of possible rays in the horizontal plane. To construct the trajectories of these 'horizontal' rays we can solve the eikonal equation in the usual way, by using, for example, ray coordinates and subsequently solving ordinary differential equations. By introducing the variable s, the arc length of a ray ($ds = \sqrt{dx^2 + dy^2}$), we can obtain equations, determining the trajectory of the ray in the horizontal plane $\vec{r} = \vec{r}(s)$

$$\frac{d\vec{r}_l}{ds} = \vec{\tau}_l \quad \frac{d\vec{\tau}_l(\vec{r})}{ds} = \frac{1}{q_l(\vec{r})} \nabla_\perp q_l(\vec{r}) \tag{3.37}$$

where $\vec{\tau}_l(\vec{r}) = \nabla_{\vec{r}} \theta_l(\vec{r}) / |\theta_l(\vec{r})|$ is the unit vector, tangent to the ray, and the operator $\nabla_\perp = \nabla_{\vec{r}} - \vec{\tau}(\vec{\tau} \nabla_{\vec{r}})$ denotes the gradient in the direction perpendicular to the direction of the ray, i.e. perpendicular to $\vec{\tau}$. Solving equation (3.37) and thereby determining the ray trajectory, by integrating along this trajectory, we can find the eikonal

$$\theta_l(s) = \int_0^s q_l(\vec{r}_l(s))\,ds \tag{3.38}$$

and also the amplitude $A_l(s)$ from the transport equation:

$$A_l^2(s) = A_l^2(0) \exp\left[-\int_0^s \gamma_l(\vec{r}(s))\,ds\right] \tag{3.39}$$

Barridge (1977) and Brekhovskikh and Lysanov (1991) constructed horizontal rays for a point source in a wedge-shape region. In this case, the horizontal rays are hyperbolas. They have caustics, which are also hyperbolas. It makes sense to solve the equations for the trajectory of a horizontal ray in the general case of a shallow-water waveguide using perturbation theory, assuming the smallness of horizontal refraction. The perturbation theory for rays (Kravtsov *et al.* (1984), Kravtsov and Orlov (1990)) for our model of shallow water is constructed as follows. Suppose the source is located at the point with coordinates $(0, 0, z_0)$, and the receiver is at the point with coordinates $(x_1, 0, z_1)$ (see Figure 3.2). We shall assume that the unperturbed horizontal ray is a straight line along the x-axis. The field of the point source, neglecting horizontal refraction, at the point of observation, is determined by the

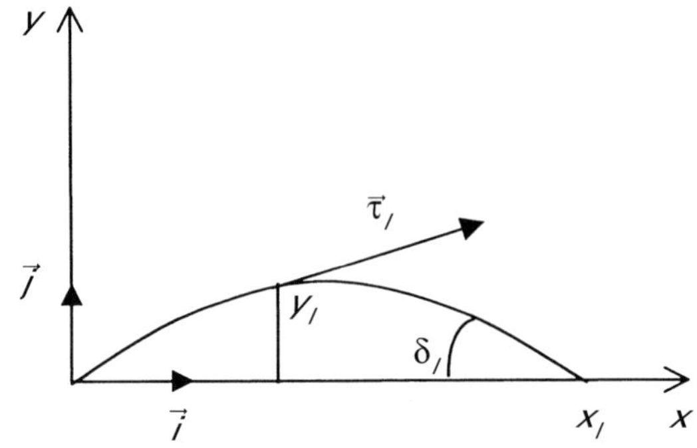

Figure 3.2. Horizontal ray in the (x, y)-plane.

expression (3.21):

$$\psi(x_1, 0, z) = \sum_n A_l(0,0)\psi_l(x_1, 0; z)\exp\left\{\int_0^{x_1}\left[iq_l(x,0) - \frac{\gamma_l(x,0)}{2}\right]dx\right\} \tag{3.40}$$

corresponding to a straight-line horizontal ray $\vec{r}_l^{\,0}(s) = \vec{r}_l^{\,0}(x) = \vec{i}x$, where $\vec{i}$ is a unit vector in the x-direction, $\vec{\tau}_l^{\,0}(x) = \vec{i}$, and the length of the ray is equal to the coordinate $s = x$. We shall determine first-order corrections, due to presence of a dependence on the transverse coordinate y, to the parameters of the problem, the sound speed and the depth of the waveguide. We do this based on calculation of corrections to the function q_l in the integrand and integration of (3.37) and (3.38) along the unperturbed ray (Kravtsov and Orlov (1990)), which is possible, assuming smallness of the displacement of the ray in the horizontal plane. In those cases where the equations obtained are used, this condition is satisfied. In the case when the horizontal displacements are very large, a more accurate form of perturbation theory has to be used (Kravtsov and Orlov (1990)), when the corrections are found from previous integration along the constructed 'perturbed' ray.

Thus, the corrections to $q_l(\vec{r})$ are determined by the equation:

$$q_l(\vec{r}) = q_l^0(\vec{r}) + q_l^1(\vec{r}) = q_l(x,0) + \left.\frac{\partial q_l}{\partial y}\right|_{y=0} y = q_l(x) + q_{ly}(x)y \tag{3.41}$$

where we have denoted

$$q_{ly}(x) = \left.\frac{\partial q_l(x,y)}{\partial y}\right|_{y=0} \qquad q_l(x) = q_l(x,0) \tag{3.42}$$

and corresponding corrections to the eikonal and the ray trajectory will be sought in the form:

$$\begin{aligned}
\theta_l(\vec{r}) &= \theta_l^0(\vec{r}) + \theta_l^1(\vec{r}) \\
\vec{r}_l(s) &= \vec{i}x + \vec{r}_l^{\,1}(x) \\
\vec{\tau}_l(s) &= \vec{i} + \vec{\tau}_l^{\,1}(x) \\
\theta_l^0(\vec{r}) &= \int_0^x q_l(x', 0)dx'
\end{aligned} \tag{3.43}$$

We substitute these expressions into the ray equation, taking into account the fact that the unperturbed ray coincides with the x-axis and the transverse gradient is determined by the partial derivative with respect to y.

For first-order corrections, we have the equation:

$$\frac{d^2\vec{r}_l^{\,1}(x)}{dx^2} = \vec{j}\nu_l(x) \tag{3.44}$$

where $\nu_l(x) = q_{ly}(x)/q_l(x)$ ($\vec{j}$ is a unit vector along the y-axis), whence, after integrating twice we obtain

$$\vec{r}_l^{\,1}(x) = \vec{r}_l^{\,1}(0) + \vec{C}x + \vec{j}\int_0^x dx' \int_0^{x'} dx''\nu_l(x'') = \vec{r}_l^{\,1}(0) + \vec{C}x + \vec{j}\int_0^x (x - x')\nu_l(x')dx' \tag{3.45}$$

We determine the constants $\vec{r}_n^{\,1}(0)$ and $\vec{C}$ from the conditions that our ray must start at the point $(0, 0)$ in the (x, y)-plane and arrive at the point of observation $(x_1, 0)$; that is, $\vec{r}_l^{\,1}(0) = \vec{r}_l^{\,1}(x_1) = 0$. Then

$$\vec{r}_l^{\,1}(x) = -\vec{j}\frac{x}{x_1}\int_0^{x_1} (x_1 - x')\nu_l(x')dx' + \vec{j}\int_0^x (x - x')\nu_l(x')dx' \tag{3.46}$$

Using this expression, we can find the first-order correction to the eikonal at the point of observation. This correction is obtained after substituting in (3.38) the expression for the horizontal ray, obtained taking into account the correction (3.46):

$$\theta_l(x_1, 0) = \int_0^{x_1} q_l(\vec{r}_l^{\,0} + \vec{r}_l^{\,1})dx = \int_0^{x_1} q_l(x)dx + \int_0^{x_1} q_{ly}(x)|\vec{r}_l^{\,1}|dx \tag{3.47}$$

After substituting the value of $\vec{r}_n^{\,1}$ into the second integral, we obtain for the correction to the eikonal and the correction to the tangent vector:

$$\theta_l^1(x_1) = \int_0^{x_1} q_{ly}(x)\left[-\frac{x}{x_1}\int_0^{x_1} (x_1 - x')\nu_l(x')dx' + \int_0^x (x - x')\nu_l(x')dx'\right]dx \tag{3.48}$$

$$\vec{\tau}_l^{\,1}(x_1) = \left.\frac{d\vec{r}_l^{\,1}(x)}{dx}\right|_{x=x_1} = -\vec{j}\frac{1}{x_1}\int_0^{x_1} (x_1 - x)\nu_l(x)dx \tag{3.49}$$

The value of the last vector determines the angle of refraction of the horizontal ray at the point of observation $(x_1, 0)$:

$$\delta_l = \frac{1}{x_1}\int_0^{x_1}(x_1 - x)\nu_l(x)dx \tag{3.50}$$

From these formulae it is clear that the corrections are determined by integrals along the unperturbed ray of a quantity of the type of a transverse derivative of waveguide parameters. Of course, if the waveguide does not have a predominant slope of the bottom, for example, or change in sound speed, then these perturbations will be to some extent compensated on integration along a ray and the corrections at the point of receipt will be insignificant. If the same waveguide had predominant transverse gradients (for example, if the track passes along a coastline, where there is a wedge, or along a temperature front) then the perturbations will accumulate and the corrections to the eikonal may be significant. In particular, if we have a soliton-like internal wave, with a long wavefront, then as sound propagates along the front, such refraction may even lead to the formation of a waveguide in the horizontal plane (Katsnelson and Pereselkov (2000)).

We can estimate the characteristic values of the horizontal refractive indices for typical shallow-water conditions. Suppose that along the whole track the transverse gradient of the sound speed profile and the depth of the water are more or less constant. In this case, we shall estimate the values of the angle of horizontal refraction, the values to be added to the eikonal (additional phase incursion) and also of the maximum horizontal deflection of a ray from the track for modes with index l.

If the values of $\nu_l(x) \approx \bar{\nu}_l = \text{constant}$ due to transverse variations of the waveguide parameters are taken outside the integrals of (3.48)–(3.50), estimates of the lateral deflection of the ray $y_l(x)$ can be obtained:

$$y_l(x) \approx \frac{\bar{\nu}_l}{2}x(x - x_1) \tag{3.51}$$

In this approximation, a horizontal ray has a parabola shape. A ray in the horizontal plane is deflected from a straight line (along the x-axis) towards decreasing values of q_l as a function of y; in other words, in the opposite direction to the transverse gradient of q_l. Since the change in an eigenvalue as the sound speed profile or the depth varies depends on the index and may be both positive and negative for different indices, horizontal rays corresponding to different mode indices may be deflected from a straight line in different directions. The maximum deflection of a ray trajectory from the direction of the track is equal to

$$\bar{y}_l = \frac{\bar{\nu}_l x_1^2}{8} \tag{3.52}$$

The additional phase incursion (of the eikonal) due to horizontal refraction can be estimated from the formula

$$\theta_l \approx \frac{q_{ly}^2}{12q_l}x_1^3 \tag{3.53}$$

We note that the supplement to the eikonal depends on the mode index, which in a real waveguide may lead to a change in the field interference pattern at the receiving array in comparison with the case when there is no horizontal refraction. We can also estimate the angle in the horizontal plane between a ray reaching the point of receipt and the direction of the sensing track:

$$\delta_l = -\frac{1}{x_1}\int_0^{x_1}(x_l - x)\nu_l(x)dx \approx -\frac{q_{ly}x_1}{2q_l} \tag{3.54}$$

The effect of variations in the waveguide depth and the sound speed profile can be estimated separately using a number of models. For example, assuming that waveguide depth varies ($H = H(y)$) for a constant sound speed $q_l = \sqrt{k^2 - \sigma_l^2}$, $\sigma_l = l^*\pi/H$, where $q_{ly} = (\sigma_l^2/q_l)(H_y/H)$, $(H_y = \partial H/\partial y)$. Taking into account what was said earlier about the direction of deflection of a ray, it is clear that a ray (emanated from the point $x = 0$ and reached $x = x_1$) in the horizontal plane is deflected in the direction of decreasing depth. Substituting the transverse gradient of the eigenvalues in formulae (3.52)–(3.54) we obtain

$$\bar{y}_l \approx \frac{\pi^2 l^2}{8k^2H^2\tilde{L}_H}x_1^2 \quad \theta_l \approx \frac{x_1^3}{12k}\left(\frac{\pi^2 l^2}{H^2k\tilde{L}_H}\right)^2 \quad \delta_l \approx \frac{x_1}{2\tilde{L}_H}\frac{\sigma_l^2}{q_l^2} \tag{3.55}$$

where we have denoted $\tilde{L}_H = H/H_y$, being the characteristic scale of the change in depth. We can estimate the values of the parameters obtained in typical shallow-water conditions. For example, if $H = 100\,\text{m}$, $f = 300\,\text{Hz}$, $H_y = 10^{-3}$, $x_1 = 30\,\text{km}$, we obtain, that the deflection of a horizontal ray with index l from a straight line is approximately equal to $\bar{y}_l \approx l^2$ (m), the phase increment is $\theta_l \approx 2 \times 10^{-4} l^4$ (rad), and the angle at which the horizontal ray reaches the point of observation is $\delta_l \approx 10^{-4} l^2$ (rad). If we introduce the grazing angle of a vertical ray with the bottom $\chi_l \approx \sigma_l/q_l$ into the ray approximation, then the estimates for the given parameters of horizontal refraction can be expressed in terms of this angle $\delta_l = (x_1/2L_H)\chi_l^2$.

Clearly, horizontal rays corresponding to higher modes experience greater horizontal refraction, where the dependence on the mode index in this example is very sharp. Let us now consider a waveguide with a negative gradient of the sound speed and calculate the horizontal refraction for rays corresponding to near-bottom modes. Suppose we have a sound speed profile $c(y; z) = c_S[1 - \varepsilon(y)z]$ or $k^2(y, z) = k_S^2[1 + 2\varepsilon(y)z]$, where c_S is the speed on the surface and c_H is the speed near the bottom. The eigenvalues of near-bottom rays (which do not graze the surface) are determined from the WKB formulae

$$q_l = \sqrt{k_H^2 - \sigma_l^2} \quad \sigma_l = \left[3\left(l + \frac{1}{2}\right)\pi\varepsilon(y)k_S^2\right]^{1/3} \tag{3.56}$$

Hence $q_{ly} = -(\sigma_l^2/3q_l)(\varepsilon_y/\varepsilon)$. If, analogously to earlier, we introduce the spatial scale of the change of sound speed in the transverse direction (it is actually the spatial scale of the change in gradient) $\tilde{L}_c = c/c_y = a/a_y$ and the grazing angle of a ray with the bottom, then for the horizontal angle of refraction we obtain the estimate $\delta_l \approx -(x_1/3\tilde{L}_c)\chi_l^2$. It is clear that, as in the previous case, modes with high indices deviate further, the direction of deviation of a horizontal ray coincides with the direction of increasing gradient; that is, where rays are more strongly held to the bottom.

In conclusion, we note that, as the above estimates show, for typical shallow-water conditions, for low (weakly decaying) modes on tracks of several tens of kilometres, the horizontal angle of refraction may reach several degrees. The sound speed profile (gradient) may vary markedly at these distances, although it is true that when these quantities vary simultaneously their mutual compensation or reinforcement may play a role. In Doolitle *et al.* (1988) horizontal refraction was measured experimentally in a coastal wedge region. The inclination was very significant, with $H_y \approx 10^{-2}$, so that the horizontal angle of refraction amounted to several degrees. Let us now consider the interesting case when horizontal refraction may be significant in conditions when a single internal wave propagates along a wavefront. The displacement of the surface of constant density $\zeta(\vec{r}, t)$, due to oscillations in the vertical plane, leads to a corresponding change in the sound speed profile. For soliton-like internal waves, this displacement, in a first approximation, can be described by a function $\zeta(y - \nu t)$ corresponding to the propagation of a planar front in the direction y with velocity ν. In real conditions, we may have $|\zeta| \sim 5\,\mathrm{m}$, and the velocity itself is very small, $\sim 1\,\mathrm{m/s}$, which enables us to consider the sound propagation problem in a 'frozen time' approximation, fixing the value of t (for example $t = 0$). If the function $\zeta(y)$ is periodic (or quasi-periodic) then regions of space corresponding to neighbourhoods of the minimum ($y_{\min}$) or maximum ($y_{\max}$) of the function $\zeta(y)$ represent poles in the horizontal plane, parallel to the x-axis. In the first case, we have conditions, when, because of horizontal refraction, horizontal rays are deflected towards the axis $y = y_{\min}$; that is, there is a focussing of horizontal rays, which for a sufficiently long wavefront, may lead to the formation of the waveguide in the horizontal plane. The region in the neighbourhood of $y = y_{\max}$ corresponds to an anti-waveguide in the horizontal plane. Figure 3.3 shows the results of calculations of the ray pattern in the horizontal plane, corresponding to displacements of the surface of constant density in the region of the thermocline, also shown on the figure, in accordance with the results of the work of Katsnelson and Pereselkov (2000).

To conclude this section, we note that this method has been developed further in Collins (1993), where the calculation of the field in the horizontal plane is performed using a parabolic equation technique, which provides an increase in accuracy and eliminates the caustic problem. For the application of this approach, see also Katsnelson and Pereselkov (2000).

3.4 CORRECTIONS TO THE ADIABATIC APPROXIMATION; GENERALIZED MODES OF AN IRREGULAR WAVEGUIDE

The system of equations (3.18) or the analogous system obtained in another approximation, determines the behaviour of sound in an irregular waveguide. The field,

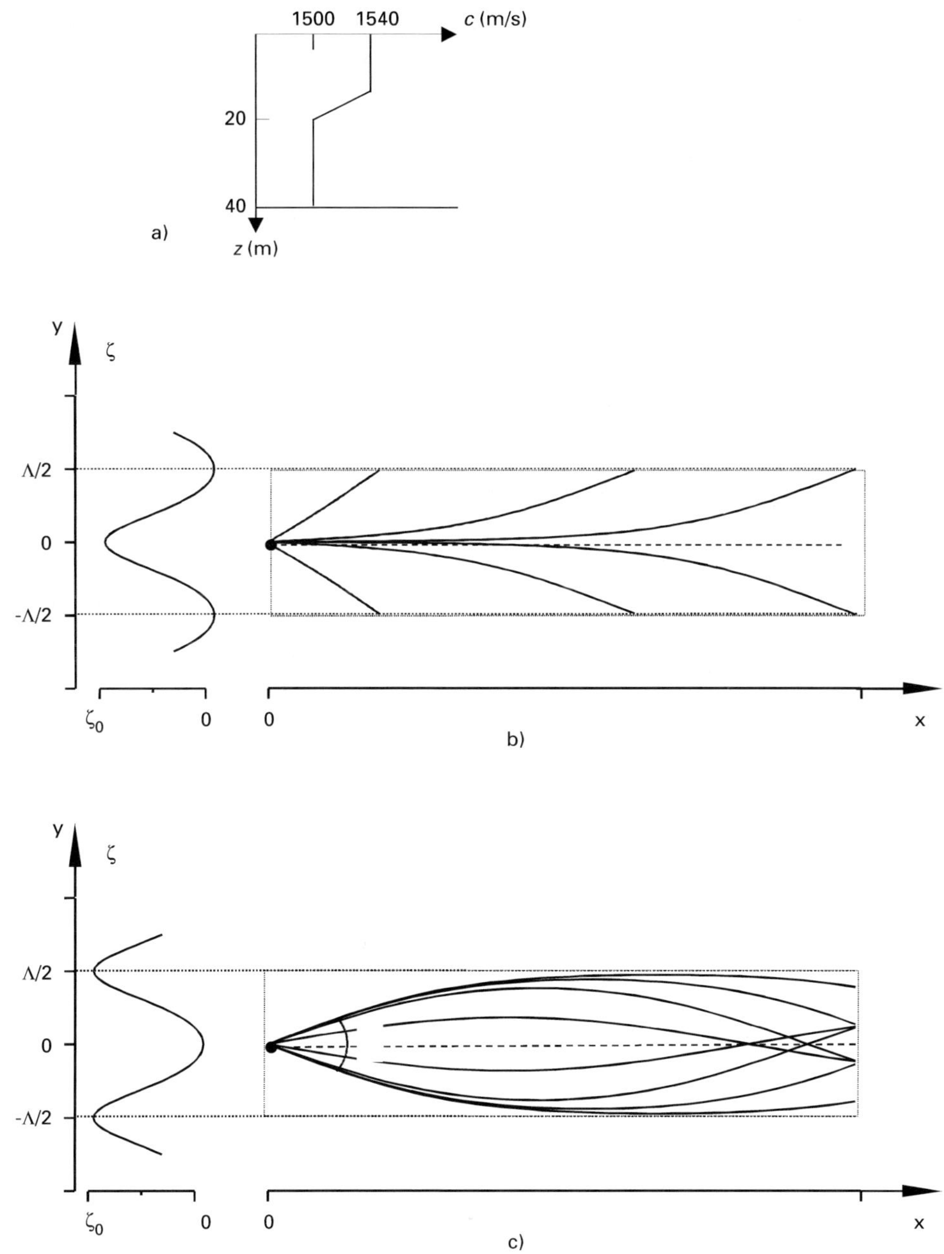

Figure 3.3. Pattern of horizontal rays corresponding to the first vertical mode of the waveguide in the presence of perturbations due to an internal wave with amplitude $\zeta(y) = (\xi_0/2)[1 + \cos(2\pi(y - y_0)/\Lambda)]$. Parameter values $\Lambda = 400\,\text{m}$, $\zeta_0 = 5\,\text{m}$, frequency $f = 300\,\text{Hz}$. The sound speed profile of the unperturbed waveguide is shown in (*a*). (*b*) corresponds to the value $y_0 = 0$, i.e. to the anti-waveguide (defocussing) in the horizontal plane. (*c*) corresponds to $y_0 = \pi$, i.e. focussing (waveguide formation) in the horizontal plane.

constructed based on solution of this system in the form (3.21) corresponds to neglecting mode couplings. However, if the non-adiabaticity parameter $\tilde{\kappa}$ ((3.23) and (3.24)), which determines the value of mode couplings, is not small, then in a particular approximation, it will be necessary to solve the system (3.18) or a similar system. This has been done by many authors (Graves *et al.* (1975), McDaniel (1978), Nagl *et al.* (1978), Pierce (1965, 1982), Williams (1980)), using various methods. If, based on physical considerations, one restricts oneself to a finite number of coupling modes (McDaniel (1978), Parsons *et al.* (1996)), then the system of differential equations obtained can be solved numerically. In McDaniel (1978), for example, in a channel with constant sound speed and linearly varying depth, the coupling of three modes was considered numerically, for parameter values $\tilde{\kappa} \sim 1$–2. It was shown that even at distances of ~ 1 km, the mode amplitudes $|C_l|$ vary considerably. These results were used to analyse the passage of waves through an irregular transitional section between two regular waveguides.

Another technique involves solving the system (3.18) asymptotically, taking into account the coupling of all modes, in some approximation. Such a calculation was performed, for example, in Gurevich and Tsedilina (1973) and Katsnelson *et al.* (1986), based on the construction of so-called generalized adiabatic modes or generalized modes (GM). These modes are linear combinations of local modes in the irregular section and coincide with the local modes in the regular section. GM are solutions of an equation of parabolic type (not to be confused with the parabolic equation technique) and can be constructed in analytic form for model waveguides. This was done, for example, in Gurevich and Tsedilina (1973), where GM were constructed for a characteristic channel for the ionosphere, with a parabolic profile of the dielectric constant in the transverse direction and a randomly varying profile in the longitudinal direction. In Katsnelson *et al.* (1986), GM were constructed for characteristic channels for oceanic waveguides; with a constant sound speed in a transverse section and a channel depth varying along the track, and also with a linearly varying sound speed in a transverse section and a variable channel depth.

Below, we study the theory of GM and provide examples of a number of definitive results from their calculation.

Let us again consider equation (3.13) with the corresponding boundary conditions. We shall search for a solution of this equation in a form analogous to (3.16), neglecting at once reflected waves:

$$\Psi(r,z) = \sum_{l=1}^{M} \frac{a_l(r)}{\sqrt{r\eta_l(r)}} \Phi_l(r,z) \exp\left[i\int_0^r \eta_l(r')dr'\right] \tag{3.57}$$

where the form of the function $\Phi_l(r,z)$ and the value of $\eta_l(r)$ are not yet firmly established. Substituting (3.57) in (3.13), for $a_l(r), \Phi_l(r,z), \eta_l(r)$, we obtain the equation

$$\sum_{l=1}^{M}\frac{\exp(i\int\eta_l dr)}{\sqrt{\eta_l r}}\left\{\left(2i\eta_l\frac{\partial\Phi_l}{\partial r}+\frac{\partial^2\Phi_l}{\partial z^2}+[k^2(r,z)-\eta_l^2(r)]\Phi_l\right)a_l+2i\eta_l\Phi_l(r,z)\frac{da_l}{dr}\right\}$$
$$+\sum_{l=1}^{M}\frac{\exp(i\int\eta_l dr)}{\sqrt{\eta_l(r)}}\left\{\frac{d^2a_l}{dr^2}\Phi_l-\frac{d\eta_l}{dr}\frac{da_l}{dr}\frac{\Phi_l}{\eta_l}+\frac{2da_l}{dr}\frac{\partial\Phi_l}{\partial r}\right.$$
$$\left.-\sqrt{\eta_l}\frac{d}{dr}\left(\frac{1}{2(\eta_l)^{3/2}}\right)a_l\Phi_l-\frac{d\ln\eta_l}{dr}\frac{\partial\Phi_l}{\partial r}a_l+\frac{\partial^2\Phi_l}{\partial r^2}a_l\right\}=0 \tag{3.58}$$

We exclude the second sum from consideration in this expression, since it is of a higher order of smallness than the first. This approximation will be analysed below. Let us now choose the function $\Phi_l(\tau,z)$ and the quantities $\eta_l(r)$ such that the following equation is satisfied

$$2i\eta_l\frac{\partial\Phi_l}{\partial r}+\frac{\partial^2\Phi_l}{\partial z^2}+[k^2(r,z)-\eta_l^2(r)]\Phi_l=0 \tag{3.59}$$

setting the expression in curly brackets in the first sum of (3.58) to zero. Since at this stage there is some degree of freedom associated with the arbitrary function $\eta_l(r)$, we shall construct the function $\Phi_l(\tau,z)$ such that the normalisation condition is satisfied:

$$\int_0^\infty\frac{\rho_S}{\rho(z)}\Phi_l\Phi_n^* dz=\delta_{nl} \tag{3.60}$$

The solutions of equation (3.59) for different l will be called generalized modes and the quantities $\eta_l(r)$ generalized eigenvalues. In this case, the sum of (3.57) gives a decomposition of the field into generalized modes, where the quantities $a_l(r)$ serve as the coefficients of this decomposition. We note that if for the functions $\Phi_l(r,z)$ and $\eta_l(r)$ we choose normal waves and eigenvalues of equation (3.13), then setting the first sum in (3.58) to zero leads to equations (3.18) of the transverse sections method for $a_l(r)$. In our case in this approximation, i.e. neglecting the second sum in (3.58), we obtain equations $da_l/dr=0$ for $a_l(r)$. Hence we have $a_l(r)=a_l(0)=\text{constant}$, i.e. in this approximation, generalized modes do not couple with each other, and

$$\Psi(r,z)=\sum_{l=1}^{N}\frac{a_l(0)}{\sqrt{r\eta_l(r)}}\Phi_l(r,z)\exp\left[i\int_0^r\eta_l(r')dr'\right] \tag{3.61}$$

In order to establish the connection between local modes $\psi_l(r,z)$ and generalized modes, the function $\Phi_l(r,z)$ should be represented in the form of a decomposition into an orthonormalised system of local modes $\psi_l(r,z)$ with unknown coefficients:

$$\Phi_l(r,z)=\sum_m\sqrt{\frac{\eta_l(r)}{\xi_m(r)}}C_m^{(l)}(\tau)\psi_m(r;z)\exp\left(i\int_0^r(\xi_m-\eta_l))dr\right) \tag{3.62}$$

The coefficients of this decomposition are the quantities $C_m(r)$ introduced earlier, which are governed by the system of equations (3.18). Therefore, one can say that generalized modes are solutions, which take into account all normal local modes. The 'width' of the decomposition (3.62), i.e. the number of coefficients $C_m(r)$, with an appreciable weight present in the decomposition (3.62), depends on the parameter $\tilde{\kappa}$ and on the mode index. The width grows as the parameter $\tilde{\kappa}$ increases, i.e. as the 'irregularity' increases and as the mode index l grows, since higher local modes couple more strongly with one another. This can be seen from the value of the mode coupling coefficient V_{ml} (3.19) and in particular from the specific case (3.26).

To solve equation (3.59) or to construct generalized modes based on a solution of the system, it is necessary to assign conditions at $r = 0$. We shall assume that at the start of the track, an GM coincides with a local mode: $\Phi_l(0, z) = \psi_l(0; z)$; which corresponds to the 'initial' condition $C_m^{(l)}(0) = \delta_{lm}$. If the parameters of the waveguide remain unchanged, then in the regular section GM coincide with local modes, since $\partial\Phi_l/\partial r = 0$ where the value $C_m^{(l)}(r) = \delta_{ml}$ remains as before. The coefficients of the decomposition a_l in (3.57) and (3.61) coincide with the coefficients of the decomposition of the field in the regular section into normal modes, and if we consider the field after passing through the irregular section, (3.61) again becomes a decomposition into normal modes, and although the corresponding coefficients $C_m^{(l)}$ again become constant, their value changes in the irregular section (unlike the coefficients a_l). In this connection we note the following.

While a mode excited in the regular (left-hand) section of the waveguide passes through the irregular section, at the right-hand end (also regular) of the waveguide the field is already a combination of waveguide modes, characterised as mode conversion. This phenomenon is widely studied in the literature (McDaniel (1978), Parsons (1994), Williams (1980)). A local mode, which coincides with a waveguide mode in the left-hand section, also coincides with this same waveguide mode to the right of the irregular section.

This means that actual conversion of modes is not taken into account in the construction of local modes, hence their name.

We further note that when constructing GM (i.e. solutions of equation (3.59)) as the functions $\eta_l(r)$ we can take the eigenvalues $\xi_l(r)$, obtained from the solution of equation (3.14), although the function $\Phi_l(r, z)$ obtained then does not even necessarily satisfy the normalisation condition.

Conditions for the applicability of the expression (3.21) have already been discussed: these are the inequalities (3.23) and (3.24). The terms which are rejected when expression (3.21) is used have order $\tilde{\kappa} \sim D|V_{ml}|$. Terms of this order of magnitude are taken into account in the construction of the generalized modes.

Let us now determine the condition for applicability of equation (3.61), for which we shall estimate the order of magnitude of the terms to be rejected. The contribution of the rejected sum in (3.58) to the value of the coefficient $a_l(r)$ can be calculated if we introduce a system of equations for $a_l(r)$ from (3.58), similarly to what was done for the coefficients C_l in (3.18). Performing the calculations, we find

$$2i\dot{a}_l\left(\eta_l - \frac{i\dot{\eta}_l}{2\eta_l}\right) + \dot{a}_l + \eta_l\left(\dot{\eta}_l/2(\eta_l)^{3/2}\right)a_l$$

$$= \sum_m \sqrt{\frac{\eta_m}{\eta_l}} \exp\left(i\int_0^r (\eta_m - \eta_l)\,dr\right)\left[2V_{ml}\dot{a}_l + \left(F_{ml} - \frac{\dot{\eta}_m}{\eta_m}V_{ml}\right)a_l\right] \quad (3.63)$$

$$V_{ml} = \int \Phi_l \frac{\partial\Phi_m}{\partial r}\frac{\rho_0}{\rho(z)}\,dr, \quad F_{ml} = \int \Phi_l \frac{\partial^2\Phi_m}{\partial r^2}\frac{\rho_0}{\rho(z)}\,dz$$

where a dot above a symbol denotes differentiation with respect to r.

Integrating the system (3.63) approximately, taking into account the most significant terms, we obtain an estimate for the error due to the presence of the right-hand side of (3.63)

$$|\delta_{a_l}| \sim \frac{1}{\eta_l}\frac{1}{|\eta_m - \eta_l|}\frac{\dot{\eta}_l}{\eta_l}V_{ml} \sim \left(\frac{D}{\tilde{L}_k}\right)^2 \quad (3.64)$$

The condition for applicability of the GM approximation takes the form

$$\nu \sim \left(\frac{D}{\tilde{L}_k}\right)^2 \ll 1 \quad (3.65)$$

In accordance with this estimate, for a channel with constant speed and variable depth we have

$$|\delta a_l| \sim (lH'/kH)^2 \quad (3.66)$$

where, H' denotes derivative of H with respect to r.

Comparing this expression with $\tilde{\kappa} \sim kHH'/2$, we see that corrections to the solution for the adiabatic approximation are very much greater than the corrections to the GM. This corresponds to the known fact that mode transformations have an exponentially small non-adiabaticity parameter $\tilde{\kappa}$.

For concrete parameters ($H \sim 100$ m, bottom slope $\delta \sim 5^\circ$, frequency 100 Hz), the parameter $\tilde{\kappa}$ is equal to 5, whereas ν amounts to only 10^{-3}. Consequently, the generalized modes in this case describe the field very well, whereas the decomposition into local modes (3.21) admits a very large error. Similar considerations also hold for the conditions of the experiment of Parsons *et al.* (1994).

For some models of an underwater sound channel with parameters varying along the track, the generalized modes can be constructed explicitly (Gurevich and Tsedilina (1973), Katsnelson *et al.* (1986)).

As an example, let us construct GM for a model of a waveguide with a constant sound speed $c = \text{constant}$, and depth $H(r)$ varying along the track. For this, we need to solve the equation (3.59) with boundary conditions for the function $\Phi(r,z)$:

$$\Phi|_{z=0} = \partial\Phi/\partial n|_{z=H(r)} = 0 \quad (3.67)$$

where $\partial\Phi/\partial n$ is the derivative with respect to the normal to the surface $H(r)$. Making the replacement of variables in equation (3.59) $\zeta = z/H(r)$, $0 \leq \zeta \leq 1$, we can obtain, modulo terms $\sim HH'k$

$$\Phi_l = \frac{1}{\sqrt{H(r)}} \exp(i\kappa_l(r)\zeta^2)\tilde{\Phi}_l \tag{3.68}$$

where $\kappa_l(r) = \eta_l H' H/2$, and the function $\tilde{\Phi}_l$ satisfies the equation:

$$\frac{d^2\tilde{\Phi}_l}{d\zeta^2} + (k_0^2 - \eta_l^2)\tilde{\Phi}_l = 0 \tag{3.69}$$

Choosing the function $\eta_l(r)$ in the form $\eta_l(r) = \sqrt{k^2 - (l^*\pi/H(r))^2}$, we obtain for $\tilde{\Phi}_l$ the equation

$$\frac{d^2\tilde{\Phi}_l}{d\zeta^2} + (l^*\pi)^2\tilde{\Phi}_l = 0 \tag{3.70}$$

with boundary conditions $\tilde{\Phi}_l(0) = \partial\Phi_l/\partial\zeta|_{\zeta=1} = 0$. In deriving the last expression, $\partial/\partial n$ is expressed in the form:

$$\frac{\partial}{\partial n} = (n\nabla) = \frac{\partial}{\partial r} - \frac{H'}{H}\zeta\frac{\partial}{\partial\zeta} - \frac{1}{HH'}\frac{\partial}{\partial\zeta} \tag{3.71}$$

and the slope angle is assumed to be small, $\delta \approx H'$. Then, for the function $\tilde{\Phi}_l$, we shall have:

$$\tilde{\Phi}_l(\zeta) = \sin(\pi l^*\zeta) \quad l^* = l - 1/2 \quad l = 1, 2, 3, \ldots \tag{3.72}$$

and, finally,

$$\Phi_l = \frac{\sin(\pi(l - 1/2)\zeta)}{\sqrt{H(r)}} \exp(i\kappa_l(r)\zeta^2) \tag{3.73}$$

A generalized mode is a packet of local modes, where the width of the packet in this case is determined by the parameter κ_l.

GM can also be constructed for a more complex model of the waveguide (near-bottom sound channel) with a linear profile $c(r, z) = c_0(r)(1 - \varepsilon(r)z)$ and variable depth $H = H(r)$. In this case in the equation of parabolic type (3.59), we make the replacement of variables

$$\zeta = (k_0^2\varepsilon)^{1/3}(z - H(r)) = (z - H)/H_0, \quad H_0 = (k_0^2\varepsilon)^{-1/3} \tag{3.74}$$

Then, one can obtain that

$$\Phi_l(r, \zeta) = \varepsilon^{1/6} \exp[i\nu_l\zeta + i\kappa_l\zeta^2]\tilde{\Phi}_l(\zeta) \tag{3.75}$$

where $\tilde{\Phi}_l(\zeta)$ satisfies the Airy equation

$$\frac{d^2\tilde{\Phi}_l}{d\zeta^2} + (\sigma_l^2 - \zeta)\tilde{\Phi}_l = 0 \tag{3.76}$$

where σ_l is the transverse wavenumber, $\eta_l(r) = \sqrt{k_0^2(1 - \varepsilon H) - \sigma_l(k_0^2\varepsilon)^{1/3}}$ is an longitudinal wavenumber, $\kappa_l(r) = \eta_l H' H_0$, $\nu_l(r) = -\eta_l\varepsilon' H_0^2/6\varepsilon$.

3.5 INTERFERENCE STRUCTURE AND AVERAGED DESCRIPTION OF THE SOUND FIELD IN A WAVEGUIDE

As can be seen from the results obtained above, the sound field in a waveguide is a complex interference structure characterised by many spatial scales. A typical pattern obtained as a result of numerical modelling is shown in Figure 3.4 (see colour plate section), which shows the sound field from an acoustic point source, lying on the bottom in a model waveguide. The sound intensity is shown in gradations of colour in the characteristic case of a 'summer' profile of sound speed with a negative gradient towards the bottom (near-bottom sound channel). It is interesting to note that the result of interference between modes corresponds to the ray pattern (shown in Figures 3.4(*b*)), in the sense that the ray pattern gives the spatial scales of beatings, determined by the cycles of different rays in the waveguide. It is further notable that as one moves away from the source, the sound field pattern becomes less regular, taking on a rather blurred structure, which may appear random, despite the fact that this pattern is completely deterministic.

In this connection, we note that when one studies the propagation of sound in an underwater sound channel, it is often not the fine interference structure of the field (it may be very difficult to compare measured and calculated values for this) but certain average characteristics, relating mostly to the square of the sound pressure averaged over some region (modulo a sound intensity constant), which are of interest (Brekhovskikh (1965), Denham (1986), Brekhovskikh and Lysanov (1991), Borodin (1992)). By the intensity, we mean the amount of sound energy passing through a unit area in unit time. The spatial scale of the averaging is chosen to be greater than the largest scale of the interference of normal waves, which is the longest cycle distance of rays in the channel. As a result of this averaging, interference (with spatial scales of the order of the wavelength λ) and zonal (with scale $\approx D$) variations of the sound field vanish, leaving a comparatively smooth dependence of the intensity on the distance from the sound source.

This averaging will be effective if a very large number of normal waves propagate in the channel (the waveguide is highly multimodal). Precise parameters characterising the extent to which such an approximation is applicable will be indicated below.

In ray theory, averaging corresponds to summing the energy of ray tubes, having the cycle distance of rays coming from the source. An analogous averaged pattern is obtained if one averages not over distance at fixed frequency, but over frequency at a fixed distance; in other words, if one considers the field of a wideband source. When one averages over a spatial interval $\Lambda_l \approx 2\pi|q_{l+1} - q_l|^{-1}$, beatings due to the presence of terms of the type $\exp[i(q_{l+1} - q_l)r]$ vanish. Similarly, when one averages over frequency, the characteristic interval will be $\Delta\omega \approx (2\pi/r)[d(q_{l+1} - q_l)/d\omega]^{-1}$, and if we take

$$\frac{dq_l}{d\omega} \approx \frac{d}{d\omega}\sqrt{\frac{\omega^2}{c^2} - \sigma_l^2} \approx \frac{\omega}{q_l c^2}$$

then we obtain that $\Delta\omega/\omega \approx \Lambda_l/r$ (Borodin (1992)). For a number of model underwater sound channels the dependence of the averaged intensity on the distance r (average decay law) can be obtained in analytic form (Brekhovskikh and Lysanov (1992), Weston (1976, 1980)). In particular, Brekhovskikh (1965) established that for the Pekeris model with an absorbing bottom, the intensity abates with distance as $r^{-3/2}$. The nature of this abatement, which is as it were intermediate between cylindrical and spherical laws, is due to the considerable role of the absorbing bottom, on which rays are multiply reflected.

The case of an irregular bottom was considered by Weston (1976), where a number of useful equations for $I(r)$ were obtained. Focke and Weston (1978) considered the role of caustics in the averaging of the sound field, while Katsnelson *et al.* (1985) compared the results of calculation based on averaged equations for a definitive track with experimental results.

Below, we consider a generalised method for obtaining an averaged description of the intensity of sound, in conditions where rays undergo multiple reflection from absorbing boundaries. This theory is applicable for waveguides with depth varying along the track and a variable sound speed profile.

Suppose that in a waveguide with parameters (the depth $H(r)$, the bottom characteristics and the sound speed profile) varying sufficiently smoothly along the track a source is located, as usual, at the point with coordinates $r = 0$, $z = z_0$ and has a total radiated power W_0. We represent the sound field in an irregular channel similarly to equation (2.41), taking into account the dependence of the eigenvalues and functions on distance as in equation (3.21). We shall write down an expression for the sound pressure at the observation point $P(r, z)$, separating, in what follows, the real and imaginary parts of the eigenvalues. In the expression within the square root in the denominator, the imaginary part of the eigenvalue is small compared with the real part and is neglected

$$P(r,z) = \sqrt{\rho_0 c_0 W_0} \sum_{l=1}^{N} \frac{\psi_l(0;z_0)\psi_l(r;z)}{\sqrt{r q_l(r)}} \exp\left[i \int_0^r q_l(r')dr' - \int_0^r \frac{\gamma_l(r')}{2} dr' + i\frac{\pi}{4}\right] \quad (3.77)$$

Here. c_0 and ρ_0 denote the sound speed and density at the source point. We shall calculate the 'local' intensity at the observation point

$$I = \frac{1}{\rho c}\left\langle \left\{ \mathrm{Re}\left[p(r,z)\mathrm{e}^{-i\omega t}\right]\right\}^2 \right\rangle \quad (3.78)$$

Here $\langle\,\rangle$ denotes averaging over an intermediate time interval, greater than the period of the sound oscillations, the values of the density ρ and the sound speed c are taken at the observation point, and we assume that the inhomogeneity of the medium is sufficiently smooth that the wavefront is locally planar, i.e. $|\nabla p| \sim k$. The averaging over the spatial interval is carried out in two steps. First it is performed for a longitudinal scale D (in the neighbourhood of the distance r) greater than the interval of the interference beatings, as a result of which we obtain

$$I = \frac{W_0}{2r}\frac{\rho_0 c_0}{\rho c}\sum_l \frac{\psi_l^2(0;z_0)\psi_l^2(r;z)}{q_l(r)}\exp\left[-\int_0^r \gamma_l(r')dr'\right] \tag{3.79}$$

Then we average over a transverse scale $\approx H$ in the neighbourhood of the receiver and the source, for which we use the WKB method (equations (2.73), (2.75), (2.81)) to represent the functions $\psi_l(0;z_0), \psi_l(r;z)$. We obtain

$$\bar{I} = \frac{2W_0}{r}\sum_{l=1}^{M}\frac{q_l(0)\exp(-\int_0^r \gamma_l(r')dr')}{\sigma_l(0,z_1)\sigma_l(r,z)D_l(0)D_l(r)} \tag{3.80}$$

where we neglect the differences in density and sound speed at the source point and the observation point in the factor in front of the intensity, since the error given by neglecting these is extremely small.

Expression (3.80) can be used to compute the average intensity; however, to obtain analytic expressions we can move to a ray representation if we replace the sum over the discrete variable l by integration over a continuous variable. This variable might, for example, be the angle from the horizontal axis χ_0 at which a ray emerges from the source. To do this, we shall use a representation of l in terms of the ray invariant J (Brekhovskikh and Lysanov (1992)):

$$dl = \frac{\omega}{2\pi}dJ \tag{3.81}$$

Here

$$J = \oint \frac{\sin\chi(r,z)}{c(r,z)}dz = \oint \left[c^{-2}(r,z) - c^{-2}(r,z_0)\cos^2\chi(r)\right]^{1/2}dz \tag{3.82}$$

where the integration is performed over a waveguide section. Here $\chi(r)$ is the ray grazing angle on the source horizon, which naturally varies with distance. The angle formed by the ray with the horizontal plane (grazing angle) $\chi(r,z)$ depends, generally speaking, on z and r. In what follows, we shall use the above notation $\chi(r) = \chi(r,z_0)$, $\chi_0 = \chi(0,z_0)$ and also the grazing angle on the bottom: $\chi_H(r) = \chi(r,H(r))$ (see Figure 3.4).

By virtue of (3.82) we have

$$dJ = \frac{\sin\chi_0}{c(0,0)}D\,d\chi_0 \tag{3.83}$$

and as a result

$$\bar{I} = \frac{2}{r}\int_{-\pi/2}^{\pi/2}\frac{E(\chi_0,r)\cos\chi_0}{D(\chi_0,r)\sin\chi(\chi_0,r)}d\chi_0. \tag{3.84}$$

Here, and later in this section, we have normalised the total power of the source $W_0 = 4\pi$, i.e. we have assumed that the source radiates unit power in a unit solid angle. The attenuation factor $E(\chi_0,r) = \exp(-\int\gamma_l dr)$ which thus emerges can also be replaced by a ray equation if we use equation (2.84):

$$E(\chi_0, r) = \exp\left[-\int_0^r \gamma_l(r')dr'\right] = \exp\left[-2\int_0^r \frac{\ln|V(\chi_H, r')|}{D(\chi_0, r')}dr'\right] \tag{3.85}$$

Instead of the mode index in (3.83) we have the angle at which a ray leaves the source χ_0, which determines the grazing angle of that ray on the bottom χ_H, on which, in turn, the attenuation coefficient in equation (3.84) depends. Equation (3.84) can be initially obtained in ray theory.

Thus, to compute the dependence $\overline{I(r)}$ (the so-called average decay law) from equation (3.84), we need to know the form of the functions $D(\chi_0, r), \chi(\chi_0, r), \chi_H(\chi_0, r), E(\chi_0, r)$. Below, we shall find these in an adiabatic approximation, based on the notion of the ray invariant (3.82).

Because the value of J is constant along the track, we can express J explicitly in terms of the above quantities, and compute these dependences explicitly. The ray cycle distance, according to Brekhovskikh and Lysanov (1992), is equal to

$$D(\chi_0, r) = \oint \frac{dz}{\tan\chi(\chi_0, r, z)} \tag{3.86}$$

It can also be computed if we know the above dependence $\chi(\chi_0, r, z)$. After the quantities χ_H, D have been found, they should be substituted in expressions (3.84) and (3.85) and integration should be performed.

3.5.1 Isovelocity channel with variable depth

Let us consider a simple underwater sound channel in which the sound speed is constant ($c = \text{constant}$) and the absorbing bottom has a variable depth (Figure 3.5). For such a channel, within a single ray cycle the angle χ does not depend on z, so that the ray invariant J is equal to

$$J = 2\frac{H(r)\sin\chi(r)}{c} = 2\frac{H(0)\sin\chi(0)}{c} = 2\frac{H_0\sin\chi_0}{c} \tag{3.87}$$

Hence, we can express $\chi(r)$ in terms of χ_0, $H(r)$ and $H_0 = H(0)$. The grazing angle

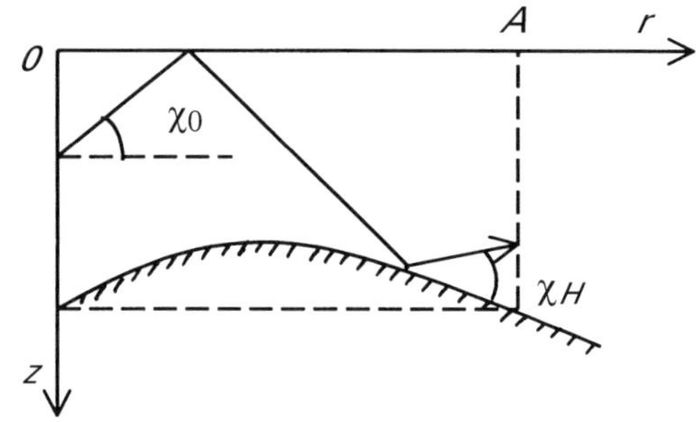

Figure 3.5. Channel with constant sound speed, bounded by an absorbing bottom of variable depth.

$\chi_H(r)$ in the adiabatic approximation is equal to $\chi(r)$. Furthermore, we can compute the ray cycle distance

$$D(\chi_0, r) = \frac{2H(r)}{\chi(\chi_0, r)} = \frac{2H^2(r)}{H_0 \sin \chi_0} \sqrt{1 - \frac{H_0^2 \sin^2 \chi_0}{H^2(r)}} \tag{3.88}$$

To compute the attenuation coefficient of a ray we use equation (2.87) in which we take χ_H for χ.

Restricting our analysis solely to the contribution of rays with small grazing angles, we have

$$\gamma = sH_0\chi_0^2/H(r) \tag{3.89}$$

$$D(\chi_0, r) = 2H^2(r)/H_0\chi_0 \tag{3.90}$$

Thus, after substituting equations (3.85), (3.89) and (3.90) in expression (3.84), we obtain

$$\bar{I}(r) = \frac{1}{rH(r)} \int_{-\pi/2}^{\pi/2} d\chi_0 \exp\left[-sH_0^2 \int_0^r \frac{dr'}{H^3(r')} \chi_0^2\right] = \frac{\sqrt{\pi/s}}{r_0 H_0 H(r)} \left[\int_0^r H^{-3}(r')dr'\right]^{-1/2} \tag{3.91}$$

For a constant depth ($H(r) = H_0 =$ constant) equation (3.91) becomes the expression known from Brekhovskikh (1965) and Brekhovskikh and Lysanov (1992):

$$\bar{I}_0 = \sqrt{\pi/sH_0} r^{-3/2} \tag{3.92}$$

Significantly, in equation (3.91), the intensity at the point r is determined integrally by the whole of the bottom profile. Using equation (3.91), one can consider the behaviour of the intensity for different bottom profiles. Results for a number of profiles were obtained in Katsnelson and Kulapin (1984). For example, in a sound channel with linearly varying depth $H(r) = H_0 \pm \varepsilon r$, calculation according to equation (3.91) produces the expression

$$\bar{I}(r) = \sqrt{\frac{\pi}{sH_0}} \frac{r^{-3/2}}{\sqrt{1 \pm \varepsilon r/2H_0}} \tag{3.93}$$

which differs from the law for the transmission loss in a regular channel by the factor $(1 \pm \varepsilon r/2H_0)^{-1/2}$. Let us consider the role of this factor. If, for example, the channel narrows with distance (minus sign), then the intensity of the sound field at distance r from the source will, naturally, be greater than $\bar{I}_0$, but less than when a conventional adiabatic factor $(1 - \varepsilon r/H_0)^{-1}$ is used, which is inversely proportional to the local depth of the waveguide. This is because the absorption coefficient (2.87) increases when the waveguide narrows, so that the grazing angle χ grows and the ray cycle distance D decreases.

In the same way, one can consider a widening waveguide. Here, the absorption coefficient is less than in a regular waveguide, and, due to the decreased loss in comparison with the regular waveguide, the intensity decreases more slowly as the depth grows.

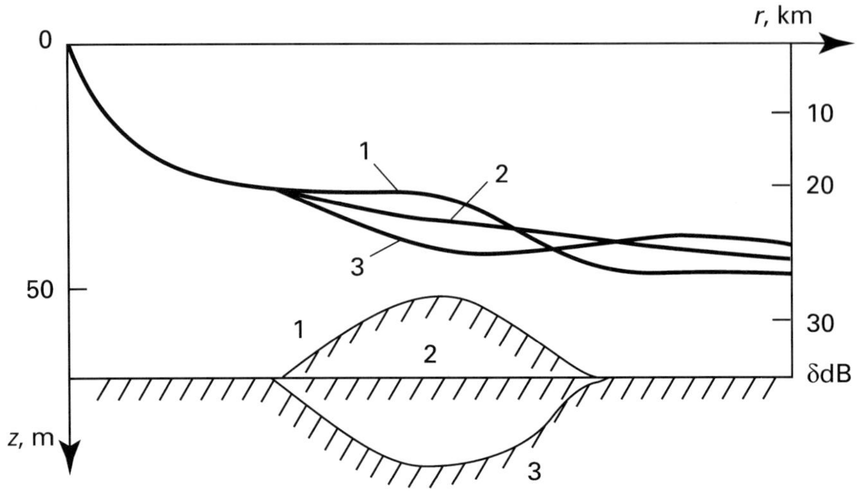

Figure 3.6. Graph of $\bar{I}(r)$ in the presence of a 'hill' and a 'trough'. 1–3 – profiles of the bottom and the corresponding averaged intensity curves.

Thus, equation (3.91), obtained in the adiabatic approximation, indicates that the average field at a given point is determined by the whole of the bottom speed profile along the area of the track passed through. This conclusion can be illustrated by the following example. Suppose the water depth in the section A (Figure 3.5) is the same as at $r = 0$. What change in the field in the section A is induced by the presence of a hill between the source and section A? In the adiabatic approximation, when the field characteristics are determined by local values of the waveguide parameters and there is no absorption, the field in section A does not change. Meanwhile, it follows from (3.91) that the integral in the presence of a hill increases and, consequently, the average intensity decreases. Similarly, if there is a trough between sections O and A the field in section A increases in comparison with the case of a regular sound channel. Figure 3.5 illustrates these arguments and shows the behaviour of the curves $\bar{I}(r)$ for the above channels.

Figure 3.7 shows an example of the calculation of the averaged intensity for a channel of variable depth ($c =$ constant). This figure shows the results of calculation using the summation of modes, studied in Section 3.2. From the figure, it is clear that in the region of applicability of equation (3.91), where the number of propagating modes is large, there is a good agreement between the curves. The distance r^*, determining the boundary of applicability of equation (3.84) (respectively of equation (3.91)), is computed from the equation

$$\int_0^{r^*} H^{-3} dr \leq 16\pi^{-2} s^{-1} \lambda^{-2} \tag{3.94}$$

This distance is shown on Figure 3.7.

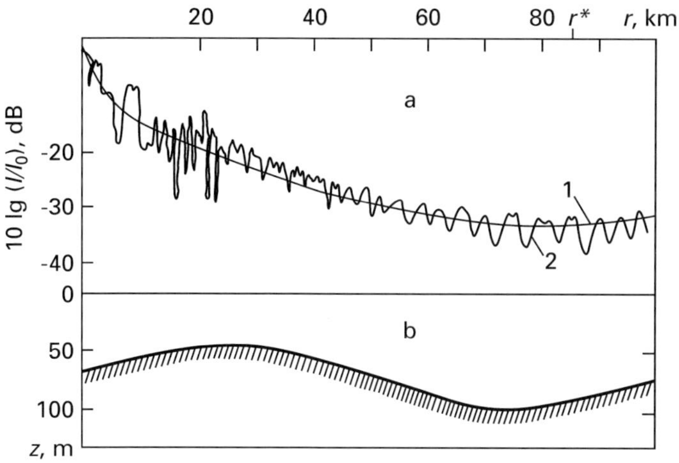

Figure 3.7. Calculated values of the relative sound intensity $10\lg(I/I_0)$ from a point source (a) and bottom profile (b). 1 – Based on averaged equation (3.91); 2 – using summation of modes; parameter values for the calculation $f = 100\,\text{Hz}$, $c = 1500\,\text{m/s}$, $c_1 = 1580\,\text{m/s}$, $s = 0.1$, $z_0 = z = 50\,\text{m}$.

3.5.2 Irregular near-bottom sound channel

Let us now consider an irregular near-bottom sound channel (Figure 3.8). Here the sound propagation takes place by successive reflections of refracted rays from the bottom. We shall assume that the sound speed varies linearly with depth and that the waveguide parameters vary smoothly along the track: $c(r,z) = c_S(r)(1 - \varepsilon(r)z)$, $H = H(r)$. The ray cycle distance is easily found from the formula:

$$D(\chi_H, r) = \frac{2c_H}{\varepsilon c_S}\tan\chi_H \tag{3.95}$$

where $c_H = c(r, H(r)) = c_S(1 - \varepsilon H)$. To calculate the averaged intensity using equation (3.84), we need to know the dependence of the grazing angle $\chi_H(r)$ on

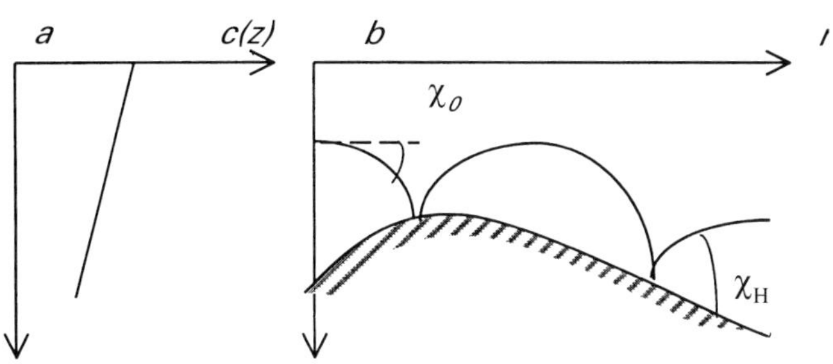

Figure 3.8. Irregular near-bottom sound channel with an homogeneous absorbing fluid bottom and variable depth. (a) – sound speed profile; (b) – bottom profile.

the angle χ_0. This dependence is determined using the ray invariant (3.82)

$$J = \frac{1}{\varepsilon c_S}\left(\ln\frac{1+\sin\chi_H}{1-\sin\chi_H} - 2\sin\chi_H\right) \tag{3.96}$$

In real conditions the main contribution to the resulting field comes from rays with small grazing angles; thus, in expressions (3.95), (3.96), we can restrict ourselves to terms of the first non-vanishing order in χ_H:

$$J \approx \frac{2}{3\varepsilon c_S}\chi_H^3 \approx \frac{2}{3\varepsilon_0 c_{S0}}\chi_{H0}^3 \tag{3.97}$$

Here, $\varepsilon_0 = \varepsilon(0)$, $c_{S0} = c_S(0)$, $\chi_{H0} = \chi_H(0)$. The dependence of $\chi_H(r)$ on the angle at which the ray leaves the source can be determined using Snellius's law and the value of J in the plane $r = 0$:

$$\chi_H = (\varepsilon c_S/\varepsilon_0 c_{S0})^{1/3} \arccos(n_S \cos\chi_0) \quad n_S = c_{H0}/c_{S0} \tag{3.98}$$

Assuming, in addition that the angle χ_0 is small, we can decompose expression (3.98) in a series in χ_0:

$$\chi_H(\chi_0, r) \approx \left(\frac{\varepsilon c_S}{\varepsilon_0 c_{S0}}\right)^{1/3}\left(\arccos n_s + \frac{n_S}{2\sqrt{1-n_S^2}}\chi_0^2\right) \tag{3.99}$$

From equations (3.98) and (3.99) it is clear that $\chi_0 = 0$ corresponds to the minimum grazing angle $\chi_{H\min} = (\varepsilon c_S/\varepsilon_0 c_{S0})^{1/3} \arccos n_S$. The factor $E(\chi_0, r)$, determining the sound loss on reflection, has the form (3.85), where $D(\chi_H, r)$ is determined by expression (3.95) and for $V(\chi_H)$ we take the Fresnel reflection coefficient (2.86) with local grazing angle $\chi_H(r)$. We proceed in the usual way, decomposing the integrand in equation (3.85) into a power series in χ_H:

$$2\frac{\ln|V(\chi_H)|}{D(\chi_H)} = \frac{\varepsilon c_S}{c_H}(-s + \nu\chi_H^2) \tag{3.100}$$

where

$$s = -\frac{\ln|V(\chi)|}{\tan\chi}\bigg|_{\chi=0} = \mathrm{Re}\frac{2m_1}{\sqrt{n_1^2(1+i\alpha)-1}} \qquad \frac{d}{d\chi}\left[\frac{\ln|V(\chi)|}{\tan\chi}\right]\bigg|_{\chi=0} = 0$$

$$\nu = \frac{1}{2}\frac{d^2}{d\chi^2}\left[\frac{\ln|V(\chi)|}{\tan\chi}\right]\bigg|_{\chi=0} = \mathrm{Re}\frac{[3n_1^2(1+i\alpha)-2m_1^2]}{3[n_1^2(1+i\alpha)-1]^{3/2}} \tag{3.101}$$

Using equations (3.98) and (3.99), we can obtain a further decomposition of equation (3.100) in a series in χ_0:

$$2\frac{\ln|V(\chi_H)|}{D(\chi_H)} = -u(r) - w(r)\chi_0^2 \tag{3.102}$$

Here

$$u(r) = -\frac{\varepsilon c_S}{c_H}\left[-s + \nu\left(\frac{\varepsilon c_S}{\varepsilon_0 c_{S0}}\right)^{2/3} \arccos^2 n_S\right]$$

$$w(r) = -\frac{\varepsilon c_S \nu n_S}{c_H}\left(\frac{\varepsilon c_S}{\varepsilon_0 c_{S0}}\right)^{2/3} \frac{\arccos n_S}{\sqrt{1 - n_S^2}} \tag{3.103}$$

Substituting expressions (3.95), (3.102), (3.85) in equation (3.84), we finally have:

$$\begin{aligned}\bar{I}(r) &= \frac{2}{r}\int_{-\pi/2}^{\pi/2} \frac{\cos\chi_0 \exp[-\int_0^r (u(r') + w(r')\chi_0^2)dr']}{D(\chi_H)\sin\chi(\chi_H, r)} d\chi_0 \\ &= \frac{\varepsilon c_S\sqrt{\pi}}{c_H r}\left(\tan\chi_{H\min}\sqrt{1 - \frac{c_S^2\cos^2\chi_{H\min}}{c_H^2}}\right)^{-1}\left[\int_0^r w(r')dr'\right]^{-1/2} \\ &\quad\times \exp\left(-\int_0^r u(r')dr'\right)\end{aligned} \tag{3.104}$$

Equation (3.104) is similar to equation (3.91). As in the latter, the averaged intensity in the section r is not determined locally but by overall parameters of the channel, characterizing all the previous track up to the point r. Using equations (3.101) and (3.102), we can relate the parameters u, w, s, ν to the values of the refractive index n and the density ratio m.

In the case of a regular waveguide (constant values of c_H, c_S, ε in equations (3.103)), instead of functions we have constants $u(r) = u$, $w(r) = w$ and (3.104) becomes

$$\bar{I}(r) = \frac{\varepsilon c_S}{c_H r^{3/2}}\sqrt{\frac{\pi}{w}}\left(\tan\chi_{H\min}\sqrt{1 - \frac{c_S^2\cos^2\chi_{H\min}}{c_H^2}}\right)^{-1} \exp(-ur) \tag{5.105}$$

This is a known result, amounting to the fact that an exponential damping $\exp(-ur)$ is imposed upon the 'three over two' law $I \approx r^{-3/2}$ (Brekhovskikh and Lysanov (1991), Katsnelson *et al.* (1986)). As in the case of a channel with constant sound speed, the criterion for applicability of the averaged equations is the existence of a large number of propagating modes; however, for the near-bottom channel, there is a small difference, namely that the set of all 'near-bottom' modes (whose number is determined by equation (2.90)), having the same attenuation coefficient determines the exponential factor in equation (3.104), while the other factor of the decrease in intensity, of type $r^{-3/2}$, is determined by the width $(\int_0^r w dr')^{-1/2}$ of the Gaussian angular distribution in the integrand in equation (3.104). Thus, the distance r^*, determining the boundaries of applicability of equation (3.104), is determined by the condition that a sufficiently large number of modes should exist at this width, It takes the form:

$$\int_0^{r^*} w dr \leq \left(\frac{2kH}{\pi}\right)^2$$

The next model is a channel with variable depth $H = H(r)$, having a linear sound speed profile $c = c_S(r)[1 - \varepsilon(r)z]$ and bounded by the surface $z = 0$. The difference from the previous case is that reflection from the surface is taken into account. The bottom is an homogeneous fluid half-space with density ρ_H, and a sound speed c_H which is complex, owing to absorption.

For this model, the dependence of the absorption coefficient on the grazing angle is non-monotonic; that is, up to some angle $\chi_{0\,\mathrm{min}}$, corresponding to reflection of the ray from the surface, absorption decreases, and then grows.

The absorption coefficient $\gamma/2 = -\ln|V|/D$ behaves correspondingly, as the graphs in Figure 2.8 show.

To find the average decay law in this model, requires the computation of the integral (3.84) with the non-monotonic dependence $\gamma(\chi_0)$. Of course, at large distances the angular spectrum of the radiation will be narrowed, and concentrated in the neighbourhood of the angle $\chi_{0\,\mathrm{min}}$ corresponding to minimum absorption.

Based on the above analysis, we conclude that for different regions corresponding to different distances from the source the averaged dependence of the loss with distance will be different in nature. Namely, at small distances from the source, propagation will be of a spherical nature, and correspondingly, the law for the decrease in intensity will be spherical. Furthermore, in the presence of a sufficiently large number of excited modes, the average decay law will be of type $r^{-3/2}$ (or a modification of that in the irregular case), with the possible imposition of exponential damping upon that if there are near-bottom modes in the channel. The spatial interval of this dependence is determined by the distance r^*, estimates of which were obtained in the Pekeris model or that of a near-bottom channel. For example, for a typical case with $H = 100\,\mathrm{m}$, $f = 200\,\mathrm{Hz}$, $s = 0.2$, we obtain a value for the characteristic distance of $r^* \approx 80\,\mathrm{km}$. At distances greater than or equal to r^* the field contains one or more modes, having the least attenuation coefficient, and the decrease in the field with distance is exponential. Of course, in addition to this, one must also take into account the attenuation of sound due to absorption in the water, which is proportional to the square of the frequency, but this attenuation is small at the frequencies with which we are dealing (see the remarks at the end of Section 2.3). We note in conclusion that the nature of the decrease in the average intensity also depends on the positions of the receiver and the source. For example, for a source located near the surface and a receiver sufficiently far away from it, the average decay law is given by a function of type $r^{-5.2}$, which is explained by the change in the modal distribution of radiated energy near the boundary (dipole radiation of the sound). These questions were considered in detail in Grachev (1983), Denham (1986) and Weston (1978, 1980, 1994).

In this chapter, we have formulated a number of methods which permit the approximate calculation of the characteristics of the sound field formed in real shallow-water waveguide. These methods will be used in what follows both to assess and interpret the experimental results, and as a first approximation for precise numerical calculations. The method(s) which should be applied depends on the properties of the waveguide in question (more precisely, on the smallness parameter in the problem), and on the characteristics to be computed. Great value is attached to so-called averaged decay laws, which enable us to estimate the intensity of the received field as a function of the distance to the source or, in other words, to estimate an important characteristic of the waveguide as a whole, such as the transmission loss (TL), the main measured quantity in experiments on long tracks (Chapter 6).

4

Sound field in shallow water with random inhomogeneities

4.1 STRUCTURE AND MODELS OF RANDOM INHOMOGENEITIES OF DIFFERENT KINDS

Until now we have considered the main approaches to the description of the sound propagation in shallow water, as a deterministic inhomogeneous medium. However, it is well known that the oceanic medium contains random inhomogeneities of various kinds, which have a considerable effect on the propagation of sound. Moreover, we know that the random inhomogeneities are primarily concentrated in the near-surface layer (Babii (1983), Shvachko (1982)), which leads to a particularly strong effect of random inhomogeneities precisely in shallow water. Recently, great attention has been paid, both experimentally (Zhou *et al.* (1996), Headrick *et al.* (2000)) and theoretically (Dozier and Tappert (1978), Gorskaya and Raevskii (1984), Ashley *et al.* (1987), Kuznetsova (1988), Derevyagina (1995)), to the study of shallow water as a randomly inhomogeneous medium. We note that the propagation of waves of different kinds in a randomly inhomogeneous medium has been studied for a long time (beginning with Rayleigh) (see, for example, the monographs Chernov (1975), Rytov *et al.* (1989), Ishimaru (1978), Flatte (1979), etc.), although the oceanic medium in general and the shallow-water medium in particular are very specific objects of that kind and, accordingly, require the development of highly specific techniques for their description.

The oceanic medium, if it is considered as varying in space and in time, is characterised by various space–time scales of variability (a detailed analysis was performed in Dickey (1991)). These scales range from dimensions of the order of 1 cm and 1 second in time for molecular processes, to several hundreds and thousands of kilometres and several months in time (synoptic storms and seasonal changes). Of course, there is a certain amount of coarseness about the division of the space–time variability into 'deterministic' and 'random' parts, associated with the

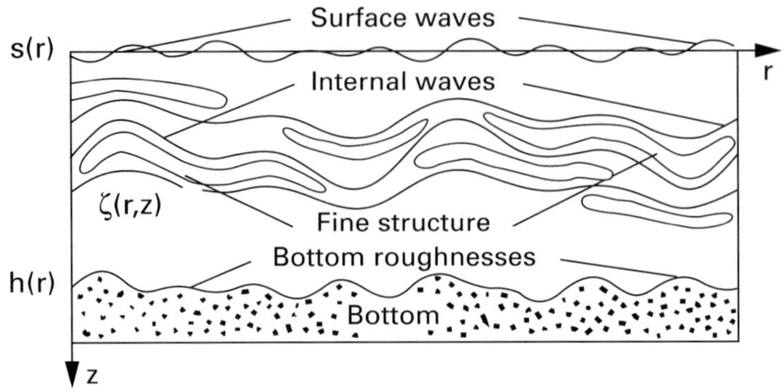

Figure 4.1. Random inhomogeneities in shallow water.

scales of averaging. In our case, the choice of those scales is dictated by the scales of the space–time variability of the low-frequency sound field, i.e. we deem phenomena with characteristic spatial scales of from one metre to several tens of metres to be randomly inhomogeneous. As far as temporal characteristics are concerned, in connection with the very small period of sound oscillations, we can, as a rule, work in an approximation of a 'frozen' medium. Thus, the main phenomena in shallow water giving rise to random inhomogeneities are taken to include the following (see Figure 4.1).

- The microstructure (thermohaline) (Fedorov (1976, 1978)) which was discussed in Section 1.2. It gives rise to fluctuations of the sound speed in the water layer $\delta c(z)$ with amplitude $\delta c_0 \sim 1\,\mathrm{m/s}$, horizontal scales of variability $L_r \sim 100$–$500\,\mathrm{m}$, vertical scales of variability $L_z \sim 0.1$–$10\,\mathrm{m}$, and a variance of the fluctuations which varies as a function of depth in the range 10^{-2}–$1\,\mathrm{m^2/s^2}$. For the calculations of the fluctuations with propagating sound, the following correlation function was used in the author's works

$$B_c(r, z : r'z') = \langle \delta c(r, z)\delta c(r'z')\rangle = \delta c_0^2 B_d(z, z')\exp(-|r - r'|/L_r) \tag{4.1}$$

  where $B_d(z, z')$ is the correlation function in the vertical direction with radius at most 10 m.
- Internal waves, displacing surfaces of constant density $\zeta(\vec{r}, z)$. One can show (see, for example, Flatte (1979)) that these displacements lead to a change in the sound speed profile of the form

$$\delta c(\vec{r}, z) = c(z)\tilde{Q}\tilde{N}^2(z)\zeta(\vec{r}, z) \tag{4.2}$$

  where $\tilde{Q} \approx 2.4\,\mathrm{s^2/m}$ is a constant, determined by the water properties, and $\tilde{N}(z) = (g\rho^{-1}(d\rho/dz))^{1/2}$ is the buoyancy frequency, determined by the density stratification. Equation (4.2) allows us to associate the correlation characteristics (for example, the spatial spectrum) of the displacements with the correlation characteristics of the sound speed field. We introduce the spectral density of the energy of internal waves $B(\vec{\kappa}, m) = |S_m(\vec{\kappa})|^2$, where the displacement is represented in the form of a series in the vertical modes of the internal waves $\tilde{\Phi}_{m\vec{\kappa}}(z)$:

$$\zeta(\vec{r}, z) = \mathrm{Re}\left\{\sum_m \int S_m(\vec{\kappa})\tilde{\Phi}_{m\vec{\kappa}}(z)\exp(i\vec{\kappa}\vec{r})\,d\vec{\kappa}\right\} \tag{4.3}$$

where m is the index of the gravitational mode of the internal waves, and $\vec{\kappa}$ is its wave vector in the horizontal plane. For $B(\vec{\kappa}, m)$ in deep water, there exists an expression, proposed by Garrett and Munk (1975), giving the energy distributed according to gravitational modes and isotropic in the horizontal plane. As mentioned earlier in Chapter 1, in shallow water this spectrum is not suitable because of the presence of considerable anisotropy and a predominance of the first gravitational mode. This corresponds to the exclusion of the sum from expression (4.3) (only the gravitational mode $\tilde{\Phi}_{1k}(z)$ remains) and integration over the angle in the horizontal plane. Using the experimentally observable fact that the energy density of internal waves of comparatively small amplitude (background internal waves) in shallow water decreases as a function of κ approximately as κ^{-p} $(p \sim 2)$[1], for background waves in shallow water the authors used the spectrum $B_{bi}(\kappa) = 2B_0\langle\zeta_0^2\rangle\kappa^{-p}$. Here B_0 is a normalisation constant, and $\langle\zeta_0^2\rangle$ is the variance at some fixed depth. Estimates of the longitudinal interval of correlations based on the given theory give a value from several hundred metres to a kilometre for background internal waves.

- Bottom roughnesses. According to the results of experimental studies (Kuperman and Jensen (*ed.*) (1980), Hampton (1973)) the surface of the ocean bed is highly smooth, the depth at comparatively large distances differing by at most several tens of metres from some average level. In addition, the characteristic longitudinal dimensions of roughnesses vary in broad ranges. It was customary to distinguish the following characteristic scales: 1) general relief, characterising the main geomorphological formations (ridges, valleys, troughs etc.), with longitudinal dimensions of 10^4– 10^5 m; 2) mesorelief, including relief forms of length 10^2–10^3 m; 3) microrelief, formed by ripple marks, boulders, hillocks, etc., with spatial scales of 10^{-2}–10 m. In this approach, general relief corresponds to regular variations of the parameters of a waveguide $H(r)$, and the other two components – mesorelief and microrelief – correspond to random fluctuations of the depth of the waveguide $h(r)$. This random function is also characterised by a spatial wave vector spectral density $B_h(k)$, associated with the correlation function for bottom roughness

$$\langle h(r)h(r')\rangle = \int_0^\infty B_h(k)\cos(|r - r'|\kappa)\,d\kappa \tag{4.4}$$

At the present time, there are not too many experimental data relating to the statistical properties of the sea bed as a randomly rough surface. According to our studies in one region of the Barents Sea, the mesoscale relief, captured by

[1] The value $p \sim 3/2$ was observed in Lynch *et al.* (1996).

high-precision echo sounding, is characterised by a spectral density of deviations well approximated by the expression:

$$B_h(\kappa) = 2\langle h_0^2 \rangle \frac{\kappa}{a^2} \exp(-\kappa^2/a^2) \tag{4.5}$$

where $\langle h_0^2 \rangle$ is the variance of the vertical deviations from the average level, and a is the spectral width, with an associated longitudinal correlation interval $L_r \sim 1/2a$. In our case, we obtained values $\langle h_0^2 \rangle \sim 2\,\mathrm{m}^2$, $a \sim 2 \times 10^3\,\mathrm{m}^{-1}$, $L_r \sim 10^3\,\mathrm{m}$. (See also Goff (1988) and Goff (1995).)

- Surface disturbance, transforming the upper boundary of the waveguide into a randomly rough surface, with a displacement described by the random function $s(\vec{r}) = s(r, \theta)$ (where θ is the azimuthal angle in the horizontal plane). This phenomenon is very well studied (Phillips (1977)), the spatial spectrum of a surface disturbance for a fixed direction, determined by the azimuthal angle θ $B_s(\kappa, \theta)$, and defined by the formula

$$B_s(r, \theta; r'\theta) = \langle s(r, \theta) s(r', \theta) \rangle = \frac{1}{2} \int_0^\infty B_s(\kappa, \theta) \cos(|r - r'|\kappa) d\kappa \tag{4.6}$$

is associated based on the dispersion relation for surface waves $\omega(\kappa)$ with the Pierson–Moskovitz (Pierson and Moskovitz (1969)) frequency spectrum:

$$B_s(\omega) = ag^2\omega^{-5} \exp[-b(g/\omega u)^4] \tag{4.7}$$

where $a = 8.1 \times 10^{-3}$, $b = 0.74$, $g = 9.8\,\mathrm{m/s}^2$, and u is the wind speed in m/s. In typical shallow-water conditions ($u \leq 10$–$12\,\mathrm{m/s}$, sea depth $H \sim 100$–$200\,\mathrm{m}$) the dispersion equation $\omega^2 = g\kappa$ can be used for surface waves. In particular, for wind speed $u \sim 10\,\mathrm{m/s}$, $\theta \sim 0$, the variance of the vertical displacements is $\langle s^2 \rangle \sim 0.25\,\mathrm{m}^2$.

4.2 COUPLED MODE EQUATIONS

To calculate the sound field in an irregular waveguide sound pressure is represented in the form of a decomposition into local modes (3.16) where the coefficients of the decomposition $C_l(r)$ in the case of smooth variation of the parameters, may be assumed to be constant. The smoothness criterion (possibility of neglecting mode coupling) was formulated for deterministic variation of these parameters (for example, the inequalities (3.23), (3.24), (3.30)). In the case when the coupling cannot be neglected, we must look for $C_l(r)$ as functions of distance, which describe a mode conversion, i.e. a transfer of energy from one mode to another. In this case, the transformation is due to deterministic variations of the waveguide parameters. In addition, when sound propagates in a waveguide, the mode coupling may be due to stochastic variations of the parameters – the random inhomogeneities in the water layer or boundary (roughnesses of the bottom or the surface) listed in the previous section. This coupling means that the coefficients of the decomposition $C_l(r)$ become random functions of the coordinate r and, accordingly, the sound field itself (3.16) becomes a random function of the coordinates. In order to determine the characteristics of the sound field, we have to construct its various statistical moments. In this chapter we shall work mainly with the second moment – the probability of the distribution of intensity across waveguide modes. Equations determining the variation of these moments with distance can be obtained in the form of a system of so-called transfer equations (master equations) and, moreover, under a number of assumptions, may be brought to closed form, in the guise of a so-called diffusion approximation. This approach, apparently first proposed in Marcuse (1972) for optical waveguides, and for acoustic waveguides in Kohler and Papanikolau (1977) and Dozier and Tappert (1978), was used for the analysis of propagation in deep water in Artelnyi and Raevskii (1984, 1987). Scattering on random inhomogeneities in deep-water waveguides was also considered in Flatte (1979) and Beran and Whitman (1978).

The specifics of shallow water, taking into account the effect of random homogeneities, involve both quantitative (different characteristic values of the waveguide parameters from in the case of deep water) and qualitative differences. As already indicated on a number of occasions, when sound propagates in shallow water, absorption in the bottom, which is taken into account in our description in terms of mode attenuation coefficients, plays a crucial role. Of course, variation of the mode structure of the field due to scattering (mode transformations) on random inhomogeneities is accompanied by its variation due to non-uniform (over modes) absorption, which naturally complicates, generally speaking, the pattern of the reconstruction of the spatial structure of the field in shallow water in comparison with deep water. Using the technique of the transfer equations (kinetic equations) and (or) the subsequent diffusion approach, we can include absorption in the bottom in a natural manner in the equation describing the variation of the field, and thereby obtain an averaged picture of the variation of the spatial distribution of the statistical characteristics of the sound field as it propagates in the waveguide.

Formally, the presence of random inhomogeneities with the parameters listed above means that the values of $k^2(r,z)$ in equation (2.2) gain fluctuating parts (or the boundary conditions gain a fluctuating part[2]). In our further study, we shall use the random inhomogeneous refractive index $n_R(r,z)$, whose square has average part $n^2(r,z)$ and fluctuating part $\mu(r,z)$ (we assume that $\mu = 0$ for $z > H$, i.e. that random inhomogeneities occur only in the water layer):

$$k^2(r,z) = k^2 n_R^2(r,z) = k^2[n^2(r,z) + \mu(r,z)] \quad |\mu| \ll \langle n_R^2 \rangle = n^2 \quad \langle \mu \rangle = 0 \tag{4.8}$$

The above facts lead to values of $\approx 10^{-2}$–10^{-3} for $k^2(r,z)$ in real situations. Despite the smallness of μ, the effects due to the presence of random inhomogeneities may not be small on sufficiently long tracks. The random part of the refractive index is described by the correlation function:

$$B(r,z;r',z') = \langle \mu(r,z)\mu(r',z') \rangle \tag{4.9}$$

[2] Stochastic boundary conditions also lead to similar coupled mode equations and will not be considered separately in what follows.

As far as the randomly inhomogeneous field of the refractive index is concerned, we shall suppose that the following relation between the spatial scales of the random inhomogeneities holds. Suppose the correlation length of the random inhomogeneities in the longitudinal direction is $R_{\parallel}$ and $R_{\perp}$ in the transverse direction, and that the characteristic scale of variations of the waveguide parameters in the longitudinal direction is $\tilde{L}$ and the characteristic distances of observation are r. We shall assume that

$$R_{\perp} \ll R_{\parallel} \ll \tilde{L}, r \tag{4.10}$$

We obtain equations describing the variation of the sound field from the source, propagating in the waveguide. We shall suppose, as in Section 3.2, that the sound propagates in a waveguide whose depth $H(r)$ varies smoothly with the distance, and that the deterministic (average) part of the refractive index also varies smoothly with distance. The sound field satisfies an equation similar to (2.6) with corresponding boundary conditions on the surface (pressure-release boundary) and on the bottom (condition of glueing together with the lower half-space, which may be chosen to be stratified). As in the previous section, we shall be interested in the intensity of the sound field, generated by an acoustic source with frequency ω, and power W_0; therefore, we shall be dealing with a sound pressure with complex amplitude, similar to (2.41) or (3.77) (the temporal dependence $\sim \mathrm{e}^{-i\omega t}$ is omitted)

$$\left[\frac{1}{r}\frac{\partial}{\partial r}\left(r\frac{\partial}{\partial r}\right) + \frac{\partial^2}{\partial z^2} + k^2(n^2(r,z) + \mu(r,z))\right] P(r,z) = 0 \tag{4.11}$$

As before we represent the complex amplitude of the sound field in the form of a decomposition into the modes of an unperturbed (without random inhomogeneities) waveguide:

$$P(r,z) = \sqrt{\rho_0 c_0 W_0} \sum_l \frac{\psi_l(r;z)}{\sqrt{r q_l(r)}} \left[C_l^+(r) \exp\left[i \int_0^r q_l(\rho) d\rho \right] \right.$$
$$\left. + C_l^-(r) \exp\left[-i \int_0^r q_l(\rho) d\rho \right] \right] \tag{4.12}$$

where $\psi_l(r;z)$ are the eigenfunctions of the matched waveguide, determined earlier (3.14). The difference from the analogous equation (3.16) is that here the coefficients of the decomposition $C_l^{\pm}(r)$ are random functions. Moreover, in the exponential exponents the integrands contain the real part of an eigenvalue. We note that both the real and the imaginary parts of the eigenvalues are smooth functions of the coordinate r. Substituting the decomposition (4.12) in equation (4.11), we obtain a system of stochastic differential equations for the random functions $C_l^{\pm}(r)$. Neglecting back-scattering terms containing coefficients C_l^- are dropped from the system of equations and the remaining coefficients C_l^+ will be denoted by C_l. In addition, the following approximations are made. In obtaining the derived quantities, we neglect adiabatic dependences on r: $\partial q_l/\partial r$, $\partial \psi_l/\partial r$ and we also suppose that

$|d^2C_l/dr^2| \ll |q_l dC_l/dr|$. Taking this into account we obtain a system of stochastic differential equations:

$$\frac{dC_l(r)}{dr} = -\frac{\gamma_l(r)}{2}C_l(r) + i\sum_m V_{lm}(r)\exp\left[i\int_0^r q_{ml}(\rho)d\rho\right]C_m(r) \tag{4.13}$$

where the matrix elements of the mode coupling V_{lm} are determined by the equation:

$$V_{lm}(r) = \frac{k^2}{2\sqrt{q_l q_m}}\int_0^\infty \frac{\rho}{\rho(r,z)}\psi_l(r;z)\mu(r,z)\psi_m(r;z)dz \tag{4.14}$$

and are random functions of r, $q_{ml}(r) = q_m(r) - q_l(r)$.[3]

The derivation of the equations for the average characteristics of the sound field is performed in the standard way. Instead of (4.13), we write down the equivalent integral equation:

$$C_l(r) = C_l(0) - \int_0^r \frac{\gamma_l(r')}{2}C_l(r')dr' + i\sum_m \int_0^r V_{lm}(r')\exp\left[i\int_0^{r'} q_{ml}(\rho)d\rho\right]C_m(r')dr' \tag{4.15}$$

After one iteration

$$\begin{aligned} C_l(r) = C_l(0) &- \int_0^r \frac{\gamma_l(r')}{2}C_l(r')dr' \\ &+ i\sum_m \int_0^r V_{lm}(r')\exp\left[i\int_0^{r'} q_{ml}(\rho)d\rho\right]C_m(0)dr' \\ &- i\sum_m \int_0^r dr' \int_0^{r'} dr'' \frac{\gamma_m(r'')}{2} V_{lm}(r')\exp\left[i\int_0^{r'} q_{ml}d\rho\right]C_m(r'') \\ &- \sum_{m,n}\int_0^r dr' \int_0^{r'} dr'' V_{lm}(r')V_{mn}(r'') \\ &\times \exp\left[i\int_0^{r'} q_{ml}(\rho)d\rho + i\int_0^{r''} q_{nm}(\rho)d\rho\right]C_n(r'') \end{aligned} \tag{4.16}$$

and, with ensemble averaging, we obtain the following equation:

$$\begin{aligned} \langle C_l(r)\rangle = \langle C_l(0)\rangle &- \int_0^r \frac{\gamma_l(r')}{2}\langle C_l(r')\rangle dr' \\ &- i\sum_m \int_0^r dr' \int_0^{r'} dr'' \frac{\gamma_m(r'')}{2}\langle V_{lm}(r')C_m(r'')\rangle \exp\left[i\int_0^{r'} q_{ml}d\rho\right] \\ &- \sum_{m,n}\int_0^r dr' \int_0^{r'} dr'' \langle V_{lm}(r')V_{mn}(r'')C_n(r'')\rangle \\ &\exp\left[i\int_0^{r'} q_{ml}(\rho)d\rho + i\int_0^{r''} q_{nm}(\rho)d\rho\right] \end{aligned} \tag{4.17}$$

[3] Integration in (4.14) in fact takes place within the water layer.

in which in approximation of the smoothing (Kohler and Papanicolaou (1977)), we separate the average expression from that in the integrand

$$\langle V_{lm}(r')V_{mn}(r'')C_n(r'')\rangle = \langle V_{lm}(r')V_{mn}(r'')\rangle\langle C_n(r'')\rangle$$
$$\langle V_{lm}(r')C_l(r'')\rangle = \langle V_{lm}(r')\rangle\langle C_l(r'')\rangle = 0 \tag{4.18}$$

The correlator of the random perturbation function in (4.18) can be represented using the correlation function (4.9)

$$\langle V_{lm}(r')V_{mn}(r'')\rangle = \frac{k^4}{4q_m\sqrt{q_lq_n}}\int_0^\infty dz'\int_0^\infty dz''\frac{\rho^2}{\rho(r',z')\rho(r'',z'')} \times \psi_l(r';z')\psi_m(r';z')B(r',z';r'',z'')\psi_m(r'';z'')\psi_n(r'';z'') \tag{4.19}$$

We shall further use the assumption that the correlation function B is a sharper function of its arguments $r'-r''$ than the dependence of the eigenfunctions on the longitudinal coordinate. Consequently, in the integral (4.19) for the eigenfunctions we can set the value of the longitudinal coordinate equal to r'. By virtue of the slowness of the density variations, the same can be done for the function ρ. This means that the correlator (4.19) can be represented in the form $\langle V_{lm}(r')V_{mn}(r'')\rangle = B_{lm,mn}(r';r'-r'')$, where the function $B_{lm,mn}$ is a smooth function of the first argument and a sharp function of the second argument. After differentiating (4.17) with respect to r, we obtain

$$\frac{d\langle C_l(r)\rangle}{dr} = -\frac{\gamma_l(r)}{2}\langle C_l(r)\rangle - \sum_{m,n}\int_0^r B_{lm,mn}(r;r-r') \times \exp\left[i\int_0^r q_{ml}(\rho)d\rho + i\int_0^{r'} q_{nm}(\rho)d\rho\right]\langle C_n(r')\rangle dr' \tag{4.20}$$

Using the smoothness of the function $\langle C_n(r')\rangle$ in comparison with the correlator, we take it out from under integral for values $r=r'$. For sufficiently large values of r' the correlation function is very small and the integral in formula (4.19) can be extended to infinity. We then obtain

$$\frac{d\langle C_l(r)\rangle}{dr} = -\frac{\gamma_l(r)}{2}\langle C_l(r)\rangle - \sum_{m,n}\left[\int_0^\infty B_{lm,mn}(r;r-r')\exp\left[i\int_r^{r'} q_{nm}(\rho)d\rho\right]dr'\right] \times \langle C_n(r)\rangle\exp\left[i\int_0^r q_{nl}(\rho)d\rho\right] \tag{4.21}$$

With the notation

$$A_{ln}(r) = \sum_m\int_0^\infty B_{lm,mn}(r;r-r')\exp\left[i\int_r^{r'} q_{nm}(\rho)d\rho\right]dr' \tag{4.22}$$

we finally obtain a system of equations for the average modal amplitudes

$$\frac{d\langle C_l(r)\rangle}{dr} = -\frac{\gamma_l(r)}{2}\langle C_l(r)\rangle - A_{ll}\langle C_l(r)\rangle - \sum_{n\neq l} A_{ln}(r)\exp\left[i\int_0^r q_{nl}(\rho)d\rho\right]\langle C_n(r)\rangle \quad (4.23)$$

The right-hand side of the system of equations describes the transfer of energy from the mode with index n to the mode with index l as a result of mode coupling (scattering on random inhomogeneities). The system (4.23) describes the change in the coherent component of the field (of the average amplitude). In the sum on the right-hand side of equation (4.23), we separate out the diagonal term corresponding to scattering at angle zero. In the language of modes, this corresponds to the coupling of the mode with index l with the same mode of index l as a result of random inhomogeneities.

The coefficient $A_{ll}(r)$ is, generally speaking, complex; this corresponds to attenuation of the coherent part of the amplitude and to the change in its phase under the effect of statistical perturbation. Thus, we can introduce the additional coefficient Γ_l of the attenuation of the coherent component of the mode with index l due to scattering on random inhomogeneities

$$\Gamma_l(r) = 2\mathrm{Re}\, A_{ll}(r) = \sum_m a_{lm}(r) \quad (4.24)$$

where we have denoted

$$a_{lm}(r) = 2\mathrm{Re}\int_0^\infty B_{lm}(r,r')\exp\left[i\int_r^{r+r'} q_{lm}(\rho)d\rho\right]dr' \quad (4.25)$$

$$B_{lm}(r,r') = B_{lm,ml}(r,r') = \langle V_{lm}(r)V_{ml}(r-r')\rangle \quad (4.26)$$

The sum on the right-hand side of equation (4.23) can be omitted if we consider sufficiently large distances, since it contains rapidly oscillating terms. Then for the amplitudes of the average (coherent) component, we obtain the following equation

$$\frac{d|\langle C_l(r)\rangle|}{dr} = -\frac{\gamma_l(r)+\Gamma_l(r)}{2}|\langle C_l(r)\rangle| \quad (4.27)$$

4.3 COUPLED INTENSITY EQUATIONS

In this section, we obtain equations for averaged square quantities in the field. As a preliminary, we note that since the sound field $P(r,z)$ in (4.12) is random, like the amplitudes $C_l(r)$ in the decomposition into modes, then when averaging squared quantities it is sensible to identify mean (coherent) and fluctuating (incoherent) components of these. Thus,

$$C_l(r) = \langle C_l(r)\rangle + \delta C_l(r) \quad \langle \delta C_l(r)\rangle = 0$$

$$P(r,z) = \langle P(r,z)\rangle + \delta P(r,z) \quad \langle \delta P(r,z)\rangle = 0 \quad (4.28)$$

respectively, for the squares (of the mode intensities)

$$\langle |C_l(r)|^2\rangle = |\langle C_l(r)\rangle|^2 + \langle |\delta C_l(r)|^2\rangle \quad (4.29.1)$$

and also for the square of the field, proportional to the total (summed over the modes) intensity

$$\langle |P(r,z)|^2 \rangle = |\langle P(r,z)\rangle|^2 + \langle |\delta P(r,z)|^2 \rangle \tag{4.29.2}$$

If we define the square of the modulus of the mean amplitude to be a quantity proportional to the intensity of the coherent component, delivered by the mode with index l,

$$|\langle C_l(r)\rangle|^2 = U_l(r) \tag{4.30}$$

then from (4.27), we immediately have the equation

$$\frac{dU_l(r)}{dr} = -[\gamma_l(r) + \Gamma_l(r)]U_l(r) \tag{4.31.1}$$

which must be supplemented by the 'initial' conditions, determined by the source

$$U_l(0) = |C_l(0)|^2 \tag{4.31.2}$$

Let us now consider the variation of the averaged square of the mode amplitudes; in other words, the total intensity of the sound, delivered by the mode with index l along the waveguide. For this, we shall integrate equation (4.15) from $r + \Delta r$ and iterate it once, thus obtaining the integral equation

$$\begin{aligned} C_l(r+\Delta r) = C_l(r) &- \int_r^{r+\Delta r} \frac{\gamma_l(r')}{2} C_l(r')dr' \\ &+ i\sum_m \int_r^{r+\Delta r} V_{lm}(r') \exp\left[i\int_r^{r'} q_{ml}(\rho)d\rho\right] C_m(r')dr' \\ &- \sum_{m,n} \int_r^{r+\Delta r} dr' \int_r^{r'} dr'' V_{lm}(r'')V_{mn}(r'') \\ &\times \exp\left[i\int_r^{r''} q_{ml}(\rho)d\rho + i\int_r^{r''} q_{nm}(\rho)d\rho\right] C_n(r'') \end{aligned} \tag{4.32}$$

We multiply this expression by the complex conjugate and leave only quadratic terms in the perturbation of the variable. After averaging over instances, we obtain

$$\begin{aligned} \langle |C_l(r+\Delta r)|^2\rangle = \langle |C_l(r)|^2\rangle &- \frac{1}{2}\int_r^{r+\Delta r} \gamma_l(r')\left[\langle C_l(r')C_l^*(r)\rangle + \langle C_l(r)C_l^*(r')\rangle\right]dr' \\ &+ \sum_{m,m'} \int_r^{r+\Delta r} \chi' \int_r^{r+\Delta r} dr'' \langle V_{lm}(r')V_{lm'}(r'')\rangle \langle C_m(r')C_{m'}^*(r'')\rangle \\ &\times \exp\left[i\int_r^{r'} q_{ml}(\rho)d\rho - i\int_r^{r''} q_{m'l}(\rho)d\rho\right] \\ &- 2\mathrm{Re}\left\{\sum_{m,n} \int_r^{r+\Delta r} dr' \int_r^{r'} dr'' \langle V_{lm}(r')V_{mn}(r'')\rangle \langle C_n(r'')C_l^*(r)\rangle \right. \\ &\left. \times \exp\left[i\int_r^{r'} q_{ml}(\rho)d\rho + i\int_r^{r''} q_{nm}(\rho)d\rho\right]\right\} \end{aligned} \tag{4.33}$$

The following simplifies the exponentials in the first and second sums in equation (4.33) to oscillate rapidly for $m' \neq m$ and $n \neq l$, respectively. Discarding terms with rapidly oscillating exponentials, we determine that in the sums only the quantities with $m' = m$ and $n = l$ remain. Let us denote

$$\langle |C_l(r)|^2 \rangle = W_l(r) \tag{4.34}$$

The quantity $W_l(r)$ on the right-hand side of (4.34) has the sense of the weight of the mode with index l (the term 'population density' used in quantum mechanics is more appropriate) in the decomposition of the field into normal modes and determines the intensity delivered by the mode with index l.

We shall also assume that the variation of the mode amplitudes is comparatively slow (in comparison with the scale of the correlation of the inhomogeneities). Subsequent transformations of expression (4.33) are based on assumptions about the 'discretisation step' Δr. We shall suppose that there exists a scale Δr, greater than the longitudinal radius of correlation of random inhomogeneities $R_{\|}$, but less than the characteristic scale of deterministic longitudinal variations of the waveguide parameters $\tilde{L}$ or, in more detail, $\tilde{L}_H \approx |H/H'|$, $\tilde{L}_c \approx |c/c'|$ (depths and sound speed profile) and correspondingly for the scales of variation of the eigenfunctions and eigenvalues $\tilde{L}_q \approx |q/q'|$ (the prime denotes the derivative with respect to r). The approximate extent to which these conditions are satisfied can be satisfied. As we have already said, typical values of $R_{\|}$ are 100 m, whereas $\tilde{L}_H$, $\tilde{L}_c$ and, what is more, $\tilde{L}_q$ are of the order of several kilometres, so that the inequality

$$R_{\|} \ll \Delta r \ll \tilde{L}_H, \tilde{L}_c, \tilde{L}_q, r \tag{4.35}$$

is completely possible in real conditions. This means that we can take C_l, W_l, q_{lm} out from under the integrals in the range from r to $r + \Delta r$ for values of the argument r. Thus, for the mode intensities, we obtain a system of transfer equations:

$$W_l(r + \Delta r) = W_l(r) - \gamma_l(r) W_l(r) \Delta r + \Delta r \sum_m a_{lm}(r)[W_m(r) - W_l(r)] \tag{4.36}$$

where the coefficients a_{lm} are defined in (4.25).

If we are interested in the field at a sufficiently large distance from the source, according to the inequalities (4.35), Δr may be assumed to be infinitely small. After moving to the limit $\Delta r \to 0$ in (4.36), we obtain an equation for the change of weight of the mode with index l.

$$\frac{dW_l}{dr} = -\gamma_l W_l + \sum_m a_{lm}(W_m - W_l) \tag{4.37.1}$$

$$W_l(0) = |C_l(0)|^2 \tag{4.37.2}$$

We note that all the quantities in equation (4.37.1) depend on r. Summing over l on the right- and left-hand sides of equation (4.37) we obtain the given equation determining the decrease in the total sound intensity as it propagates in the waveguide:

$$\frac{d}{dr}\sum_l W_l(r) = -\sum_l \gamma_l(r) W_l(r) \tag{4.38}$$

We note (Dozier and Tappert (1978)) that if $a_{lm} = 0$, $\gamma_l = 0$, then as $r \to \infty$, we have $dW_l/dr \to 0$, and from the equality to zero of the right-hand side of equation (4.38), we see that $W_m = W_l =$ constant, i.e. mode intensities are smoothed due to mode coupling. Using the quantity $W_l(r)$, we can find a number of characteristics of the sound field in the waveguide. For example, the vertical correlation function, averaged in the longitudinal direction over a sufficiently large interval

$$\begin{aligned} B(r; z, z') &= \frac{1}{\sqrt{\rho\rho' cc'}} \langle \mathrm{Re} p(r,z) \mathrm{e}^{-i\omega t} \mathrm{Re} p(r,z') \mathrm{e}^{-i\omega t} \rangle \\ &= \frac{\rho_0 c_0 W_0}{2r\sqrt{\rho\rho' cc'}} \sum_l \frac{W_l(r)}{q_l(r)} \psi_l(r;z)\psi_l(r;z') = \frac{W_0}{2r}\sum_l \frac{W_l(r)}{q_l(r)} \psi_l(r;z)\psi_l(r;z') \end{aligned} \tag{4.39}$$

We recall that $c = c(z)$, $c_0 = c(z_0)$, $c' = c(z')$, and similar equations for ρ. The last identity was obtained neglecting the difference in the values of the density and the sound speed at different points in the factor (as in Section 3.5). The normalisation in (4.39) was chosen so that for $z = z'$, from (4.39) one obtains the sound intensity averaged over the longitudinal interference structure

$$\bar{I}(r,z) = \frac{1}{\rho c}\langle [\mathrm{Re} p(r,z)\mathrm{e}^{-i\omega t}]^2 \rangle = \frac{W_0}{2r}\sum_l \frac{W_l(r)}{q_l(r)} \psi_l^2(r;z) \tag{4.40}$$

Comparing this equation with (3.79) we see that without random inhomogeneities, $W_l(r) = \psi_l^2(0, z_0)\exp[-\int_0^r \gamma_l(r')dr']$. The expression for the intensity, averaged over the position of the receiver, can be obtained using the WKB approximation for $\psi_l(r;z)$:

$$\begin{aligned} \bar{I} &= \frac{W_0}{r}\sum_l \frac{W_l(r)}{\sigma_l(r,z)D_l(r)} \qquad \sigma_l(r,z) = \sqrt{k^2 n^2(z) - q_l^2(r)} \\ D_l(r) &= 2q_l(r)\int_{\underline{z}}^{\bar{z}} \frac{dz}{\sigma_l(r,z)} \end{aligned} \tag{4.41}$$

Equation (4.41) also describes the intensity averaged over the position of the source, if we suppose that $W_l(0) = 2q_l(0)/[\sigma_l(0,z_0)D_l(0)]$.

Similarly to the derivation of (4.40) one can obtain the averaged intensity of the coherent component of the sound field

$$\bar{I}_c(r,z) = \frac{W_0}{2r}\sum_l \frac{U_l(r)}{q_l(r)}\psi_l^2(r,z) \tag{4.42}$$

where we define the difference between the total and the coherent components to be the incoherent (fluctuating) component of the intensity

$$\bar{I}_n(r,z) = \bar{I}(r,z) - \bar{I}_c(r,z) = \frac{W_0}{2r}\sum_l \frac{W_l(r) - U_l(r)}{q_l(r)}\psi_l^2(r;z) \tag{4.43}$$

If we use the WKB approximation for the modes, then we can obtain expressions similar to (4.41) for the coherent and the incoherent components. Thus, the mode intensities of the sound field on propagation in a randomly inhomogeneous waveguide vary according to equations (4.31), (4.37). It is clear that as the sound propagates, energy is redistributed between modes (for example, the total intensity is smoothed) and there is also redistribution between the coherent and incoherent components. More precisely, the proportion of the coherent component is decreased and the fluctuating component grows.

4.4 DIFFUSION EQUATION: AVERAGED DECAY LAWS

The nature of the solution of the system of equations (4.31), (4.37) is determined by the values of the elements a_{lm}, which determine the mode coupling. For a very broad class of random inhomogeneities, the nature of the coupling is such that only adjacent modes couple, i.e. effectively the only modes coupling with mode l are modes $l-1$ and $l+1$. In the language of rays, this corresponds to a predominance of scattering at small angles. The possibility of using this approximation is determined by the dependence of the correlation function for the random inhomogeneities on the transverse coordinates, defining the integral (4.19) and, correspondingly, (4.25). In particular, it is clear that if we have δ-correlation with respect to z then effectively all modes couple and the above approach is not suitable. At the same time, if the transverse timescale is large, then we can go beyond the limits of (4.10), so that the region of applicability of this approach lies at some intermediate values of $R_\perp$. In Sidenko (1989), based on numerical analysis, it is shown that at values ($R_\perp/H \leq 0.1$) which often arise in real situations and when there are not too many high modes, the contribution of non-adjacent modes is at most 10%. This topic is also discussed in Kohler and Papanikolau (1979).

Thus, in the system of transfer equations (4.37) for the modal intensity, we are left with the coupling of adjacent modes only. In this case, the system of equations can be written in the form:

$$\begin{aligned}
\frac{dW_1(r)}{dr} &= -\gamma_1(r)W_1(r) + a_{21}(r)[W_2(r) - W_1(r)] \\
\frac{dW_l(r)}{dr} &= -\gamma_l(r)W_l(r) + a_{l+1,l}(r)[W_{l+1}(r) - W_l(r)] \\
&\quad - a_{l,l-1}(r)[W_l(r) - W_{l-1}(r)] \\
\frac{dW_M(r)}{dr} &= -\gamma_M(r)W_M(r) + a_{M.M-1}(r)[W_M(r) - W_{M-1}(r)]
\end{aligned} \tag{4.44}$$

where M is the number of propagating modes.

The system of equations (4.37) or (4.44) can be solved numerically and then the quantities $W_l(r)$ can be used to find the characteristics of the sound field. The authors did this (Katsnelson *et al.* (1998)) to calculate the intensity of the sound propagating on an acoustic track, taking into account inhomogeneities of the bottom. The latter were measured with high precision by echo sounding and were used to construct the precise mesoscale relief of the track. Calculations of the intensity were based on numerical solution of the system (4.37) or (4.44) along the track in areas of a partition with step less than the characteristic longitudinal scale of homogeneities, which is of the order of 10 m. The intensity thus constructed, as a function of distance to the source is, strictly speaking, a deterministic quantity. In addition, this dependence is extremely irregular and gives an idea of the fluctuations; this is due to the statistical roughnesses of the bottom.

In addition, an equation can be obtained which may enable us to construct analytic and semi-analytic solutions. One step in this direction will be our assumption that the number of propagating modes is sufficiently large and that the distances between eigenvalues are sufficiently small that the discrete variable, the mode index n, can be replaced by a continuous variable p. This hypothesis corresponds to the ideology of a transition to ray formulae with an averaged description of the propagation of sound (Section 3.5) and is applicable to description of the sound field to the same extent as in Section 3.5. In this case, the system of equations (4.44) is a finite-difference approximation of some differential equation with continuous variables. Accordingly, we make the substitution

$$\Delta l = (l+1) - l = l - (l-1) \to dp$$
$$\Delta F_l = F_{l+1} - F_l \to dF \tag{4.45}$$

Denoting $a_p(r) = a_{p,p-1}(r) = a_{p-1,p}(r)$ from equation (4.44) for indices $2 \le n \le N$ we obtain a differential equation

$$\frac{\partial W_p(r)}{\partial r} = -\gamma_p W_p(r) + \frac{\partial}{\partial p}\left(a_p \frac{\partial W_p(r)}{\partial p}\right) \tag{4.46}$$

where for $a_p(r)$ we can obtain the expression

$$\begin{aligned} a_p(r) = {} & \frac{k^4}{q_p q_{p\pm 1}} \int_{-\infty}^{\infty} d\rho \int_0^{H(r)} \frac{\rho}{\rho(z)} dz \int_0^{H(r)} \frac{\rho}{\rho(z')} dz' \\ & \times \psi_p(r;z)\psi_p(r+\rho;z)B(r,r';z,z')\psi_{p\pm 1}(r;z)\psi_{p\pm 1}(r+\rho;z') \\ & \times \exp\left[i\int_r^{r+\rho} (q_{p\pm 1} - q_p)\,d\rho\right] \end{aligned} \tag{4.47}$$

The limits of the integration in (4.47) come from the fact that we are considering random inhomogeneities in the water layer only.

The first and last equations of the system (4.44) lead to boundary conditions for the parameter p, the region of permissible values for which corresponds precisely to the mode indices from 0 to some value M, determined by the sound frequency and by the waveguide parameters. Modes with indices greater than M are 'leaking modes' and determine one of the mechanisms for the attenuation of sound in the waveguide. The de-excitation of the modes when this index is reached is described in Kohler and Papanikolau (1979) as radiation losses, and corresponds to a certain term on the right-hand side of the last equation of the system (4.44). In other words, this attenuation corresponds (or will also include) the coefficient γ_M. In the case of shallow water, in the presence of marked absorption in the bottom, the role of these radiation losses is small, since as a result of growing with the mode index the coefficients γ_l may decay long before the indices corresponding to the exit of a wave from the waveguide are reached. Thus, the range of variation of the parameter p is defined by the inequality $0 \leq p \leq p_c$, where p_c is determined by the number of propagating modes. One can compare, for example, the grazing angle of a ray corresponding to a given mode, with the bed, to the continuous variable p. In particular, for a Pekeris waveguide, the continuous variable p can be determined using the relation $p = (kH/\pi)\sin\chi \approx (kH/\pi)\chi$, where χ is the grazing angle of a ray corresponding to the mode with index p.

The boundary conditions at $p = 0$ define the behaviour of the coefficient a_p in this region. From equation (4.47) one can establish that this quantity behaves as a function of the wave vector (of the sound frequency) proportionally to k^2. The condition of finiteness of the right-hand side of the expression in equation (4.46) leads to the condition

$$\left(\frac{\partial W_p}{\partial p}\right)_{p=0} = 0 \tag{4.48}$$

The second condition, by virtue of this mode attenuation in a shallow-water waveguide can be extended to $p \to \infty$, and gives

$$W_{p=\infty} = 0 \tag{4.49}$$

Then, if we need to compute any of the observed characteristics of the sound field based on $W_p(r)$, instead of summing over modes, we have to integrate over the continuous variable p. In particular, the averaged intensity (4.41) will be determined by the equation

$$\bar{I} = \frac{W_0}{r} \int dp \frac{W_p(r)}{\sigma_p(z) D_p} \tag{4.50}$$

which is similar to equation (3.84) describing the averaged law for the decay in a ray approximation. If we use the relation between the continuous variable p and the ray grazing angle χ, given earlier, these equations can be reduced to each other. If we use the ray approach from the beginning, then an equation similar to (4.50) can be obtained by considering the propagation of sound in the waveguide as a diffusion of the intensity at the ray grazing angle, as is done in free space (Chernov (1975)).

To conclude this section, we give similar equations for the coherent and incoherent components of the averaged intensity. These comprise a differential equation

$$\frac{dU_p(r)}{dr} = -(\gamma_p + \Gamma_p)U_p(r) \tag{4.51}$$

similar to (4.31.1) and integral expressions

$$\overline{I}_c = \frac{W_0}{r}\int dp \frac{U_p(r)}{\sigma_p(z)D_p} \quad \overline{I}_n = \frac{W_0}{r}\int dp \frac{W_p(r) - U_p(r)}{\sigma_p(z)D_p} \tag{4.52}$$

where the functions $U_p(r)$ and $\Gamma_p(r)$ are similar to $U_l(r)$ and $\Gamma_l(r)$ (equation (4.24)) with continuous values of the index, having the same sense as for the $W_p(r)$.

Next, let us consider a number of examples of the solution of equation (4.46) for simple models of a shallow-water sound channel. In these simple models one can obtain a generalisation of the averaged decay laws, if the channel includes random inhomogeneities, the nature of which is not important in this case. It is only important that their characteristics satisfy the conditions for applicability of the theory given earlier (see (4.35) and (4.10)).

4.5 SOME EXAMPLES

4.5.1 Waveguide of constant depth

Let us consider the case of a waveguide having a constant depth along the track ($H = \text{constant}$) and a constant profile of the average sound speed (refractive index in equation (4.8) $n = 1$), the Pekeris model with random inhomogeneities. The bottom, as in Section 2.2, is an homogeneous fluid, absorptive half-space with corresponding values of the density and sound speed and a corresponding absorption coefficient α. The modes of a 'deterministic' waveguide can be constructed in analytic form (2.52) with the corresponding formulae for the normalisation coefficients and the eigenvalues. The mode attenuation coefficient γ_l appearing in (4.44), which is very important for us, is determined by equation (2.62). According to what has already been said, instead of the discrete variable of the mode index l, we have to introduce a continuously varying p and construct the function γ_p. We do this by simplifying (2.62) if possible. The simplifying factor is the assumption that we have sufficiently many modes propagating at small grazing angles, which means that, assuming $\sigma_l \approx l\pi/H \ll q_l \approx k$, $N_l^2 \approx 2/H$, $\kappa_l \approx k\sqrt{1-n^2}$ ($n = c/c_1$) for γ_p, we have:

$$\gamma_p = \frac{s\pi^2 p^2}{k^2 H^3} \tag{4.53}$$

where $s = \alpha n^2 m(1-n^2)^{-3/2}$ or, which is the same thing for small α, it is determined by equation (2.87). The diffusion coefficient $a_p = a_{p,p-1} = a_{p,p+1}$ can be computed by substituting in expression (4.47) the function $\psi_l(z)$ according to equation (2.52) and the correlation function $B(r, z; r', z')$ chosen based on the specific model of the inhomogeneities. Similarly to Kohler and Papanikolau (1979), one can show that the dependence on p for the diffusion coefficient appears in the form of a correction $\approx p\pi/kH$, and, bearing in mind that because of absorption we are interested in the region of values $p \ll kH/\pi$ and subject to the condition that the longitudinal scale of the correlation of the random inhomogeneities $R_\parallel$ is small in comparison with the ray cycle distance D, this dependence can be neglected. Thus, we shall assume in this section that the diffusion coefficient a does not depend on the parameter p of the theory.

Thus

$$\frac{\partial W_p(r)}{\partial r} = -\frac{s\pi^2}{k^2 H^3} p^2 W_p(r) + a \frac{\partial^2 W_p(r)}{\partial p^2}$$

$$W_p(0) = W_0(p)$$

$$\left.\frac{dW_p}{dp}\right|_{p=0} = 0 \tag{4.54}$$

The 'initial' conditions $W_0(p)$ are determined by the sound source. Since, due to the growth of the absorption coefficient γ_p with index p, the distribution function decreases rapidly, as $p \to \infty$, we may suppose that the variable p ranges from 0 to ∞. Because the coefficient γ_p is an even function of p, the function W_p can be evenly extended to the region of negative values of p, where when calculating the integrals over p, we must take a factor of 1/2 in front of the integral.

Equation (4.54) has the form of a Schrödinger equation for a harmonic oscillator, whose solution can be given for arbitrary initial conditions $W_0(p)$:

$$W_p(r) = \frac{1}{\sqrt{4\pi a r_0 \sinh(r/r_0)}} \times \int_{-\infty}^{\infty} W_0(q) \exp\left\{\frac{1}{2ar_0}\left[-\frac{p^2+q^2}{2}\coth(r/r_0) + \frac{pq}{\sinh(r/r_0)}\right]\right\} dq \tag{4.55}$$

In particular, if the source is a non-directional point source, then from equation (4.39) we obtain $W_0(p) = H^{-1}$ and

$$W_p(r) = \frac{1}{H\sqrt{\cosh(r/r_0)}} \exp\left[-\frac{p^2}{4ar_0}\tanh(r/r_0)\right] \tag{4.56}$$

where

$$r_0 = \frac{kH}{2\pi}\sqrt{\frac{H}{as}}. \tag{4.57}$$

Clearly, $W_p(r)$ is the product of an amplitude factor $\cosh^{-1/2}(r/r_0)$ by a Gaussian distribution, whose width (i.e. the effective number of modes propagating) decreases as r increases, but is bounded below by the limiting width equal to $2\sqrt{ar_0}$. The characteristic spatial scale of these variations is r_0, which parameter takes into account the role of the random inhomogeneities in the waveguide.

Let us consider the limiting cases. If $r \ll r_0$, then

$$W_p(r) \approx H^{-1} \exp(-\pi^2 s p^2 r / k^2 H^3)$$

which corresponds to attenuation of an individual mode in the Pekeris waveguide. The factor in front of r in the exponential is the mode attenuation coefficient (4.53). In this limit, the diffusion coefficient a in equation (4.55) vanishes, which means that at distances less than r_0, random homogeneities are not manifest. If $r \gg r_0$ then

$W_p(r)$ takes the form $(r^{-3/2}\pi/H)\exp(-r/2r_0 - p^2/4ar_0)$. The intensity transported along the waveguide decreases exponentially, with mode attenuation coefficient $(2r_0)^{-1}$, independent of p, and the width of the mode intensity distribution (which is still Gaussian) becomes a fixed constant.

Since the width of the distribution determines the number of modes propagating in the waveguide, it is clear that there are infinitely many modes at $r = 0$, while the number decreases due to mode attenuation at large indices, but not to one as in Brekhovskikh (1980), because of the mode conversions on random inhomogeneities. The dependence of the width of the distribution on the waveguide parameters – its growth as H and a increase and its decrease as s grows – are evident. Let us now compute the average intensity according to (4.39), where the summation over modes must be replaced by integration over the continuous variable p (4.50). The normalisation of the radiated power is chosen as in Chapter 3 for the derivation of (3.84) $W_0 = 4\pi$. Substituting (4.56) in (4.50), we obtain

$$\bar{I} = \frac{2\pi}{kH^2 r}\sqrt{\frac{\pi r_0 a}{\sinh(r/r_0)}} \tag{4.58}$$

This equation gives the average decay law for the intensity along the waveguide, and generalises the known 3/2 law (Brekhovskikh, 1965). The characteristic parameter here is r_0. If $r \ll r_0$, then from (4.58) we obtain the law $r^{-3/2}$, describing the decay in the intensity of sound in an underwater sound channel without random inhomogeneities.

Let us now consider the behaviour of the averaged coherent component of the sound field intensity, which is obtained from equation (4.52), where the distribution function for the coherent component U_p is computed from equation (4.51). The attenuation coefficient for the coherent component in our approximation based on equation (4.24) is $\Gamma_p = a_{p,p-1} + a_{p,p+1} = 2a$. Substituting this and the attenuation coefficient γ_p from equation (4.53), we obtain

$$\frac{\partial U_p(r)}{\partial r} = -\left(\frac{s\pi^2 p^2}{k^2 H^3} + 2a\right) U_p(r) \tag{4.59}$$

$$U_p(0) = W_p(0) = W_0(p) = H^{-1} \tag{4.60}$$

Equation (4.59) with the initial conditions (4.60) is easily solved

$$U_p(r) = \frac{1}{H}\exp\left[-\left(\frac{s\pi^2 p^2}{k^2 H^3} + 2a\right) r\right] \tag{4.61}$$

whence one can find the averaged intensity of the coherent component

$$\bar{I}_c = \sqrt{\frac{\pi}{sH}}\, r^{-3/2}\exp(-2ar) \tag{4.62}$$

Comparing equations (4.62) and (3.92), we see that the decay of the coherent component is determined by the '3/2 law' with an exponential factor determining an additional decrease in the coherent component due to scattering on random

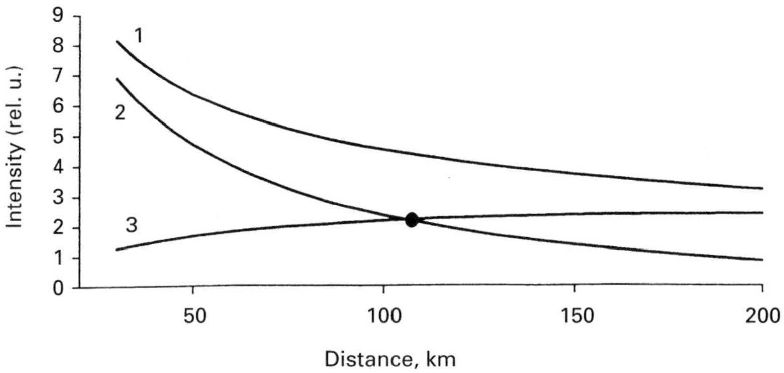

Figure 4.2. Averaged decay laws of sound with frequency $f = 300\,\text{Hz}$ in a waveguide with random inhomogeneities. Curve 1 – total intensity; 2 – coherent component; 3 – incoherent component. Waveguide parameters: $H = 150\,\text{m}$, $c = 1500\,\text{m/s}$, dimensionless coefficient of absorption in the bottom $s = 0.1$, diffusion coefficient $a = 3 \times 10^{-6}\,\text{m}^{-1}$.

inhomogeneities. Figure 4.2 shows the curves for the decay of the given components of the intensity for a model sound channel, characterised by the parameters indicated in the figure caption. As can be seen from Figure 4.2, the behaviour of the intensity of the wave field in the randomly inhomogeneous waveguide is associated with the operation of several mechanisms. These include absorption by the bottom, redistribution of the intensity between modes, and redistribution of the intensity between the coherent and the incoherent component. Near the source, when the effects of scattering on random inhomogeneities are small, the intensity of the sound field is concentrated mainly in the coherent component. (In the language of modes, this means that the fluctuations are small and a comparatively stable interference structure should be observed.) As the observer moves away from the source, considerable losses of the total intensity and its coherent component occur, whereas the incoherent component grows with the inflow of energy from the coherent component (the fluctuations grow, both in the relative sense, and in absolute value). At a certain distance from the source the incoherent component (fluctuations of intensity) reaches a maximum and then decreases because its loss to the bottom is not compensated by an inflow of energy from the coherent component due to scattering. At large distances all the components of the intensity decrease monotonically. Based on the above analysis, we can introduce the characteristic distance r_{coh} (scale of coherence) at which the coherent component and the incoherent component are equal; for $r > r_{\text{coh}}$ the radiation is dominated by the incoherent component. In the language of modes, we can say that at distances less than the scale of coherence we observe a comparatively stable interference structure, while at large distances fluctuations predominate and the field has a substantially stochastic nature. For our model case (Pekeris model with random inhomogeneities) we have the estimate $r_{\text{coh}} \sim 0.35/a$, which gives for the parameters of our waveguide, $r_{\text{coh}} \sim 100\,\text{km}$.

4.5.2 Irregular waveguide

Let us now consider an example of a very simple irregular waveguide with constant refractive index ($n = 1$) and a depth which is a function of the coordinate r: $H = H(r)$. We shall assume that this dependence is sufficiently smooth (varies little in a ray cycle distance). This problem is analysed based on a decomposition of the amplitude of the sound field $\Psi(r; z)$ into local modes $\psi_l(r, z)$, depending on r, as a parameter. In the same way, the quantities σ_p, q_p, ψ_p, D_p, γ_p exhibit a dependence on r, which can be formally obtained by substituting $H(r)$ in place of H. As noted in Sections 3.2 and 3.4, this approach can be used in the case of a sufficiently smooth variation of the waveguide depth with distance, which is determined by the non-adiabaticity parameter $\nu = kHH'/2$. If we assume that $H' \sim 10^{-3}$, $H \approx 100\,\text{m}$, $k \approx 2$, the parameter is small for several tens of modes propagating for those waveguide parameters, and the above approach is legitimate.

The dependence of the parameters upon r means that in accordance with the diffusion equation for the mode intensity $W_p(r)$ the absorption and diffusion coefficients will depend on r. For γ_p this dependence is obtained from (4.53), and we assume that a does not have such a dependence. This assumption corresponds, for example, to a scattering layer of constant width in the waveguide, or to the fact that wave transformation is caused by boundary roughnesses. On the other hand, in the presence of a dependence $a(r)$, by a simple replacement of variables it can be transferred to a coefficient in front of the first term on the right-hand side of (4.54). Thus

$$\begin{aligned} \frac{\partial W_p(r)}{\partial r} &= -\frac{s\pi^2}{k^2 H^3(r)} p^2 W_p(r) + a \frac{\partial^2 W_p(r)}{\partial p^2} \\ W_p(0) &= W_0(p) \\ \left.\frac{dW_p}{dp}\right|_{p=0} &= 0 \end{aligned} \tag{4.63}$$

Since in a smoothly irregular waveguide the index of a local mode (or in our case the parameter p) can be compared with an adiabatic invariant, equation (4.63) is interpreted as a diffusion equation in an adiabatic invariant. A similar approach was used in problems concerning the propagation of radio waves in the ionosphere by Gurevich and Tsedilina (1979).

A solution of equation (4.63) for a non-directional point source has the form:

$$W_p(r) = \frac{1}{H_0} \exp\left[-2a \int_0^r u(r')dr' - u(r)p^2\right] \tag{4.64}$$

where $H_0 = H(0)$, and u is a function satisfying the Riccati equation

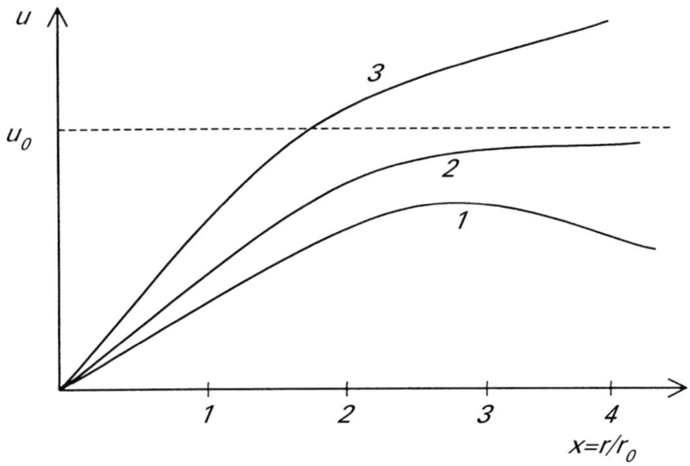

Figure 4.3. Graphs of $u(x)$ for a waveguide with a linearly varying depth: 1 – broadening waveguide ($\varepsilon = 0.1$); 2 – regular ($\varepsilon = 0$); 3 – narrowing ($\varepsilon = -0.1$); $4ar_0 = 1$; dashes – value of $u_0 = (4ar_0)^{-1} = 1$, corresponding to the asymptotic value of u in a regular channel for $H = H_0$.

$$\frac{du}{dx} + qu^2 = \frac{1}{qf^3(x)} \tag{4.65}$$

$$u(0) = 0 \; x = \frac{r}{r_0} \; r_0 = \frac{kH_0}{2\pi}\sqrt{\frac{H_0}{as}} f(x) = \frac{H(r)}{H_0} \; q = 4ar_0$$

It is clear that the distribution of the intensity by mode remains Gaussian with width $u^{-1/2}$ (as $r \to 0$ the width tends to infinity, which corresponds to all the excited modes). Let us mention some subcases. If $H = \text{constant}$, then the solution of (4.65) is a function $u = (4ar_0)^{-1}\tanh(r/r_0)$, in accordance with (4.56). For $a = 0$, $u = (s\pi^2 k^{-2} \int_0^r H^{-3}(r')dr'$ and the exponent corresponds to the attenuation coefficient of an adiabatic mode, obtained in a semi-classical approximation. Equation (4.65) is easy to solve numerically. For specificity, we assume that the waveguide depth varies linearly: $f(x) = 1 + \varepsilon x$, where ε is a dimensionless parameter and $\varepsilon > 0$ corresponds to a broadening channel and $\varepsilon < 0$ to a narrowing channel. The results of numerical calculations for equation (4.65) are shown in Figure 4.3. Let us first consider a broadening channel. Curve 1 clearly behaves non-monotonically: it grows at first, reaching a maximum at some value x, depending on q, and then falls. Recalling that $u^{-1/2}$ is the width of the Gaussian distribution W_p, the behaviour of $u(x)$ is interpreted as follows. At the point $x = 0$, $u = 0$; that is, the width of the distribution is infinite, which corresponds to the fact that all modes (rays) are present. As propagation occurs, because of absorption in the bottom, modes with high indices (steep rays) are strongly absorbed and for $r < r_0$, no random inhomogeneities are manifest, and the packet becomes narrower. For $r > r_0$ the diffusion mechanism takes effect, leading to the appearance of modes with large indices. In a channel of constant width

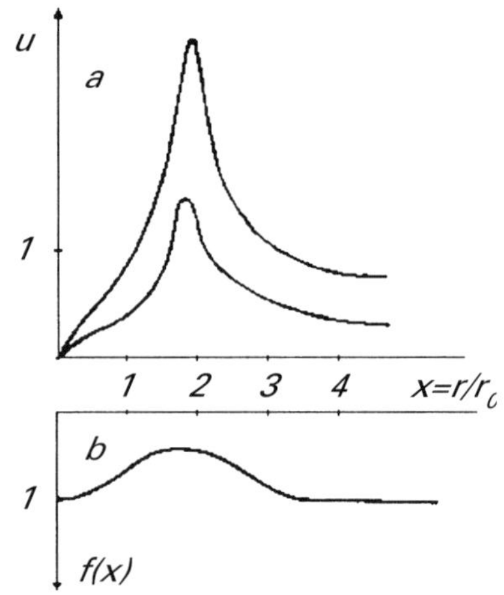

Figure 4.4. Graphs of $u(x)$ for an irregular waveguide (a) and its profile (b). 1 – $4ar_0 1$; 2 – $4ar_0 = 0.5$.

the birth and attenuation of modes cancel each other out, and at large distances the function $u(x)$ approximates asymptotically to a constant. In a broadening channel after some maximum value is reached $u(x)$ decreases as $x \to \infty$; in other words, the distribution becomes broader and new modes appear. This happens because the ray grazing angle with the boundaries in a broadening channel $\chi(r) = (H_0/H(r))\chi(0)$ decreases and, consequently, its absorption in the boundary decreases, in such a way that the appearance of new rays due to scattering on random inhomogeneities is not compensated by absorption. In the case of a narrowing channel, the angle $\chi(r)$ grows as r increases and absorption in the boundary dominates. Here, the function grows sharply as r increases; in other words, the distribution narrows.

We note that in a broadening channel, there exists a distance x_m at which the number of propagating modes is maximal. At that point $qu_{\max}^2 = 1/qf^3(x_m)$, $u_{\max} = (1/q)f^{-3/2}(x_m)$. If the waveguide has a complex profile, then by decomposing it into fragments in which it is purely narrowing or broadening, one can gain a qualitative understanding of the behaviour of the distribution of intensity. An example of such a waveguide is shown in Figure 4.4, where the behaviour of the curves can be understood from the above discussions. The average intensity has the form:

$$\bar{I} = \frac{1}{rH(r)}\sqrt{\frac{\pi}{u(r)}}\exp\left[-2a\int_0^r u(r')dr'\right] \tag{4.66}$$

In the limiting case $a = 0$, equation (3.91) follows from (4.66). For $H = H_0$, (4.66) becomes (4.58).

It is clear from (4.66) that the behaviour of the average energy density is determined by the function $u(x)$. The presence of random inhomogeneities leads to a faster exponential decay, where, even if u decreases with distance (in a broadening channel) the integral in the exponential grows. An analysis of the coherent component of the intensity in this case can be performed similarly to that in the previous section.

4.5.3 Depth dependence of the intensity

Let us now consider a reconstruction of the vertical (depth) dependence of the intensity of sound from a point source in shallow water with random inhomogeneities. Suppose that the average part of the square of the refractive index in the water layer is described by a linear dependence on depth:

$$n^2(z) = n_s^2(1 + \varepsilon z) \tag{4.67}$$

where n_s is the refractive index at the surface. We know that in the case of a negative sound speed gradient, unlike in the examples considered, the absorption coefficient depends on p, in a different way from (4.53), which has to do with the presence of groups of near-bottom modes (rays) with approximately the same attenuation coefficient. A characteristic dependence of γ_p is shown in Figure 2.8. Equation (4.54) can only be solved numerically, based on a representation of the distribution $W_p(r)$ as a sum over the eigenfunctions of the Sturm problem:

$$W_p(r) = \sum_n P_n(p) \exp(-\lambda_n r) \tag{4.68}$$

where the dependence $P_n(p)$ and λ_p are determined by

$$\frac{d}{dp}\left(a_p \frac{dP_n}{dp}\right) + (\lambda_n - \gamma_p)P_n = 0 \tag{4.69}$$

with boundary conditions $P'_n(0) = P_n(p_{\max}) = 0$.

It is clear that as r increases only one term remains in the sum (4.68); namely, that corresponding to the least value of the parameter λ. We refer to this as the asymptotic solution (in Artelnii and Raevskii (1987), the corresponding limit is called the self-similar limit; see also Whitman and Beran (1978)). Accordingly, the distribution of the field intensity with depth tends to some 'asymptotic limit':

$$\bar{I}^*(r, z) = \frac{W_0}{r} e^{-\lambda^* r} \int_0^{p_{\max}} \frac{P^*(p)}{\sigma_p(z) D_p} dp \tag{4.70}$$

where λ^* is the minimal eigenvalue and $P^*(p)$ is the corresponding eigenfunction.

Then to obtain estimates, we shall suppose that the diffusion coefficient is independent of the mode index. To estimate the distances of the self-similar regime one can use the following relation: $\min(\lambda_n - \lambda^*)r > 1$ for $n > 1$; in particular, for the Pekeris waveguide:

$$r > R_c = \frac{kH}{4\pi}\sqrt{\frac{H}{a_p s}} \tag{4.71}$$

For a waveguide with a negative sound speed gradient the estimates will be rather different. We shall suppose that the effects of scattering are so small that in the self-similar regime only 'bottom' modes remain. i.e. $16a/(s\varepsilon^2 H^3 k^2) \ll 1$. This condition is well satisfied for typical conditions $f = 300\,\text{Hz}$, $H \sim 100\,\text{m}$, $s \sim 0.1$, $a_p \sim 10^{-5}\,\text{m}^{-1}$, $\varepsilon \sim 2.5 \times 10^{-4}\,\text{m}^{-1}$, $c_S \sim 1500\,\text{m/s}$. In this case, for the estimates of the distances of the self-similar regime and of the distribution of intensity, we obtain the simple expressions:

$$r > R_c = \frac{2}{3}\frac{\varepsilon k^2 H^3}{\pi^4 a_p}, \qquad 0 \leq z < H - \left(\frac{3\pi}{2kn_s}\right)^{2/3} \varepsilon^{-1/3},$$

$$\bar{I}^*(r,z) = \frac{\tilde{W}_0}{kn_s Hr} e^{-\lambda^* r} \sqrt{\frac{\varepsilon}{H-z}} \operatorname{arccot} \sqrt{\frac{H-z}{H}} \tag{4.72}$$

where $\tilde{W}_0$ is some normalisation value for the power of source.

As one might expect, the speed of the decay in intensity is determined by the gradient of the refractive index, where the longitudinal scale of coherence grows as the gradient of the speed increases. This has to do with the fact that for large values of ε, the number of mixed modes decreases. We note that because of the need to satisfy the above inequality the transition from equation (4.72) to equation (4.71) in the limit does not take place. For typical shallow-water conditions of a drop in the sound speed between the bottom and the surface $\geq 10\,\text{m/s}$, we have $R_c \geq 70\,\text{km}$.

We note that the sense of the limiting regime is that at large distances an equilibrium distribution of $W_p(r)$ is obtained over the modes: the 'dying out' of the higher modes as a result of greater absorption than for lower modes is compensated by a transfer of energy to them as a result of scattering on inhomogeneities. As a result, the shape of the curve of the depth dependence is unchanged with distance. Figure 4.5 shows numerical calculations of the dependence of the sound field intensity on depth at different distances from the source. In the calculations, both waveguide boundaries were assumed to be pressure-release boundaries. The calculation was carried out for the values $f = 300\,\text{Hz}$, $H = 100\,\text{m}$, $s = 0.1$, $a_p = 10^{-5}\,\text{m}^{-1}$, $\varepsilon = 2.5 \times 10^{-4}\,\text{m}^{-1}$, $c_S = 1500\,\text{m/s}$. The nature of the calculated dependence of the intensity on depth can be understood to be the result of scattering on inhomogeneities of radiation energy concentrated in modes with the least attenuation; in this case, the near-bottom modes. Because of this, the graph of the field intensity becomes smoother and, as a result of the redistribution of energy over the modes, most of the radiation energy is concentrated at depths with a lower sound speed. A more important detail is the broadening of the interference structure. It is clear that at comparatively small distances the vertical interference structure is little disrupted, whereas at large distances (for our values, starting from $\sim 100\,\text{km}$) it is practically blurred. In Figure 4.5, curve 1, for a distance of 200 km, already corresponds to the limiting distribution (for the chosen values of the waveguide parameters, the longitudinal scale of coherence is $\sim 170\,\text{km}$). We note that unlike at small distances from the source, where there is a stable interference structure, at the distances of the limiting regime this constant interference field no longer exists. Thus, in sounding (or recovering the waveguide parameters) at such distances one needs to use the statistical characteristics of the field.

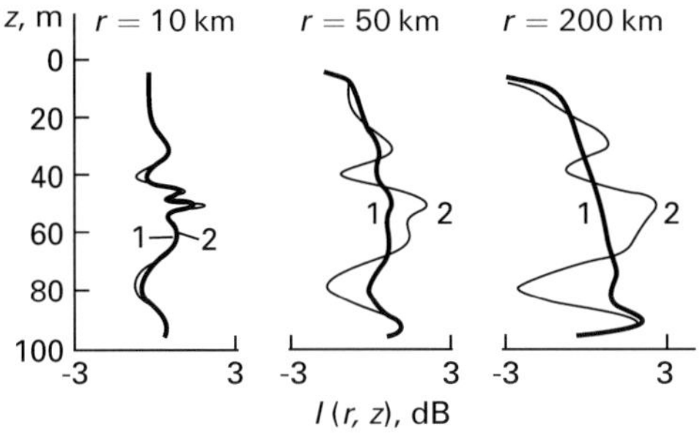

Figure 4.5. Reconstruction of the depth dependence of the intensity of sound at frequency $f = 300$ Hz, in a waveguide with a linear sound speed profile $c(z) = c_S(1 - \varepsilon z)$, $c_S = 1500$ m/s, $\varepsilon = 2.5 \times 10^{-4}\,\mathrm{m}^{-1}$, $H = 100$ m, depth of source 50 m, diffusion coefficient $a = 10^{-5}\,\mathrm{m}^{-1}$, dimensionless coefficient of absorption in the bottom $s = 0.1$. Distances from the source are shown in the Figure. Curves 1 and 2 for each distance describe intensity with and without random inhomogeneities correspondingly.

Thus, in this chapter, we have formulated the tools (techniques) for describing, calculating and interpreting data on the propagation of sound in a waveguide with random inhomogeneities of an arbitrary nature. These tools are constructed in the language of modes and take into account the properties of a shallow-water waveguide as a very specific random inhomogeneous medium. The key points of the description are various forms (in one approximation or another) of the propagation equation for the modal intensity distribution function. Within this approach, we have introduced a specific parameter of a random inhomogeneous waveguide, namely the diffusion coefficient, taking into account both parameters of the waveguide and statistical characteristics of the medium filling it. These results are of particular value for estimation of the degree of stability of the interference pattern, the measurement and analysis of which form the subject of Chapters 5 and 6, and also for certain problems associated with noise (Chapter 7).

5

Acoustic experiment in shallow water

5.1 FREQUENCY BAND USED FOR LARGE-SCALE ACOUSTIC MONITORING

The propagation of sound in different shallow-water regions has a large number of distinctive properties, associated with the specifics of difference shelf zones of the World Ocean. These specifics are due not only to the bottom relief, the structure and the acoustic properties of the bottom sediments, but also to the spatial distribution of the sound field and to the perturbations of the latter caused by characteristic hydrodynamic processes on the ocean shelf. Quantitative characteristics describing the propagation of sound are always geographically attached to a specific region and even to a time of year. This has stimulated a great many experimental studies of the long-range propagation of sound in various shallow-water regions. These studies have been directed not only towards the determination of numerical characteristics, but also towards the development of the physical foundations for acoustic monitoring of the ocean shelf. Coarsely, these studies can be divided into the following directions:

- Attenuation of sound on propagation.
- Space–time fluctuations of sound waves, associated with hydrodynamic processes in shallow water.
- Dispersion distortions of sound signals.
- Matched field signal processing in acoustic monitoring of shallow-water.

The results of studies in each of these directions and their interpretation have been presented in Chapter 4. However, their analysis enables us to identify a number of characteristic traits of the technical aspects of physical experiments associated with the acoustic sensing of shallow water at large distances.

First and foremost among these traits, we must mention the application of sound signals with frequencies less than 400 Hz. The use of such low frequencies is primarily associated with the strong attenuation of acoustic waves in shallow water due to the interaction with the bottom (see Section 6.1). For example, for a negative gradient of the sound speed profile, the transmission loss of acoustic signals ($\mathrm{TL} \equiv 10\lg(\bar{p}/\bar{p}_0)$) is described by the three terms, inside square brackets, corresponding to different decay laws for the sound field with distance

$$\mathrm{TL} = [20\lg(r_0/r_1)] + [15\lg(r_1/r^*) - \beta_1(r^* - r_1)] + [10\lg(r^*/r) - \beta(r - r^*)] \quad (5.1)$$

Here $\bar{p}$ is the mean acoustic pressure at the point of receipt, defined in terms of the mean intensity of the sound field ($\bar{p}(r) = \sqrt{I\rho c}$) (see Section 3.5)[1], and $\bar{p}_0$ is the acoustic pressure at distance $r_0 = 1$ m from the sound source ($\bar{p}_0 = \sqrt{I\rho c} = \sqrt{W_0\rho c/4\pi r_0^2}$). The first term corresponds to spherical spreading of the wave front at the start of the distance (2.3), the second to the so-called 'three over two law' with additional attenuation, and the third to cylindrical spreading with additional attenuation. The expression (5.1) is empirical. However, as shown in Section 3.5, the second term corresponds to the situation when a large number of modes are present in the waveguide propagation, and the third term corresponds to a number of near-bottom modes with the same attenuation coefficient (here, the coefficient β is given by $\beta = 10^4 u \lg e\,\mathrm{dB/km}$, where u is defined in equation (3.103)). The distances r_1 and r^* define the dissipation interval $[r_1 - r^*]$. in which the 'three over two' law is satisfied. The distance r_1 (it can be called the transitional distance) can be estimated using the following simple relation $r_1 \cong H^2/2\lambda$ (at this distance the Fresnel radius of a ray is equal to half the depth of the waveguide (Kravtsov *et al.* (1990))). For a waveguide depth of 100 m, the transitional distance is equal to 333 m at frequency 100 Hz (wavelength 15 m), and 1000 m at frequency 300 Hz (wavelength 5 m). The quantities r^*, β_1 and β depend on the length of the sound wave and the waveguide parameters, including the acoustic properties of the bottom. For example, for one region with a negative gradient of the sound speed profile (Bunkin *et al.* (1984)), these quantities were equal to $r^* = 23$ km, $\beta_1 = 0.8$ dB/km, $\beta = 0.24$ dB/km, for a waveguide depth of 100 m and a frequency of the acoustic signals of 100 Hz. At a frequency of 300 Hz, these same parameters were equal to $r^* = 13$ km, $\beta = 1.3$ dB/km, $\beta = 0.38$ dB/km. The figures show that the transmission loss increases sharply with frequency, and for an emission level of $\bar{p}_0 = 190$ dB[2] relative to 1 μPa, the sensing range for the waveguide studied in Bunkin *et al.* (1984) is $r = 235$ km at frequency 100 Hz and $r = 185$ km at frequency 300 Hz (estimates of r obtained using equation (5.1) assume that the

[1] In experiments, $\bar{p}$ is defined as $\bar{p} = \langle |p(\vec{R}, t)| \rangle$, where the pointed brackets denote averaging over a spatial interval greater than the period of the interference beats, and over a time greater than the period of the oscillations.

[2] The radiation level of 190 dB corresponds to the acoustic power ($W_0 \cong 100$ W) of typical non-directional low-frequency sound sources, used in experiments on acoustic tomography of the ocean.

signal-to-noise ratio at the point of receipt is at least 20 dB, and that the accumulation time is equal to one second).

These estimates demonstrate the orders of magnitude for the range of acoustic sensing in shallow water. These limiting ranges are significantly less than in a deep ocean, where experiments on the ultra long-range propagation of sound at distances of several thousand kilometres are known (Munk *et al.* (1994)). On the other hand, problems in acoustic monitoring of shallow water themselves usually require relatively small sensing distances.

Here, we once again stress that the significant growth in the attenuation of sound with frequency means that the use of high frequencies for acoustic sensing of shallow water at large distances holds absolutely no promise, despite the possibility of increasing the source power with frequency. In fact, in the above example, to obtain a sensing range of 235 km at frequency 300 Hz, requires an emission level of 209 dB, which is a very complex engineering problem.

5.2 SELECTION OF NORMAL WAVES

Small sensing ranges and low frequencies also lead to a spatial method for receiving and processing acoustic signals, which is characteristic for experiments in shallow water. In fact, in acoustic monitoring of the ocean, narrow-band signals are usually employed, whose spectral width $\Delta\omega$ is significantly narrower than the carrier frequency ω_0 (the ratio $\Delta\omega/\omega_0$ is usually at most 10%, which is basically associated with the use of resonant sound sources, having a very large power, unlike non-resonant wideband sources). In this case, when low frequencies are used, the value of $\Delta\omega$ is small. For short sensing ranges the main procedure of acoustic monitoring of the ocean (Spindel (1985)), involving the identification of signals corresponding to individual rays or modes, is often impossible. This is indicated in the first place by simple analytic estimates. Indeed, for a clear temporal resolution of two signals consecutively reaching the point of receipt and corresponding to two different modes of the waveguide, the difference in the arrival time τ_1 must satisfy $\tau_1 \geq \Delta\tau = 2\pi/\Delta\omega$, where the period $\Delta\tau$ is the duration of a signal at the output of an optimal receiver[3]. (Here, we have in mind a sufficiently good resolution of the two signals, enabling us not only to detect the one pulsed signal against the background of the other, but also to measure its amplitude with the requisite accuracy). On the other hand, for a Pekeris waveguide (see Section 2.2) with a rigid bottom ($m_1 \to 0$, $n_1 \to 0$), the difference in the time of arrival for the mth and the lth modes[4] is:

$$\tau_1 = \left(\frac{r}{v_m^{gr}} - \frac{r}{v_l^{gr}}\right) \cong \frac{\pi^2 rc\left[\left(m-\frac{1}{2}\right)^2 - \left(l-\frac{1}{2}\right)^2\right]}{2\omega_0^2 H^2} \tag{5.2}$$

[3] Note that the duration $\Delta\tau$ may be different from the duration T of the radiated signal. The latter determines the energy of the radiated signal, and consequently, the possibility of detecting it in a noise background.

[4] Equation (5.2) was obtained for the modes bearing the lowest energy, i.e. on the assumption that the transverse wavenumbers $\sigma_{m,l}$ are much less than k.

Thus, the signals of two adjacent modes with indices $m+1$ and m are only resolved at distances r between the sound source and receiver greater than:

$$r \geq r_{bm} = \frac{2H^2\omega_0^2}{\pi mc\,\Delta\omega} \tag{5.3}$$

For a signal frequency $f \equiv \omega_0/2\pi = 300\,\mathrm{Hz}$, and a waveguide depth $H = 100\,\mathrm{m}$, signals corresponding to the first and second modes are separated by their time of arrival only at distances greater than 80 km, even for a signal band ratio of $\Delta\omega/\omega_0 = 10\%$. Thus, the distances from which temporal separation of signals, corresponding to different modes, begins are comparable with the limiting sensing ranges, estimates of which were presented earlier. At lower distances these signals may be superimposed on each other. This same conclusion can also be reached from the results of full-scale experiments described in Section 6.3.

To distinguish signals at small distances it is advisable to use long chains of hydrophonic receivers and a special method for processing the received signals. In fact, technically it is not very difficult in shallow water to set up such a chain vertically covering the whole waveguide. The simplest means of tuning such a chain to receive just one mode involves summing the signals received at each hydrophone, taking into account a linear phase compensation law or, in other words, using the chain as a vertical linear array with directional diagram pointed in the direction of the Brillouin rays, corresponding to a received normal wave. If on reception one simultaneously forms two directional diagrams (using two phase compensators) pointing, respectively, to the upper and lower rays, and sums the signals from the two diagrams simultaneously, then, as shown in Eliseevnin *et al.* (1981) the equivalent array length is equal to twice the depth of the waveguide. The angular resolution of such an array in the vertical plane is $\lambda/2H$, which is again equal to the difference in the angles of arrival of the Brillouin rays of adjacent lower modes for the Pekeris waveguide.

Numerical modelling, performed in the same work by Eliseevnin *et al.* (1981) showed that the separation of modes using this method was still not sufficiently good and can only be used for a large signal-to-noise ratio at the point of reception. For better separation of modes, one has to increase the array angular resolution, which cannot be achieved using a vertical chain. In this case, help can be provided by a horizontal chain, extending in the direction of the sound source (Kravtsov *et al.* (1988)). Being phased, such a chain forms a so-called 'travelling-wave' array. In this case, there are no natural technical constraints on the array length L, associated with the depth of the waveguide, and its resolution is equal to $\cong 2\sqrt{\lambda/L}$. For a length $L \geq 16H^2/\lambda$ such an array can be used to distinguish signals corresponding to different modes. However, we note that in this case there are physical constraints on the length of the array. In fact, the quantity R_m, which can be called the length of the mth mode, to which the array is tuned, $R_m = r/\cos\chi_m$, must satisfy the far-zone requirement $R_m \geq L^2/\lambda$ (here, χ_m is the Brillouin angle of the mth mode). For $m = 1$ ($\cos\chi_1 \cong 1$) we obtain the estimate $L_{\max} = \sqrt{r\lambda}$. Thus, the following constraints are imposed upon the length of a horizontal linear array designed for mode identification:

$$16H^2/\lambda < L < \sqrt{rL} \tag{5.4}$$

The identification of signals corresponding to different modes can be performed in a stable manner in shallow water using a vertical chain of hydrophones, but using more complex matched field signal processing. The essence of this processing is the application of a set of matched filters, each of which is tuned to receive a signal corresponding to a single waveguide mode and has the following transfer function (Miller *et al.* (1993)).

$$Q_m(r, r', z_j, z', \omega) = \sum_j^{J_h} \delta(z' - z_j)\delta(r - r')\psi_m(r', z', \omega)S^*(\omega) \tag{5.5}$$

Here, j is the index of the receiving hydrophone, J_h is the number of receiving hydrophones, $\psi_m(r', z'; \omega)$ is an eigenfunction of the mth mode, where the dependence on frequency is given explicitly, r, z_j are the coordinates of the relative position of the hydrophones, and $S(\omega)$ is the spectrum of the radiated signal. The spectrum of the received signal at the output of such a filter is equal to:

$$Y_m(r, \omega) = \frac{1}{2\pi r}\int_V S(r', z', \omega)Q_m(r, r', z_j, z', \omega)dV' \tag{5.6}$$

where V is the waveguide volume, and $S(r', z', \omega)$ is the spectrum of the signal in the waveguide (at the input to the filter), which according to (1.12), (2.32) and (2.39) and neglecting the intramode dispersion, can be written in the following form:

$$S(r', z'; \omega) = S(\omega)\frac{1}{2}\sqrt{\frac{1}{2\pi r'}}\sum_l^M \frac{\psi_l(0, z_0; \omega)\psi_l(r', z'; \omega)}{\sqrt{\xi_l}} \exp[i(\xi_l r' + \pi/4)] \tag{5.7}$$

Here, M is the maximum number of energy-bearing modes. Because of the orthogonality of the eigenfunctions $\psi_m(r', z'; \omega)$ for a chain length equal to the waveguide depth, and a distance between hydrophones of less than half the wavelength, the spectrum γ_l (and, consequently, the output signal itself) depends only on the parameters of a normal wave:

$$Y_m(r, \omega) = C_m\sqrt{\frac{1}{\xi_m r}}\hat{N}_m|S(\omega)|^2 \exp(i\xi_m r) = C_m\sqrt{\frac{1}{\xi_m r}}\hat{N}_m|S(\omega)|^2 \exp\left(i\omega t_m - \frac{\gamma_m}{2}r\right) \tag{5.8}$$

where $\hat{N}_m$ is a coefficient, determining the gain of the chain under this signal processing technique. ($\hat{N}_m = \sum_j^{J_h} \psi_m^2(r, z_j, \omega)$), $C_m = \psi_m(0, z, \omega)e^{i\pi/4}/2\sqrt{2\pi}$, t_m is the signal propagation time corresponding to the mth mode. This signal processing technique has been successfully used in a tomographic experiment in the Barents Sea (Parsons *et al.* (1996)).

5.3 PURE-TONE SOUND SOURCES IN A SEA EXPERIMENT

In experimental studies of the attenuation of sound in shallow water, acoustic sensing is usually performed from a ship which is moving away, using both pure-tone (harmonic or narrow-band) and wideband (essentially explosive) sound sources. Both methods have their advantages and disadvantages. When pure-tone sources are used, the measurement accuracy is of course greater. This is primarily related to the possibility of performing almost continuous sensing as the ship with the towed sound source moves away. The discreteness of the measurements of the amplitude of the sound field with respect to distance, is equal to $\Delta r_d = \nu\hat{T}$, where ν is the speed of the ship and $\hat{T}$ is the signal processing (accumulation) time. For typical values of the speed of the ship $\nu \sim 1.5\,\text{m/s}$ and times $\hat{T} \sim 50\,\text{s}$, the value $\Delta r_d \sim 75\,\text{m}$ is usually less than the characteristic scale of variation of the sound field in the waveguide $\Lambda_{\min}$, determined by the minimum period of intermode interference beats. According to (2.49) and (2.50) for the Pekeris waveguide[5]

$$\Lambda_{\min} = 2\pi|q_1 - q_M|^{-1} = 2\pi\left|\sqrt{k^2 - \frac{0.56\pi^2}{H^2}} - \sqrt{k^2 - \frac{M^2\pi^2}{H^2}}\right|^{-1} \tag{5.9}$$

The estimated value of $\Lambda_{\min}$ for a waveguide of depth 100 m with constant sound speed is equal to $\Lambda_{\min} \cong 142\,\text{m}$, for a sound speed in the bottom of $c_1 = 1700\,\text{m/s}$ and wavelength $\lambda = 15\,\text{m}$ (The estimate of Λ_m is valid only for relatively small distances from the sound source. As the distance increases in shallow water only a few energy-bearing modes remain and the value of $\Lambda_{\min}$ becomes even larger. See Chapter 2 for more details.) This implies, according to the Shannon–Kotelnikov's theorem, that measurements of the amplitude of the sound field in the waveguide, and, in particular, of its changes due to the attenuation of sound, are completely correct.

The measurement errors in this case are in many respects associated with the presence of acoustic noise. For measurements against this background, the signal accumulation time should be increased to the maximum, particularly at large distances from the sound source, where the amplitude of the field is small. However, in experiments of this type, the time $\hat{T}$ is constrained by the instability of the movement of the ship towing the pure-tone source. Indeed, for random variations of ν, the Doppler effect leads not only to a frequency shift in the received harmonic signal by $\Delta f \cong f\nu/c$, but also to a 'blurring' δf of its spectrum. An example of such a spectrum for a radiated signal frequency $f = 100\,\text{Hz}$ and an average ship speed $\nu = 1.5\,\text{m/s}$ is shown in Figure 5.1 (Bunkin *et al.* (1984)). The width of the spectral line is equal[6] $\delta f = 0.01\,\text{Hz}$. Accordingly, the accumulation time in this case should be at most $\hat{T} \leq 1/\delta f \leq 100\,\text{s}$. Here, we note two things. First the quantity δf for a harmonic signal radiated from a moving ship is somewhat greater than the 'blurred' spectrum due to the hydrodynamic processes in shallow water (see

[5] For the first mode ($l = 1$) l^* is assumed to equal $l^* = 0.75$.

[6] Note that the width of the spectral line of the radiated signal here was determined by the stability of the frequency of the quartz generator and amounted to $\sim 10^{-4}$ Hz.

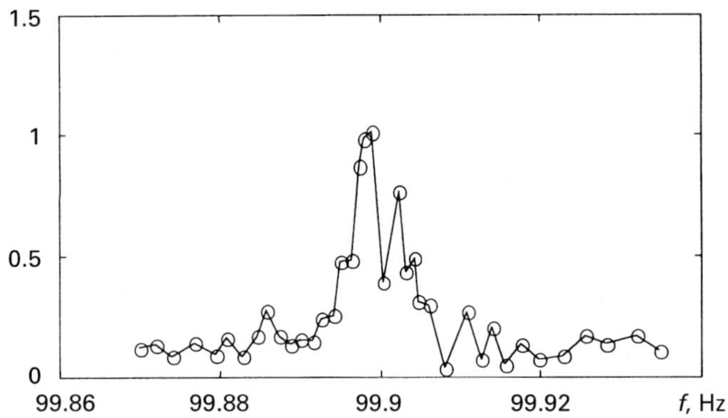

Figure 5.1. Spectrum of an acoustic signal with frequency 100 Hz, radiated from a moving ship.

Chapter 6 for more details). Second, the width of the spectral line in this case has an absolute minimum $\delta f_{\min}$ for a measurement time $\hat{T}_{\text{opt}}$ in the interval $0 \ll \hat{T}_{\text{opt}} < \tau_\nu$ ($\delta f_{\min} \cong 1/\hat{T}_{\text{opt}}$), where τ_ν is the correlation time of random variations of the ship speed (Gorskaya *et al.* (1981)). Usually, this time is of the order of $\tau_\nu \sim (1-5) \times 10^2$ s. For a longer or a shorter measurement time δf grows[7]. Thus, the accumulation time has some optimal value $\hat{T} = \hat{T}_{\text{opt}} \sim \tau_\nu/3$, corresponding to the most accurate measurement of the parameters of the acoustic field.

Sounding by harmonic signals from a ship moving away is sometimes also used to study the modal composition of the sound field in a waveguide, including the study of the attenuation of individual modes. The relatively simple methodology, from the experimental point of view, which we discussed earlier, is based on the fact that the Doppler frequency shift, strictly speaking, depends upon the phase velocity $\nu_m^{\text{ph}} = \omega_0/q_m$ of a normal wave: $\Delta f_m \cong f\nu/\nu_m^{\text{ph}}$ (Gorskaya *et al.* (1981)) (here, Δf_m is the signal frequency shift, corresponding to the mth normal wave). Spectral analysis of the received signal with a high frequency resolution, enables us to identify the spectral components corresponding to individual modes, even in the case of reception at a point receiver, without employing the long arrays, discussed earlier. Of course, that is only possible when the frequency interval between components is greater than the line width δf. Unfortunately, in practice, this condition is only satisfied at very low frequencies, where, on the one hand, δf is small, and, on the other hand, the frequency interval $\tilde{\Delta} f_m$ between the mth and the $(m+1)$th spectral components of a mode is relatively large. For the lowest modes of a Pekeris waveguide with a rigid bottom ($m_1 \to 0, n_1 \to 0$), $\tilde{\Delta} f_m$ can be estimated using the

[7] Note that in order to obtain the minimum width of the spectral line, one must not only choose an optimal measurement time, but also exclude frequency drift in the processing (see Burlakova *et al.* (1987) for more details).

following relation

$$\tilde{\Delta} f_m \equiv \Delta f_{m+1} - \Delta f_m = \frac{\nu}{2\pi}\left(\sqrt{k^2 - \sigma_{m+1}^2} - \sqrt{k^2 - \sigma_m^2}\right) \cong \frac{\pi \nu m c}{2H^2 \omega_0} \tag{5.10}$$

Here σ_m is the transverse wavenumber given by equation (2.49′) ($\sigma_m \ll k$).

For a waveguide depth $H = 100\,\text{m}$, frequency $\Lambda_{\min} = 20\,\text{Hz}$, $\nu = 1.5\,\text{m/s}$ and $m = 1$, we have $\tilde{\Delta} f_m = 0.0028\,\text{Hz}$. This value is comparable with $\delta f = 0.002\,\text{Hz}$, which one might expect at this frequency according to experimental data (Bunkin *et al.* (1984)). As equation (5.10) implies, as the frequency grows, $\tilde{\Delta} f_m$ decreases, and conversely, the value of δf itself, which is proportional to the frequency, increases, which makes it impossible to identify individual modes. Doppler selection of individual modes in shallow water has been successfully used, for example in Belov *et al.* (1996) for a sensing signal frequency of 17 Hz.

We note, however, that Doppler selection requires a large signal accumulation time and is only good for regular waveguides, whose characteristics do not vary in the interval $\Delta r = \nu \hat{T}$. In the case of weakly irregular waveguides, where in this region smooth variations of the parameters of the individual modes already occur, and the adiabatic approximation (see Chapter 3) is valid, normal waves are identified with high accuracy using the Wiener transform (Belov *et al.* (1996)).

5.4 WIDEBAND SOUND SOURCES

One evident disadvantage of the use of pure-tone signals is associated with the impossibility of determining the frequency dependence of the characteristics of the propagation of sound waves. Practically no full-scale experiments have succeeded in studying three pure-tone signals simultaneously, and signals from wideband sound sources are therefore usually used to determine the frequency dependence. The term 'wideband' covers sound sources in the form of explosions, pneumatic systems and electrical discharges. However, at low frequencies, only explosions provide sufficient power for long-range (up to several hundred kilometres) acoustic sensing. There are also low-frequency wideband sound sources of the piston type with powerful electric motors. One important advantage of these systems over the sources mentioned earlier is the possibility of studying signals with a given dependence of parameters (amplitude, phase, frequency) on time. However, these sources are extremely expensive and complicated to manufacture, and thus only exist in the form of one-off examples. As a rule, these sources are not suitable for being towed underwater. Section 6.1 contains an example of the use of such a source.

The sound source for explosions is a shock wave, followed by pulses of gas bubbles. The dependence on time of the pressure of the shock wave pulse is described by a function (Urick (1975)):

$$f(t) = p_0 e^{-t/t_0} \tag{5.11}$$

where, because of the considerable contribution of the nonlinearity of the medium at large ranges, p_0 and t_0 depend upon distance[8]

$$p_0 = 5.15 \times 10^7 \left(\frac{w^{1/3}}{r}\right)^{1.13} \quad [Pa] \tag{5.12}$$

$$t_0 = 93 \times w^{1/3} \left(\frac{w^{1/3}}{r}\right)^{-0.22} \quad [\mu s] \tag{5.13}$$

Here, z is the distance in meters, and w is the mass of the charge in kilogrammes. The sound pressure of the first gas bubble pulse depends on the depth of the explosion and is comparable in value with the pressure of the shock wave. The pressure of the second pulse is approximately a factor of 5 less, and subsequent pulses are negligibly small.

The characteristics of explosive sound sources have been described in detail; for example, in Urick (1975), and Weston (1963). Here, we shall present only a few properties of these sound sources, mentioned in Vadov (1994), which are significant for acoustic sensing in shallow water at large distances.

- The spectrum of the radiated signal up to a frequency of 3–5 kHz is characterised by strong irregularity, associated with interference between the pulse of the shock wave and the gas bubble pulses. At large (greater than 10 km) distances the irregularity is determined mainly by the first bubble pulse. The position of the first maximum on the frequency scale f_0 and the distance Δf between successive minimum in the spectrum are related to the period of the first bubble pulse T_1 by the equation: $f_0 T_1 = \Delta f T_1 = 1$. T_1 can be estimated from the equation due to Arons (1948):

$$T_1 = 0.48 \frac{K_0 w^{1/3}}{(z+10)^{5/6}} \tag{5.14}$$

  Here, K_0 is a coefficient of proportionality (the value of K_0 is weakly dependent upon the type of the explosive material and for TNT is equal to 4.36), and z is the depth of the explosion in metres. (We note that in the experiment of Vadov (1994), the beat period varied according to the law $T_1 \sim 1/z^{1.1}$; i.e. it depended more strongly on depth than in equation (5.14)).
- The decay of the spectral density of the energy of the explosive signal with frequency (based on the envelope, without taking the interference structure into account) follows a law close to $1/f$ at frequencies less than 1 kHz.
- The repetition rate of the spectral characteristics (standard deviation of the

[8] For typical explosive sound sources (where the mass of the explosive material is a few hundred grammes), this dependence occurs for distances of several hundred metres. For large distances, p_0 and t_0 are time independent and the equation of linear acoustics (1.6) holds.

levels) of explosive signals for frequency bands bounded by consecutive interference minima is 0.2–0.9 dB.

However, we note that for a charge mass of 400 g (similar masses of explosive material are usually employed in experiments on the long-range propagation of sound) the spectral density of the radiated signal in the band $\Delta f = 1/T$, even at the frequency f_0, is at least 10 dB less than the level of standard pure-tone sound sources. (Here, we have in mind the effective spectral density of an explosive signal, calculated at distance 1 m from the source according to linear laws of acoustics). This is the situation even for small depths of the explosion ~ 100 m, when the frequency f_0 is equal to ~ 100 Hz. Thus the sounding range and the accuracy of measurements of the characteristics of sound propagation when pure-tone sources are used are greater than for explosions. It is not usual to increase the mass of the charge in a physical experiment, for the following reasons:

- the low effectiveness of such an increase (level of emission proportional to $\sim w^{1/3}$);
- the possibility of destroying or injuring seriously a large number of marine organisms;
- a significant increase in the cost of the individual charge;
- the complexity and danger of working with charges of high mass.

Measurement errors associated with the use of explosive sources are often also due to the choice of the discreteness of the blasts when the distance between the individual explosions is greater than Λ_m.

5.5 EXPERIMENTS ON STATIONARY ACOUSTIC TRACKS

Studies of the space–time fluctuations of sound waves in shallow water, like acoustic tomography of hydrodynamic processes in shallow water, are usually performed on stationary acoustic tracks for a stationary sound source and receiver anchored to the bottom. In the simplest form, this is done using two ships, each of which is braced between two anchored mooring buoys, put in place beforehand using a lifting-and-mooring barge in the study region (Katsnelson *et al.* (1992)) (a sketch of the experiment is shown in Figure 5.2). From on board one of the ships the sound source is lowered to the bottom, while from on board the other a vertical chain of hydrophones is lowered. The 'receiving' ship with the chain must have a quiet mode, i.e. present a possibility of operating only on a special silent diesel generator. The main problem in stationary experiments is to minimise all displacements whatsoever of the sound receivers and sources, and to take control measurements of any such displacements which may occur. Based on this, the pure-tone source is usually anchored at the least possible distance (≤ 1 m) from the sea bed. (We note that in this case the position of the source in shallow water is most effective as far as the excitation of acoustic waves is concerned (Zhuralev *et al.*, 1987)). To this end, the cable linking the

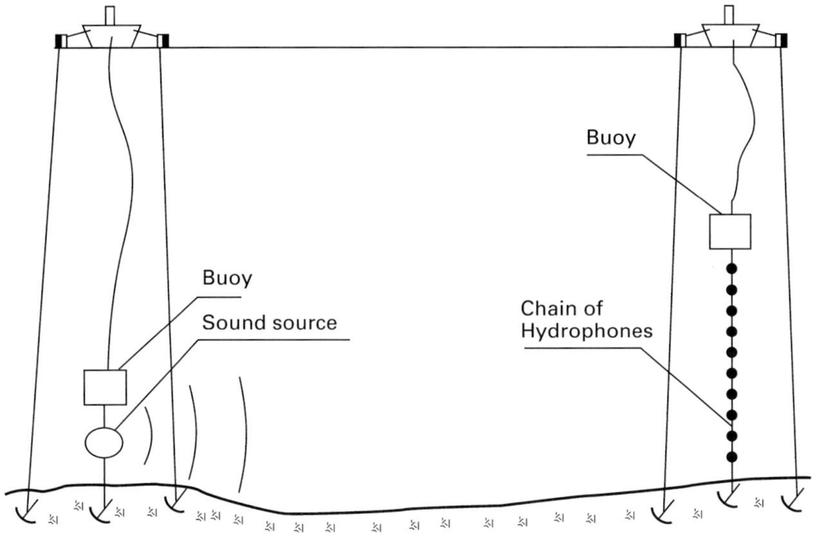

Figure 5.2. Diagram of an experiment on a stationary acoustic track.

lowering systems to the ship is not put under tension at the time of the experiment, and small drift movements of the ships do not lead to drift of the source or the chain of hydrophones. However, the chain of hydrophones may be highly inclined under the effect of currents. If the chain retains a rectilinear form (one attempts to achieve this by providing each hydrophone with a neutral buoyancy) then the horizontal drift of the jth receiver Δr_j can be estimated from the readings z of a depth sensor positioned near the upper hydrophone $\Delta r_j \cong l_j \Delta\varphi_0 \cong l_j \sqrt{1 - (H - z)/L_0)^2}$, or (more precisely) from the readings of an inclinometer positioned there, measuring the angle of inclination $\Delta\varphi_0$ of the chain. Here l_j is the distance from the jth hydrophone to the chain anchor, L_0 is the distance from the depth sensor to the anchor, and H is the depth of the sea at that point of installation, measured using an echo sounder. The direction of drift (direction of inclination) is estimated from the readings of a compass on the chain or from the direction of currents recorded on the ship using an acoustic Doppler current profiler.

The use of two ships on a stationary track greatly increases the cost of the experiments. The 'receiving' ship can be dispensed with, using an autonomous vertical array transmitting information on a radio link to the emitting vessel. One version of such an array, used in Miller *et al.* (1993), is shown in simplified form in Figure 5.3. We note, however, that the range of operation of the radio channel depends on the rough sea and is at most a few tens of kilometres.

A more promising method, in terms of cost, for long-term acoustic monitoring on stationary tracks on the ocean shelf involves the use of autonomous bottom stations, carrying both a sound source and vertical receiving arrays (Derevnin *et al.*, 1986; Boutin *et al.*, 1989). Using such systems, equipped with their own

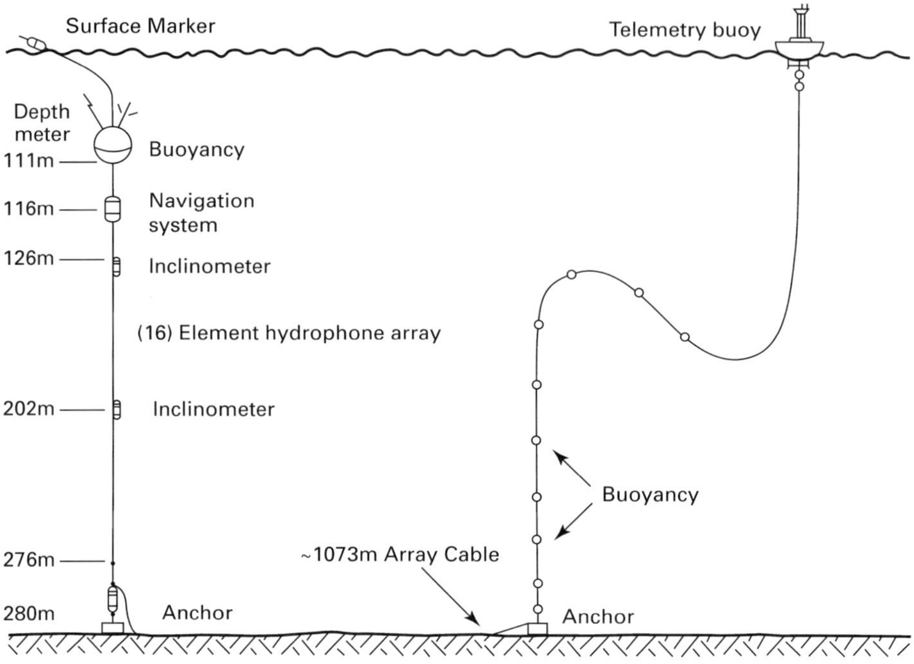

Figure 5.3. Autonomous vertical receiving array.

power supply together with a data collection and processing system one can in principle carry out tomographic experiments of several years duration. However, when carrying out long-term experiments one must have a facility for operational (real-time) monitoring and output of the information received. This problem can be solved, for example, by connecting, where possible, bottom stations to an existing submarine cable network. It is significant that, for these purposes, one can use old telephone cables and connect the cables and the stations (as noted in Lynch *et al.* (1992)) inductively without destroying the integrity of the cable.

However, attachment to communication cables which are already laid is not always expedient. The tomographic module (Webb tomography transceiver), described in Lynch *et al.* (1992) and Boutin *et al.* (1989), which combines an information collection and processing device and a satellite communications channel for periodic output of this information and control of the operation of module, is optimal. Figure 5.4 shows a diagram of such a module, tested in tomographic experiments in shallow water (Lynch *et al.*, 1996). The technical characteristics of the module are as follows: carrier frequency for the acoustic signals, 400 Hz; operational frequency band, 100 Hz; source level 180 dB relative to 1 μPa; device information storage capacity, 200 Mbyte. One important property of the module is the presence of a navigation system for the sound source and receivers using acoustic expendable transponders on the bottom, and also stable clocks with a precision of $\sim 20\,\mu$s, synchronised by signals from the GPS satellite positioning system.

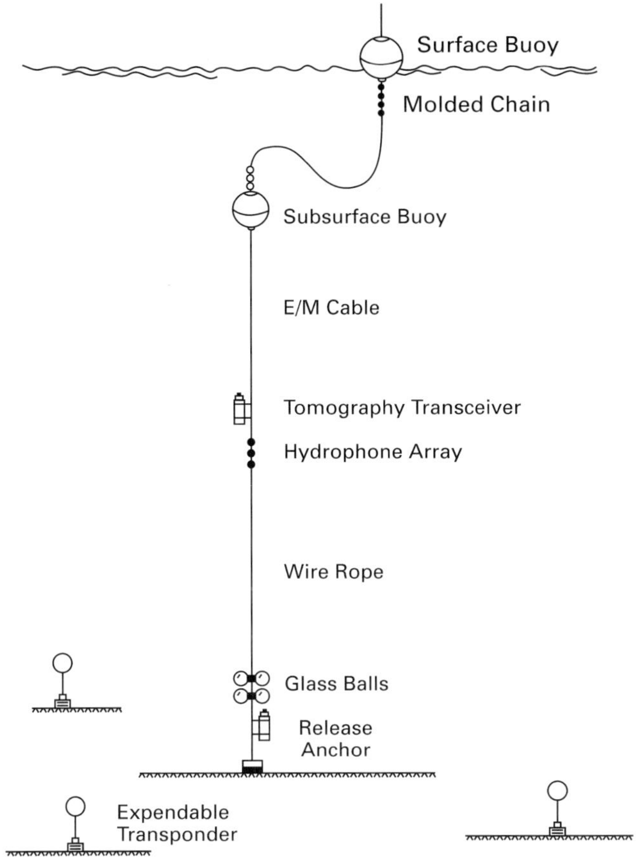

Figure 5.4. Autonomous tomography module (tomography mooring).

We also note that a navigation system using acoustic transponders on the bottom based on a depth sensor or an inclinometer and a compass forms part of most receiving arrays designed for acoustic monitoring on stationary tracks. However, these expensive systems can be dispensed with if one uses (as proposed in Burov *et al.* (1991)) four pure-tone sound sources simultaneously, where the frequencies of three of these ω_1, ω_2, ω_3 are arbitrary and the frequency of the fourth ω_4 satisfies the equation

$$\omega_1 + \omega_4 = \omega_2 + \omega_3 \tag{5.15}$$

Then, the additional phase incursion due to the small deviation of the receiving node of the array from the vertical, will be compensated if in the monitoring process one analyses not the field itself, but a combination of values of the received field in the form of a product of order four at the four frequencies[9]:

$$M = P_{\omega_1}(r, z, t) \cdot P^*_{\omega_2}(r, z, t) \cdot P^*_{\omega_3}(r, z, t) \cdot P_{\omega_4}(r, z, t) \tag{5.16}$$

where $P_\omega(r, z, t)$ is the field of a point pure-tone source at frequency ω. Here, we have in mind monitoring based on matched field signal processing (for further details, see Section 6.4), where, as the criterion for comparing th- calculated M_{th} and experimental data M_{e}, for example, the following correlation formula is used:

$$K = \frac{\sum_j^{J_h} M_{\text{e}}(z_j) M_{\text{th}}(z_j)}{\left(\sum_j^{J_h} M_{\text{e}}^2 \sum_j^{J_h} M_{\text{th}}^2\right)^{1/2}} \tag{5.17}$$

In conclusion, we note that the range of technical tools used in experiments on the sea shelf is broader than those discussed in this chapter. We have attempted to identify only the main, essential characteristics of the methods and techniques of physical experiments in shallow-water (application of low-frequency sound cources, selection of waveguide modes, organisation of stationary acoustics tracks, etc.).

[9] The possibility of this compensation follows from the expression for the likelihood function for the calculated and the recorded sound field at a number of frequencies (for further details, see Burov and Sergeev (1992)).

6

Long-range propagation of sound in shallow water and acoustic monitoring: results and possibilities

6.1 ATTENUATION OF SOUND AND ESTIMATION OF PARAMETERS OF THE SEA BED BASED ON ACOUSTIC SENSING DATA

The possibilities for large-scale acoustic monitoring on the ocean shelf are in many respects determined by the attenuation of low-frequency sound waves in long-range propagation in shallow water. The attenuation of sound is also one of the most important characteristics for other applied problems in shallow water. In this connection, the study of the attenuation of low-frequency acoustic waves on the shelf forms one of the fundamental directions of shallow-water acoustics.

The attenuation of low-frequency sound in shallow water is mainly determined by the interaction of acoustic waves with the sea bed. As mentioned earlier, on the ocean shelf the sea bed is usually composed of recent marine sediments and is an extremely inhomogeneous multilayered medium. Many parameters affect the sound field in a waveguide, including the sound speed in the fluid and in the layer of sediments, the attenuation coefficients for longitudinal and transverse waves, the density of the medium, the thickness of the layers, etc. Not all these parameters can be measured directly (or even indirectly), thus for the calculations, one resorts to modelling, i.e. in practice, to the tuning of certain parameters in order to agree with the experiment. In this connection, approaches to the description of the propagation of sound at large distances in shallow water can be loosely divided into two types. The first relates to the modelling of the sea bed as a multilayered medium (Section 2.4) with the introduction of the fullest possible description of the characteristics of each layer: h_n, c_n, ρ_n, α_n, etc., each of which may depend on distance. The model has a large number of parameters (of the order of tens or more), whose values can be refined by special methods. The other type of approach involves

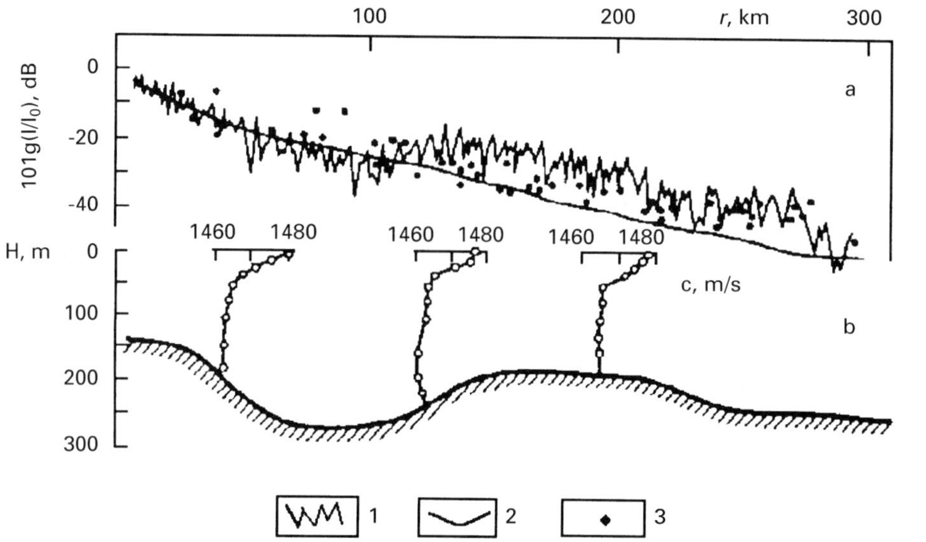

Figure 6.1. Calculated (1), (2) and experimental (3) values of the relative intensity $(10\lg(I/I_0))$ of the sound from a point source in a region of the Barents Sea (*a*), together with changes in the depth $H(r)$ and vertical profiles $c(z)$ along the sound propagation track (*b*). (1) – Based on summation of modes; (2) – from the averaged relation (3.104).

representing the sea bed as a randomly inhomogeneous medium with a number of effective parameters c_1, ρ_1, α, taking into account all the complex bottom structure in some averaged way, or more precisely, reducing the effect of the complex bottom structure on the sound field to the impedance (2.14.2). Naturally, the applicability of each of these approaches is determined by the specific situation. In particular, in the case of large (several hundred kilometres) distances between the sound source and the receiver, the second approach, despite its apparent simplicity, may be more effective. This has to do with the fact that the sediment layers in shallow water tend to taper out and be intermittent, which is impossible to describe by a reasonable deterministic multilayer model.

The possibilities for using a model of a fluid homogeneous bottom with effective parameters for calculations of the sound field intensity at frequency 300 Hz at large distances ($\sim$300 km) are demonstrated in Figure 6.1 for the Barents Sea (Katsnelson *et al.* (1987)). The calculation was carried out based on summation of modes, i.e. based on equation (2.41) and on the averaged equation (3.104). As one can see from the figure, the calculated curves for the sound field intensity are covered to a high degree of accuracy by the results of the experiment, despite the fact that the effective parameters were assumed to be unchanged along the whole track of the sound propagation[1].

[1] As one might expect, at large distances, where only a few of the lowest modes survive, the curve calculated from the averaged equation diverges from the experimental data and from the results of calculation based on summation of modes (for further details, see Section 3.5).

Here, it is pertinent to make a small detour from the main theme of this section and talk in more detail about the methodology for numerical calculations of the sound field in shallow water. For any approach to a model of the bottom, the arbitrary nature of the sound speed profile makes the channel model highly complex. This complexity leads to a transformation of the dominant role, especially of numerical calculations for determining the sound fields in real situations. In shallow-water conditions most calculations are carried out using summation of modes (mode algorithms and programs, some of which are well known; for example, the KRAKEN program (Porter (1992)), or the SNAP program from the SACLANTCEN centre (Jensen and Ferla (1979))). Let us briefly consider a number of the main features of numerical modelling, specific to shallow-water acoustics. In the cross-sections method (Section 3.2) or taking corrections into account (Section 3.4), the calculations of the sound field are based on solution of the Sturm–Liouville problem for a set of cross sections, corresponding to different values of the distance r. Numerical solution of this problem has been considered in many works both in general mathematical terms (Hall and Watt (1976)) and as applied to the acoustics of the ocean (DeSanto (*ed.*) (1979), Zavadskii (1972)). Let us note a number of features of this problem in shallow-water acoustics.

First, as already mentioned on a number of occasions, there is the non-self-conjugacy of the problem due to absorption in the bottom, as a result of which we have to determine the complex spectrum. On the other hand, the presence of absorption leads to a 'dying out' of modes with high indices, which are subject to great absorption, and also of modes of the continuous spectrum. This means that in the numerical solution of the problem, we can restrict ourselves to constructing only a quite limited number of propagating modes. However, even in this case, there is a high relative density of the mode spectrum. For example, the simplest estimate in the Pekeris model for the parameters $H = 100\,\text{m}$, $f = 200\,\text{Hz}$, $c = 1,500\,\text{m/s}$, $c_1 = 1,800\,\text{m/s}$, shows that for the eigenvalues with the lowest indices the relative spectral density satisfies

$$\left|\frac{\Delta\xi_l}{\xi_l}\right| = \left|\frac{\xi_l - \xi_{l-1}}{\xi_l}\right| \approx 0.01$$

i.e. successive eigenvalues differ in the second or third decimal place. As the frequency and the waveguide depth increase, the density grows and application of any numerical method for solving the Sturm–Liouville problem requires special methods to distinguish the eigenvalues.

Furthermore, the closeness of the eigenvalues makes it necessary to compute them with a small relative error. For example one of the main characteristics of the calculated field, the spatial period of signal beats, is determined by the difference of the real parts of the eigenvalues $\Lambda_{lm} = 2\pi|q_l - q_m|^{-1}$, and its calculation with an accuracy of 10^{-2} (the values of the beats are determined experimentally with approximately the same accuracy) for a relative spectral density of 10^{-3} requires the calculation of q_i with an accuracy of 10^{-5}–10^{-6}. The imaginary parts of the eigenvalues $\gamma_l/2$, which determine the decay of the mode intensity, have to be calculated with no less accuracy, since at distances of several tens and even hundreds of kilometres, at which we are working, a large error makes an inadmissible contribution to the total error.

The above accuracy requirements lead to the preferential use of numerical methods for the calculations, since because of the complexity of the real medium, it is practically impossible to construct a plausible model of the dependence of the waveguide parameters on the coordinates, for which there exists an analytic solution of the Sturm–Liouville problem. On the other hand the known analytic approximation of the WKB method or its modification is not sufficiently accurate.

We also note that in the present work, we shall not touch upon the analysis of 'direct' numerical methods of calculating the field in a waveguide, in the form of direct solution of the propagation equation (2.2) based on difference methods (Zavadskii (1972)) or the parabolic equation method, which is now widely used for both deep and shallow water (Keller and Papadakis (1977), Avilov (1995)).

Thus, the basis for calculations of the field of a point source is the expression

$$\Psi(r,z) = \sum_{l=1}^{M} \frac{\psi_l(0;z_0)\psi_l(r,z)}{\sqrt{\xi_l r}} \exp\left(i\int_0^r \xi_l dr\right) \tag{6.1}$$

which differs from (3.21) only in that here the value $C_l^0 = \psi_l(0;z_0)$ is explicitly substituted. Moreover, here, the constant multiplier is admitted, since the signal decay is usually calculated in decibels, i.e. one is interested in the quantity $\delta = 20\lg|\Psi/\Psi_0| = 10\lg(I/I_o)$, where Ψ_0 and I_0 are, respectively, the sound pressure and the intensity of the sound field at some reference distance r_0^*, for example, 1 km. The quantities $\psi_l(r,z)$ and $\xi_l(r)$ are solutions of the transverse wave equation

$$\frac{d^2\psi_l}{dz^2} + \left[\frac{\omega^2}{c^2(r,z)} - \xi_l^2(r)\right]\psi_l = 0 \tag{6.2}$$

$$\psi|_{Z=0} = 0, \quad \left[\psi_l(z) + g\frac{d\psi_l}{dz}\right]_{z=H(r)} = 0$$

The region of effect of equation (6.2) is bounded by the interval $0 < z < H(r)$, since we are interested in the field in the water layer only. The region $z > H(r)$ is taken into account by the introduction of a corresponding impedance boundary condition at $z = H(r)$. Here, $g = m_1\psi_l/(d\psi_l/dz)|_{z\to H+}$ is a coefficient depending on the parameters of the medium, on the frequency f and on the eigenvalue ξ_l. For example, if the bottom is an homogeneous fluid half-space with density ρ_1 and sound speed c_1, then g is given by equation (2.14.2). An expression for g can also be obtained for a layered (see Section 2.4) or inhomogeneous bottom taking into account shear waves, etc.

Thus, the greatest complexity in numerical modelling is associated with the calculation of the eigenvalues ξ_l. In what follows, we shall introduce the complex-valued function $\Phi(\xi)$, defined by the equation

$$\Phi(\xi) = \left[\psi + g(\xi)\frac{d\psi}{dz}\right]_{z=H(r)} \tag{6.3}$$

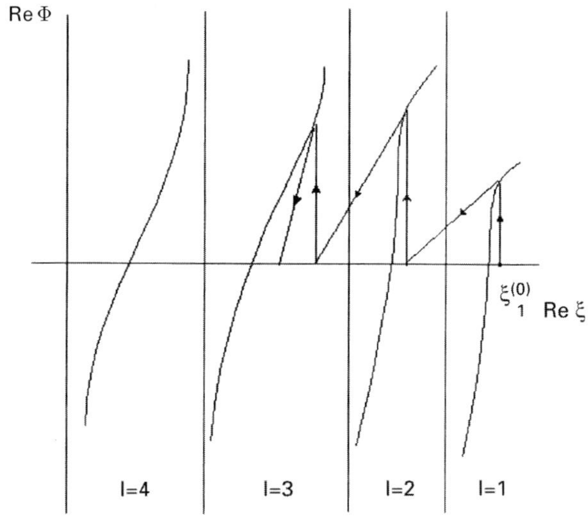

Figure 6.2. The search for roots of the function $\Phi(\xi)$ (eigenvalues) and the possibility that the roots fall outside the limits of localisation.

where $\psi(z)$ is an arbitrary continuous function, twice-differentiable for any fixed complex value ξ, satisfying equation (6.2) with the boundary condition at $z = H(r)$. In Section 2.1, we showed that the roots of the equation

$$\Phi(\xi) = 0 \tag{6.4}$$

are eigenvalues of the Sturm–Liouville problem (6.2). We immediately note that the case of additional waves, corresponding to degeneracy of the spectrum at some l, is excluded from the analysis[2]. The orthornormalised functions $\psi_l(z)$ satisfying equations (6.4) and (6.2) characterise the sound field of different modes at depth z.

Thus, the search for eigenvalues ξ_l reduces to finding the zeros of the complex function $\Phi(z)$. The function $\Phi(z)$ is computed as follows. For given $\xi = \xi' + i\xi''$ one solves the Cauchy problem for equation (6.2) with initial conditions:

$$\psi_l(0) = 0, \quad \frac{d\psi_l(0)}{dz} = \text{constant} \neq 0$$

Integration is then performed using a known method (Runge-Kutta, Numerov, etc.). Using the values of $\psi(z)$ and $d\psi/dz$ at the point $z = H$, we then find the value of $\Phi(\xi)$. The typical behaviour of the function $\Phi(\xi)$ is shown in Figure 6.2.

To search for the zeros of the function $\Phi(\xi)$ one can use iterative methods; for example, a combination of false position and secant methods for the complex case.

[2] At interesting distances, from the point of view of large-scale acoustic monitoring, of several tens of kilometres, the contribution to the total field of waves of the continuous spectrum is infinitesimally small.

For the initial approximation, we take the real value $\xi_l^{(0)} + i0$, being the solution of a transcendental equation in the WKB approximation:

$$\int_0^{H(r)} \sqrt{\frac{\omega^2}{c^2(r,z)} - (\xi_l^{(0)})^2}\, dz = \pi\left(l - \frac{1}{2}\right) \tag{6.5}$$

which is solved numerically. We then look for a zero of the function $\Phi_1 = \mathrm{Re}[\Phi(\xi + 0i)]$ for $\mathrm{Im}\xi = 0$. Because of the properties of the behaviour of the function $\Phi_1(\xi)$, particularly for small l, i.e. for lower modes, there is a danger that the solution will go into an area in which roots corresponding to another l are localised (see Figure 6.2).

To control $\xi_l^{(n)}$ (n is the number of the iteration) going beyond the boundary for the localisation of the root ξ_l, we use the property of the solution $\psi_l(z)$ that for any l, for $\xi_l^{(n)}$ in the region of localisation of the root, $\mathrm{Re}\psi_l(z)$ has l zeros, not counting the first one (theorem on the number of zeros of the eigenfunction (Titchmarsh (1958))). Note that this condition is not strict, but is well satisfied for small values of α in the neighbourhood of the root. Then $\xi_l^{(N)}$, corresponding to a zero of the function $\Phi_1(\xi)$, obtained with the necessary accuracy in the Nth step of the iteration, is taken as an initial value for equation (6.4), for example, for the secants method, generalied to the complex case. Experience shows that this choice provides fast convergence (three to six iterations) to the requisite complex value, without going beyond the boundary of the localisation of the root. At the same time, we compute the functions $\psi_l(r,z)$, which it only remains to orthonormalise taking into account the relation

$$\int_0^H \psi_l^2(z)\, dz + \frac{\psi_l^2(H)}{2\xi_l}\left(\frac{dg^{-1}(\xi)}{d\xi}\right)_{\xi=\xi_l} = 1$$

After calculating the eigenvalues ξ_l and the eigenfunctions $\psi_l(z)$ for $l = 1, 2, 3, \ldots, M$, we verify the ordering of the roots $\mathrm{Re}\xi_1 > \mathrm{Re}\xi_2 > \ldots$ and the correspondence between the number of zeros of the function $\mathrm{Re}\psi_l(z)$ and the mode index, based on the aforementioned theorem (Titchmarsh (1958)).

Here, we must say a few words about the choice of the maximum number of modes M, used in the calculations. An upper bound for M is given by equation (2.49), but at large distances modes with the highest indices decay strongly and can be neglected. Usually in practice, when calculating the sound field at large distances, we determine maximum number M in summation, as the number of modes producing a remarkable contribution (corresponding to the necessary accuracy) in the calculated sound field. That number decreases while the distance increases, and this method can therefore save computational time.

This method of finding the spectrum of normal waves for regular oceanic waveguides corresponds to that used in Zavadskii (1972), and is now in common use. One property of the method is the technique of searching for eigenvalues of the Sturm–Liouville problem, providing a resolution of the roots when their values are very close.

As an example of this method, we present the results of calculations of the

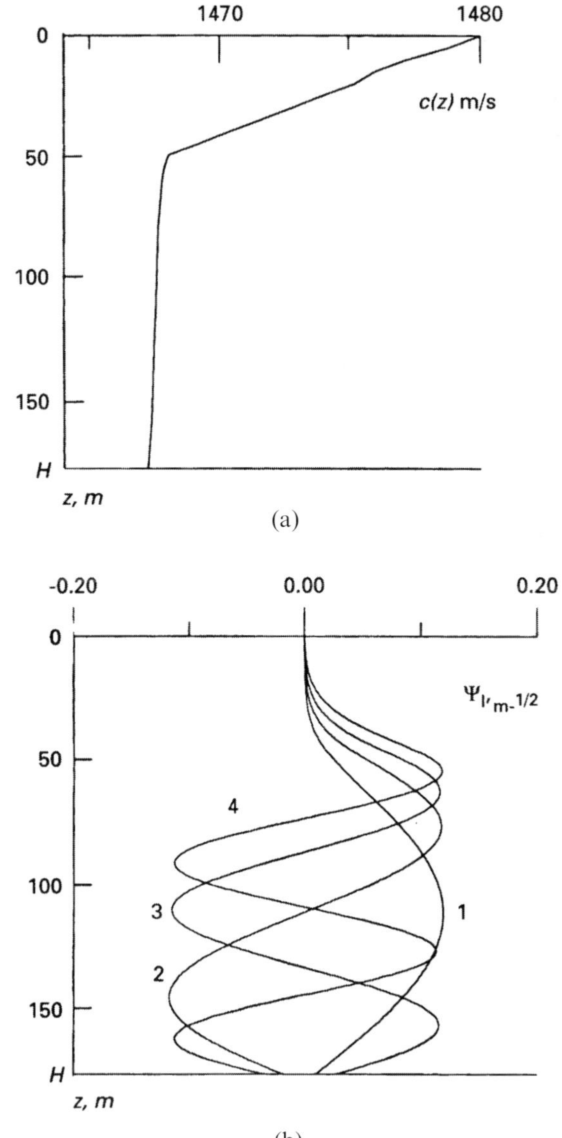

Figure 6.3. Experiment sound speed profile (*a*) and corresponding first four eigenmodes (1)–(4) at frequency 300 Hz (*b*), $c_1 = 1580\,\text{m/s}$, $\rho_1 = 1.7\,\text{g/cm}^3\text{m}$ $\alpha - 0.015$.

eigenfunctions and eigenvalues for the acoustic track shown in Figure 6.1. Figure 6.3(*a*) shows the sound speed profile in one of the sections. Figure 6.3(*b*) shows the first four modes for frequency 300 Hz. The values of the mode attenuation coefficients as a function of mode index are shown in Figure 6.4.

Turning to the main theme of this section, we again stress that the results of

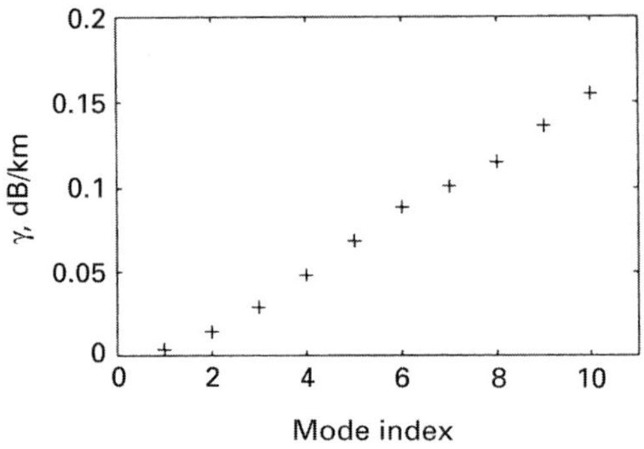

Figure 6.4. Mode attenuation coefficients at frequency 300 Hz.

calculations of the sound field in shallow water at large distances, in the framework of the model of a randomly inhomogeneous bottom with effective parameters c_1, ρ_1, α are in satisfactory agreement with experimental data[3]. The only question is the proper choice of these parameters, taking into account the geographical variability of the bottom in shallow water. For this choice, it is expedient to recommend the method studied in Bunkin *et al.* (1989). The essence of this method is a comparison of the experimental and calculated (computed for various values of the parameters) dependences of the sound field intensity on distance. By comparing these dependences based on a numerical criterion, one can choose the calculated dependence which provides the best agreement of theory and experiment and for the effective target parameters one can take the values c_1, ρ_1, α, used in calculating this last dependence.

In experiments, the dependence of the sound field intensity on distance is usually measured with a stationary receiver and a pure-tone signal source being towed away from it. Because of the reciprocity theorem, the dependence of the sound field intensity on the distance r obtained in this way is equivalent to the dependence of the intensity on the distance to a stationary source, given by equation (3.78) for a fixed receiver depth. As previously mentioned, in reality, this dependence is constructed on a logarithmic scale; in other words, when speaking about the dependence of intensity on distance, we have in mind the function $\delta(r,z) = 10\lg[I(r,z)/I(r_0^*, z)]$. The quantity δ, as a function of spatial coordinates, is very irregular, being the result of the interference of spatial beats of different scales, and to compare the experimentally measured $\delta_e(r,z)$ and the calculated functions $\delta_{th}(r,z)$, we have to use special numerical criteria. One such criterion might be the following (Bunkin *et al.* (1989)). We represent the function $\delta(r,z)$ in the form of a sum of the mean and the fluctuating parts $\delta(r,z) = \bar{\delta}(r,z) + \tilde{\delta}(r,z)$, where $\bar{\delta}$ corresponds to the intensity averaged over a longitudinal spatial interval greater than the maximum scale of the interference beats (equation (3.79)), and $\tilde{\delta}$ corresponds to the fluctuating part of the intensity

[3] We note that, formally, in the calculations, the bottom is assumed to be homogeneous with effective parameters c_1, ρ_1, α.

Colour plates

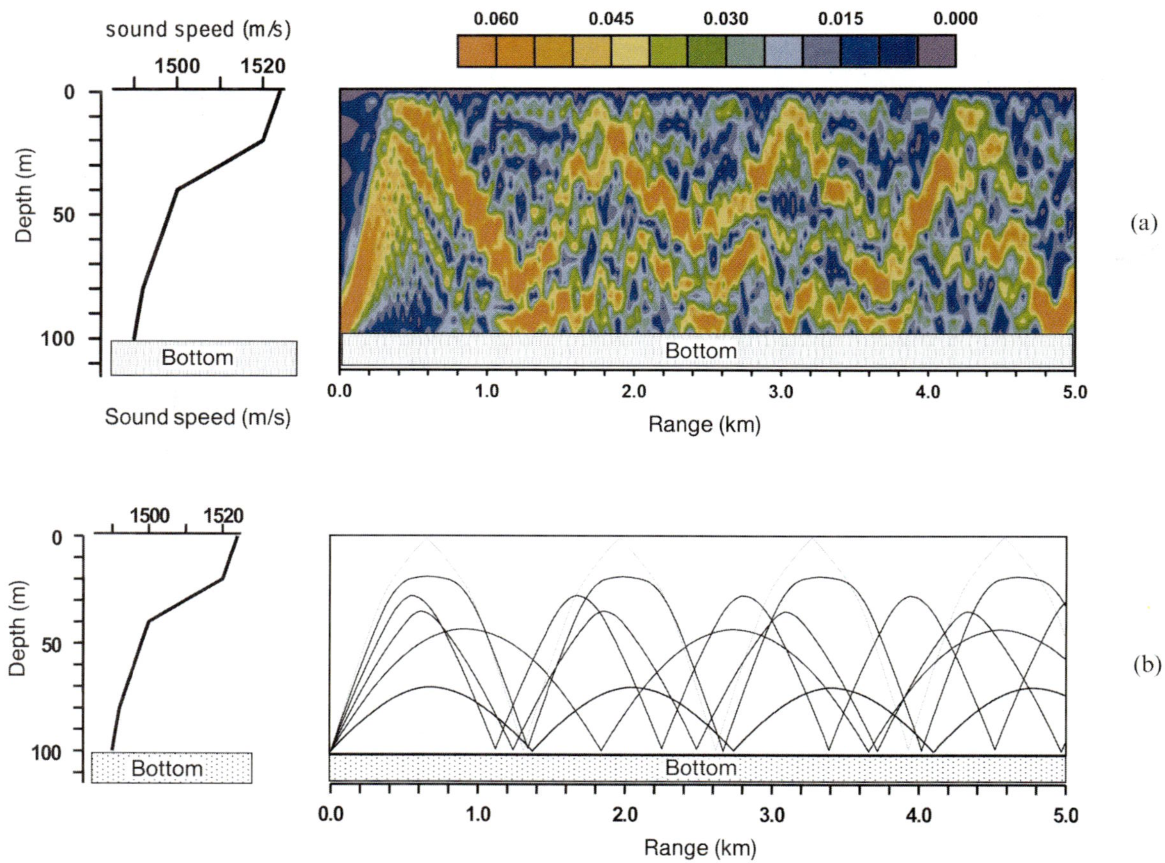

Figure 3.4. Interference (a) and ray (b) pattern for the sound field from a point source. The figure shows the sound speed profile. The source is located on the bottom, frequency $f = 250$ Hz. Bottom parameters $c_1 = 1750$ m/s, $\rho_1 = 1.8$ g/cm^3, $\alpha = 0.02$. The scale of colour gradations by amplitude is shown in (a).

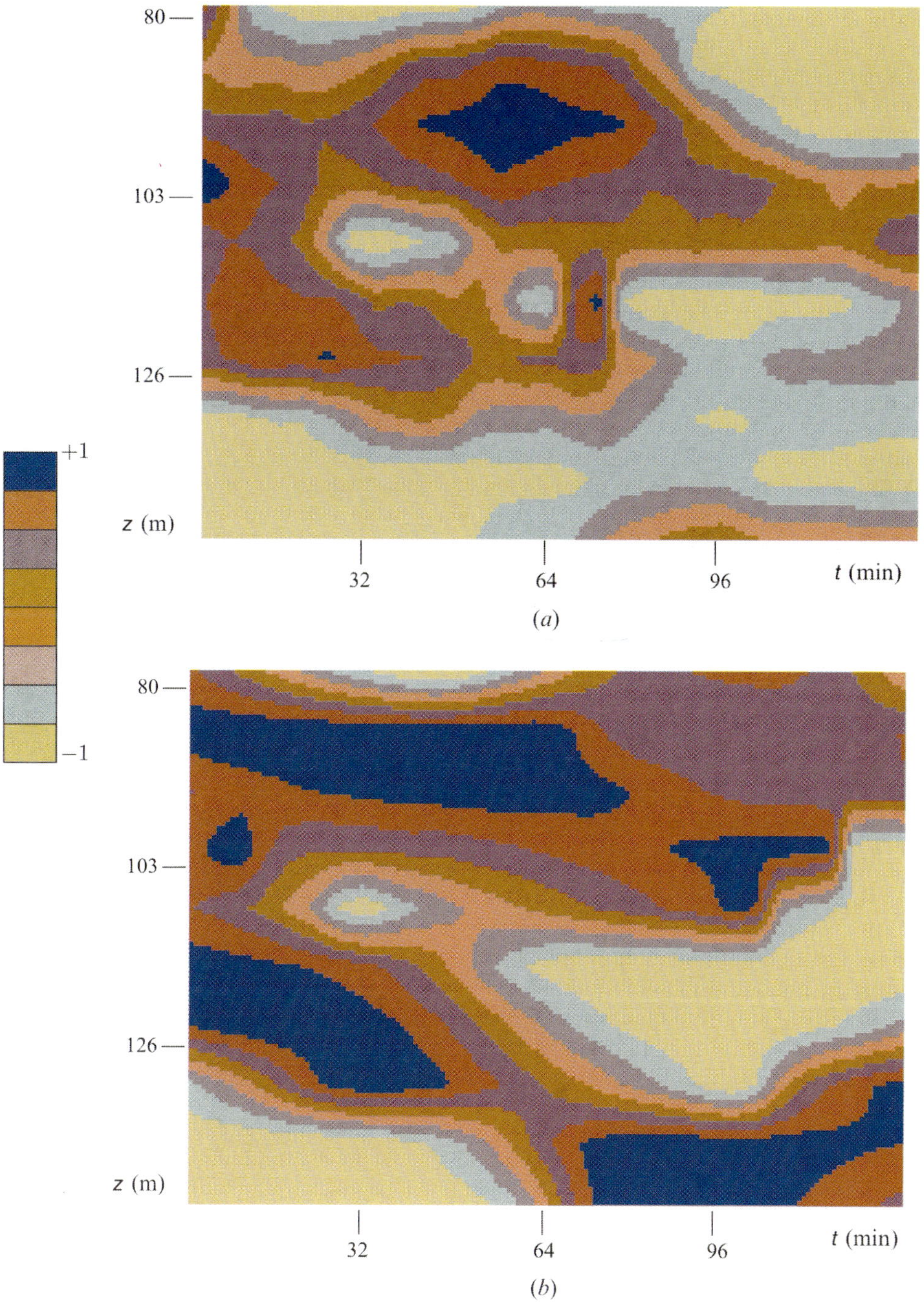

Figure 6.38. Time diagram of the variations of $\cos\varphi(t)$: (*a*) experiment, (*b*) theoretical. The scale has eight colour gradations.

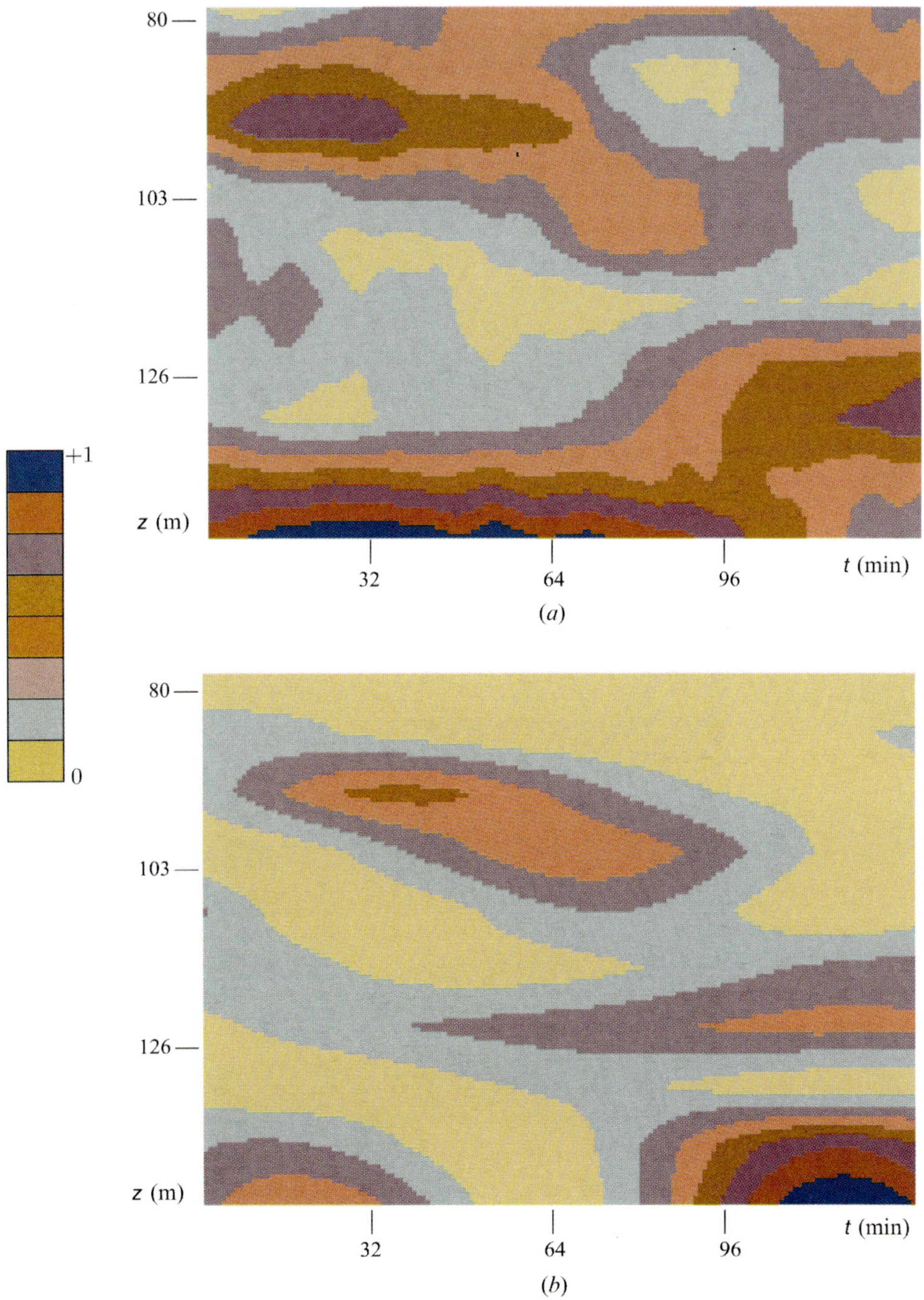

Figure 6.39. Time diagram of the vertical interference pattern of the sound field amplitude in colour gradation, analogous to Figure 6.38. (*a*) experiment, (*b*) theoretical.

$\tilde{I}(r,z) = I(r,z) - \overline{I}(r,z)$. According to this, the 'proximity coefficient' of functions also consists of two terms and is defined by the following expression:

$$K = \overline{K} + \tilde{K}, \quad \delta = \overline{\delta} + \tilde{\delta}$$

$$\overline{K} = \frac{\int_0^r \overline{\delta}_{\mathrm{e}} \overline{\delta}_{\mathrm{th}} \, dr}{\int_0^r \overline{\delta}_{\mathrm{e}}^2 \, dr + \int_0^r \overline{\delta}_{\mathrm{th}}^2 \, dr}, \qquad \tilde{K} = \frac{\int_0^r \tilde{\delta}_{\mathrm{e}} \tilde{\delta}_{\mathrm{th}} \, dr}{\int_0^r \tilde{\delta}_{\mathrm{e}}^2 \, dr + \int_0^r \tilde{\delta}_{\mathrm{th}}^2 \, dr}. \tag{6.6}$$

It is easy to see that $0 \leq K \leq 1$ and the maximum value of K (ideally 1) corresponds to the best agreement of calculation and experiment. We note that the use in this criterion for calculations of the function $\delta_{\mathrm{th}}(r,z)$ of the 'deterministic' approach (based on equations (6.1), (6.2)) is due to the fact that the spatial dependence of the sound field interference pattern, measured in experiments in shallow water at low frequencies, is sufficiently stable in time, and is smooth.

The next important matter, requiring discussion, has to do with which parameters are chosen in the first place. In the chosen model of the medium, this is determined by which parameters (of the least known) have the greatest effect on the propagation of sound. For models of a fluid homogeneous bottom, one such parameter is the coefficient α, corresponding to absorption of sound in the bottom, which, as we know (Kuperman (1980)) is the most indeterminate. According to the nature of the bottom, α may vary by a factor of 20–30, whereas the sound speed c_1 and density ρ_1 vary by 10–20% from their mean values.

To illustrate the possibilities of this method, we shall demonstrate its operation in the Barents Sea, where the dependence of the sound field intensity on distance was measured simultaneously at two frequencies: 100 and 230 Hz. Profiles of the sound speed and the water depth in the measurement region are shown in Figure 6.5. Measurements were performed with the participation of two ships, one of which towed sound sources at depth $\cong 40$ m, while from the other a vertical chain of hydrophones located at relative depths of 65, 80, 95 and 125 m was lowered. The results of the measurements and calculations[4] for a frequency of 100 Hz are shown in Figure 6.6 for a reference distance $r_0^* = 0.5$ km. In constructing the curves, the values of $I(r,z)$ were multiplied by r/r_0^*, which enabled us to correct the decrease in I with distance due to cylindrical broadening of the wavefront, and to show the interference beats more graphically. One of the calculated curves corresponds to optimal α (when K is maximal), and the other to an arbitrary α. (The other parameters of the bottom in the calculations were taken to be $\rho_1 = 2\,\mathrm{g/cm^3}$, $c_1 = 1630$ m/s, which corresponded to typical values of the density and sound speed in the region studied). One can clearly see that for non-optimal values of α the experimental and numerical curves diverge strongly. From this and analogous graphs, one can construct dependences $K(\alpha)$ for difference values of the depth of the sound receivers. Such graphs for frequencies 100 and 230 Hz are shown in Figure 6.7. Evidently, most of the dependences $K(\alpha)$ have a very clearly expressed maximum, where this maximum is found

[4] The method used for the calculations, based on the summation of normal modes, was studied earlier in this section.

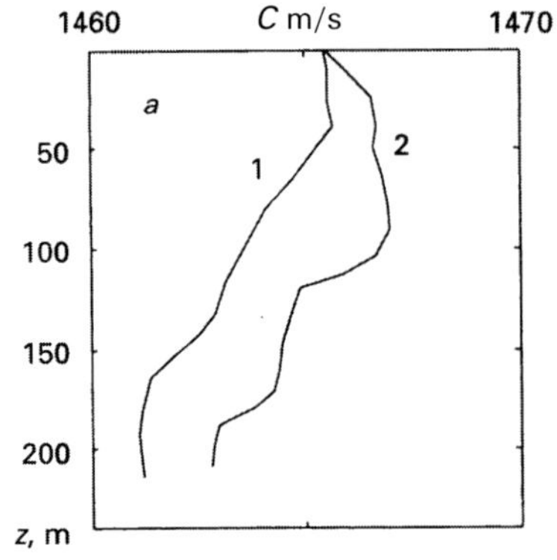

(a)

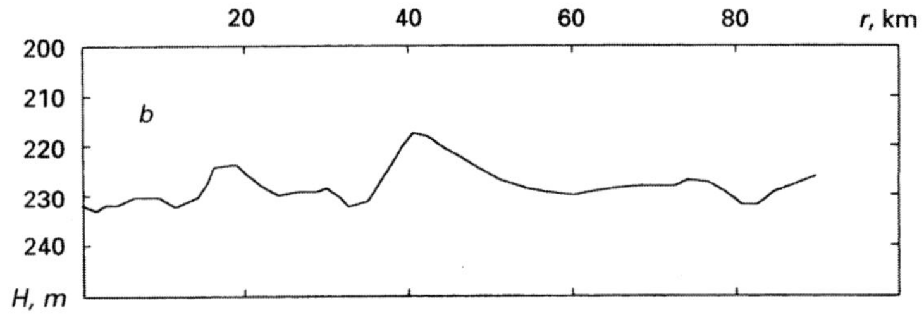

(b)

Figure 6.5. Sound speed profiles $c(z)$ at the start (1) and the end (2) of an acoustic track (a), changes in depth of the waveguide $H(r)$ along the track (b).

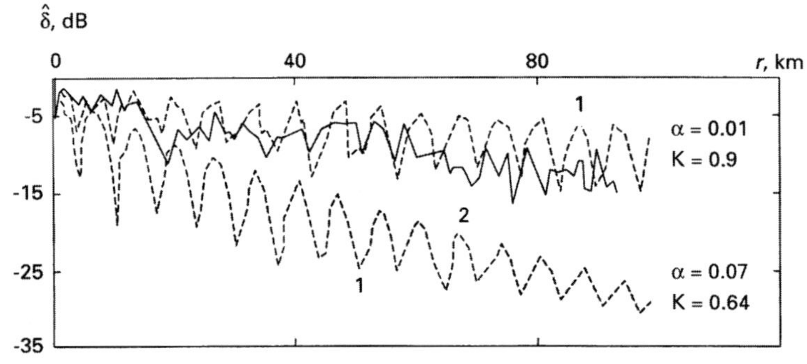

Figure 6.6. Calculated (1) and experimental (2) dependences of the relative intensity of a point sound source on distance.

approximately in a single range of values of $\alpha \approx 0.015$–0.03 for different depths and even for different frequencies.

The closeness of the values for different frequencies shows that the chosen model corresponds to a linear dependence of absorption on frequency. Naturally, such a

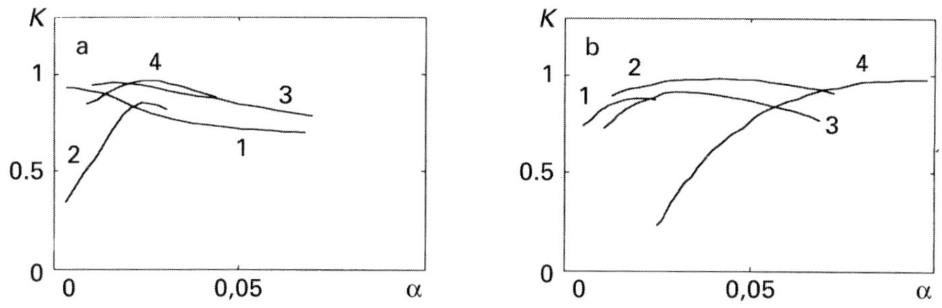

Figure 6.7. Dependence of the values of the numerical proximity criterion K on the parameter α for frequency 100 (*a*) and 230 Hz (*b*). 1–4 – receiver depth in metres: 1 – 125 m, 2 – 95 m, 3 – 80 m, 4 – 65 m.

dependence does not necessarily occur in other shallow-water regions and in other frequency bands. As mentioned earlier in Section 1.2, modern theories of the propagation of sound in sediments (Hampton (1973), Biot (1956)) assume a multi-component structure of the sedimentary rocks and identify at least three possible mechanisms for the absorption of sound in an homogeneous bottom:

- inelastic losses in the sediment matrix;
- viscous losses, associated with the relative movement of the interstitial fluid and the matrix;
- losses, due to volume scattering.

The different mechanisms give different dependences of the absorption coefficient on frequency: from proportionality to the square root of the frequency for the model of viscous losses to proportionality to the fourth power of the frequency, associated with volume scattering, when the size of the scatterers is much less than the length of a sound wave. In this connection, special studies are needed to explain the frequency dependence of the absorption of sound in the bottom for a specific shallow-water region and for a given frequency band.

The method considered earlier for estimating the effective parameters of the bottom from acoustic sensing data can be used, with certain modifications, for these purposes. Here, it is not necessary to carry out multiple preliminary towings of sound sources to large distances with radiation at different frequencies. (We note that the powerful wideband low-frequency sound sources, adapted for towing, which are required for this are extremely complex to produce and do not exist for all frequency bands). Estimates of the frequency dependence can also be obtained from the results of sensing on a stationary acoustic track with a stationary (and therefore simpler to produce) wideband noise source. Let us demonstrate this possibility on the example of a stationary track in the Barents Sea (Grigoryev *et al.* (1996)).

The sensing was carried out using a piston-type source, generating wideband pulsed signals with linear frequency modulation (LFM) in the band 25–95 Hz. In other words, the spectrum $S(\omega)$ of such a signal is characterised by a practically

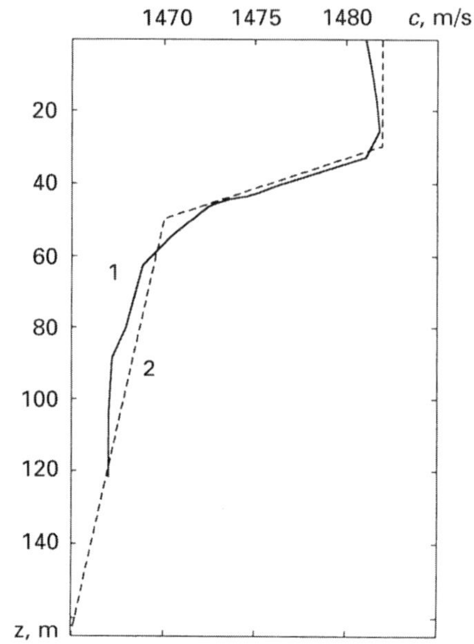

Figure 6.8. Vertical sound speed profiles: 1 – experimental, 2 – used in the calculations.

constant value in this frequency range. The duration of each pulse was $T = 40$ s and the repetition period was equal to 60 s. The sound source was lowered to a depth of 70 m from on board a scientific research ship, braced between two anchored buoys. The sound was received by an anchored hydrophone, located at a distance of 18.5 m from the bottom (the hydrophone was lowered at a distance of 13.82 km from the source from on board another ship, also braced between buoys). The depth of the water varied weakly along the track and was equal to $H \approx 170$ m. Sensing of the upper layer of the bottom sediments, carried out using a 'Parasound' echo sounder, did not reveal a stable layered structure in the studied region, so that it was then possible to use a model of an homogeneous bottom. The sound speed profile measured at the point of radiation is shown in Figure 6.8.

The essence of this method for estimating the frequency dependence of the sound absorption coefficient comprises the following:

- measurement of the spectral modulus $|S_e(r, z; \omega)|$ of the received signal, as a function of frequency, for fixed values of distance and depth;
- calculation of the analogous spectrum $|S_{th}(r, z; \omega)|$ for a model of a waveguide with a fluid absorbing bottom and coefficient α, whose dependence on frequency is given (see Section 1.2) by the exponential function

$$\alpha = \alpha_0 (f/f_0)^q \tag{6.7}$$

- the choice (based on a numerical criterion) of model parameters and, in the first

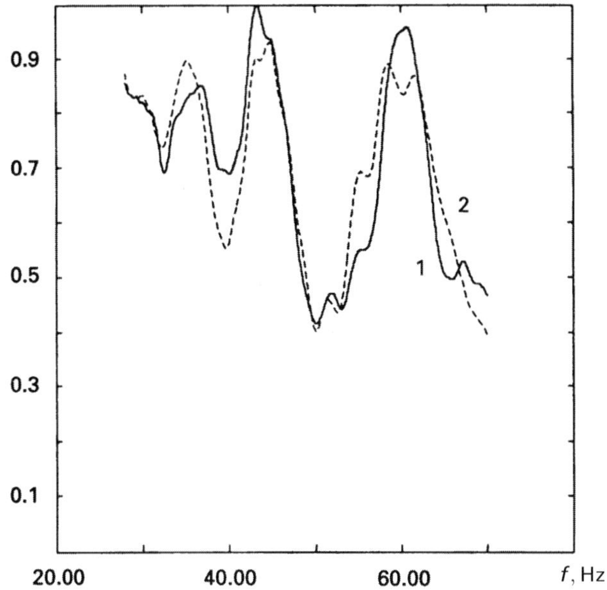

Figure 6.9. Experimental (1) and calculated (2) spectral moduli of the signal.

place, of the coefficient q, providing the best agreement of the theoretical and experimental spectra as a function of frequency.

Following this method, in the processing of the experimental data the spectral moduli of the received signals were obtained with frequency averaging in a 5 Hz band. The spectra obtained in this way were practically unchanged throughout the whole measurement period $\cong$ 100 min. In Figure 6.9, the continuous curve shows the frequency dependence of the spectral modulus of the received signal, normalised so that its maximum value is 1.

The calculated value used was

$$|S_{\text{th}}(r, z; \omega)| = C|\Psi(r, z, O, z_0; \omega)| \tag{6.8}$$

which takes explicit account of the dependence of the eigenfunctions ψ_l and the eigenvalues ξ_l on frequency. Here, $C = |S(\omega)| = \text{constant}$, since in the given frequency band the source produces a uniform spectrum. In connection with the inadequate accuracy of absolute calibration of the source and the receiving hydrophone, the parameter C, together with the exponent etc., is a tunable parameter of the theory. For the computation of (6.8), equation (6.2) was solved numerically for fixed values of r, z, z_0, but for various frequency values in the band 25–95 Hz. The parameters of the bottom in the calculation were taken to be the following: $\rho_1 = 2$ g/cm^2, $c_1 = 1780$ m/s (these parameter values were obtained as a result of comparing calculated and experimental perturbations of the sound field due to tides in this region of the Barents Sea (see Section 6.4)). We note that, according to the estimate (2.49), the number of modes giving a marked contribution to the field and, accordingly, taken into account at different frequencies, varied. Furthermore, the calculated and experimental spectra were compared based on the following numerical criterion:

$$\tilde{\sigma} = \frac{1}{J_\omega}\sqrt{\sum_{j=1}^{J_\omega}\left[\left|S_{\mathrm{e}}(r,z;\omega_j)\right| - \left|S_{\mathrm{th}}(r,z;\omega_j)\right|\right]^2} \tag{6.9}$$

Here, J_ω is the number of frequency readings in the measurement and computation of the spectrum, j is the index of a frequency reading. Under the condition of normalisation of the functions $\max_j\{|S_{\mathrm{e}}(r,z;\omega_j)|, |S_{\mathrm{th}}(r,z;\omega_j)|\} = 1$, the parameter $0 \le \tilde{\sigma} \le 1$ and the minimum of $\tilde{\sigma}$ (ideally 0) provides the best agreement of theory and experiment. Thus, the choice of optimal parameters was made based on minimisation of $\tilde{\sigma}$.

The realisation of this method for the experiment described above gave the following results. The best agreement of theory and experiment was observed for the following effective parameters of the waveguide: $\alpha_0 = 0.005$, $q = 2.2$ in the band 28–70 Hz (the value of f_0 was taken to be 28 Hz). The calculated spectral modulus corresponding to these parameters is shown by dashes in Figure 6.9.

It is true that $\tilde{\sigma}$ is practically independent of the value of α_0 (the value $\alpha_0 = 0.005$, like the effective sound speeds and density values, was determined by comparing calculated and experimental perturbations of the sound field (see Section 6.4)). However, a dependence on the coefficient q was obtained in explicit form (see Figure 6.10). The effective value obtained of $q = 2.2$ evidently points to a very strong effect of volume scattering on the attenuation of sound in the frequency range analysed, 28–70 Hz. Here, we note that a satisfactory correspondence between experiment and calculation, and also a smooth dependence $\tilde{\sigma}(q)$, only occurred in the above frequency range. When the width of the frequency band was increased to 28–95 Hz, such a correspondence was not observed. Evidently, this has to do with the fact that the proposed dependence (6.7) is not satisfied for unchanged values of α_0 and q. In the experiment, described earlier, to study the attenuation of sound using a towed source (carried out, it is true, in another region) no difference in the values of the coefficient α at frequencies 100 and 230 Hz was generally observed.

As mentioned earlier, at small (at most several tens of kilometres) distances a model of a multilayered bottom in shallow water may turn out to be more realistic than a model of an homogeneous bottom. In the first place, this will be the case for regions where, based on the results of seismography a stable layered structure is actually observed. Here, we note that a layered structure may destroy the monotonic dependence of the attenuation of sound on frequency in shallow water, which is characteristic for a model of an homogeneous bottom with an exponential dependence of the coefficient α as in (6.7). The resonance increase of the mode absorption coefficients, mentioned in Section 2.4, may give rise to a corresponding increase in the attenuation of the whole sound field in general at a specific frequency. However, such an anomalous attenuation will only be marked at large distances

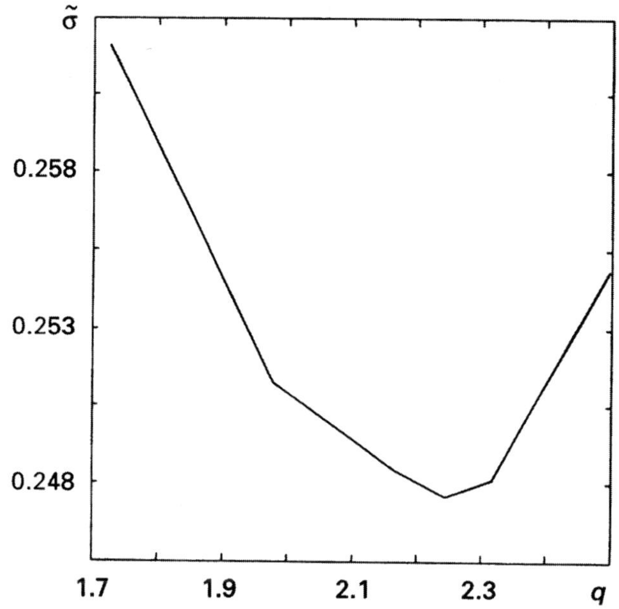

Figure 6.10. Dependence of $\tilde{\sigma}$ on q.

from the sound source, when only a few energy-bearing modes[5] remain in the waveguide, for one of which beyond the index l a resonance increase of the mode absorption coefficient takes place. The depth of the sound source position in this case must also be such that this mode is excited in the waveguide in the optimal way (the function $\psi_l(z)$ must have a maximum at the point $z = z_0$ (see (2.41)).

The sound attenuation coefficients β at different frequencies, shown in Figure 6.11, were obtained under just such conditions. The experiment was performed in the Barents Sea using explosive sound sources. The explosions were carried out at depths of 50 m and 100 m every 8 km along the track, where the starting point was at a distance of 150 km from the sound receiver and the end point at 300 km. The water depth was $\cong$ 200 m. The sound speed profile was characterised by a negative gradient. The received signals were filtered in a 2.5 Hz band at specific frequencies, separated by an interval of 10 Hz, and lying in the range 40–200 Hz. Then, for each specific frequency, the dependence $\hat{\delta}_e(r) = \delta_e(r) + 10\lg(r/r_0^*)$ was constructed which, as one can easily deduce from (5.1), when the interference beats are averaged, must be described by the expression $\beta(r - r_0^*)$ (r_0^* was taken to be 150 km). The attenuation coefficient given in Figure 6.11 was defined as the slope of the straight line interpolating the real dependence $\hat{\delta}_e(r)$ based on the least squares

[5] The presence of only a few energy-bearing modes far from the source is characteristic for a waveguide with a negative gradient of the sound speed profile (see Section 2.9).

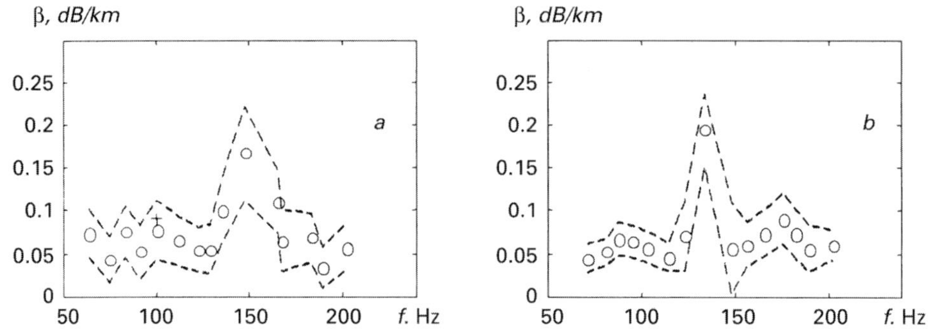

Figure 6.11. Dependence of the sound attenuation coefficient on frequency. (*a*) Explosion depth 50 m, *b*) explosion depth 100 m. The cross denotes the value of β obtained on the same track in experiments with pure-tone signals; the dots denote experimental values; the curves bound the area of values of β with fiducial probability 90%.

method. As can be seen from the figure, an anomalously high absorption is observed at frequency 150 Hz and 130 Hz for explosion depths of 50 m and 100 m, respectively. The difference in frequencies is apparently related to the excitation of different energy-bearing modes in the explosions at depths 50 m and 100 m, which accordingly experience resonance absorption at different frequencies.

Unfortunately, profile shooting only allows us to estimate the thickness of a layer h_n, whereas other parameters of the layer (sound speed, density, attenuation coefficient) remain unknown. Other geological methods, based on drilling in the bottom, can be used to determine these. However, those methods are expensive, and only allow one to obtain data for a small range of distances, In this connection, the parameter estimation methods, similar to those considered earlier, and based on a comparison of experimental and calculated dependences of changes in the sound speed in a waveguide, appear more promising. In principle, these methods include the technique proposed by Ageeva *et al.* (1994) for estimating the parameters of a multilayered bottom, based on comparison of experimental dependences $\bar{\delta}_{\mathrm{e}}(r, z, \omega)$ (obtained using explosive sound sources) and calculated dependences $\bar{\delta}_{\mathrm{th}}(r, z, \omega)$ for a number of frequencies in the range 50–630 Hz. However, Ageeva *et al.* (1994) does not present a numerical criterion for comparing the proximity of these dependences. In general, these methods are referred to as ocean tomography, based on matched field signal processing. (They are discussed in more detail in Section 6.4)

We further note that roughnesses of the bottom may also have a considerable effect on the attenuation of sound in shallow water. From the physical point of view, this is related to transformation of the modal composition as the sound propagates, and to different (see Section 2.3) attenuation coefficients of the individual modes. The effect of roughnesses of the bottom and of the surface on the sound field in an oceanic underwater sound channel (USC) has been studied (with a number of constraints) in McDaniel and McCammon (1987), and in Gorskaya and Raevskii (1990), assuming a constant sound speed and no absorption, and in Krazhev and Kudrashov (1984) with impedance boundary conditions on the bottom. In our further study of the attenuation of sound, let us consider the more general case of shallow-water USC, constructed (in the r, z-plane) in accordance with Chapter 4, in the form of a water layer, with depth bounded by the free surface $s(r)$ and the bottom $H(r) + h(r)$. The density and the sound speed profile in the water layer (refractive index : $n(r, z)$) and in the bottom are assigned as in Section 2.1. The boundaries of the USC are assumed to be random functions of the longitudinal coordinate: $s(r)$ is the deviation of the free surface from the equilibrium position, $h(r)$ is the mesoscale relief of the bottom that describes the roughness with horizontal scale ($\sim 10^2$– 10^3 m), $H(r)$ is the general relief of the bottom.

The sound field generated by a non-directional point source of frequency ω is sought in the form (4.12); and accordingly, the mode amplitudes $C_l(r)$ are determined from equation (4.13), where for the matrix elements of the mode coupling due to scattering on roughness of the surface and the bottom, one must take

$$V_{lm}(r) = \frac{s(r)}{2\sqrt{q_l q_m}} \psi_l'(r, 0)\psi_m'(r, 0) \tag{6.10}$$

$$\begin{aligned} V_{lm}(r) = \frac{h(r)}{2\sqrt{q_l q_m}} \Bigg\{ \left(\frac{\rho_1}{\rho} - 1\right) \Bigg) \psi_l(r, H)\psi_m'(r, H) - \Bigg[(k^2 n^2(r, H) - q_m^2) \\ - \frac{\rho}{\rho_1}(k^2 n_1^2 - q_m^2) \Bigg] \psi_l(r, H)\psi_m(r; H) \Bigg\} \\ + \frac{ih' q_m (1 - \eta)}{2\sqrt{q_l q_m}} \psi_l(r, H)\psi_m(r, H) \end{aligned} \tag{6.11}$$

In the expressions (6.10), (6.11), we have used the notation: $\psi_l' = d\psi_l/dz$, $H' = dH/dr$. Calculation of the sound field, taking into account scattering on random USC boundaries reduces to numerical solution of the Sturm–Liouville problem and to numerical integration of a system of equations of the form (4.13) along the acoustic track, which, for these calculations, is divided into a sequence with steps less than the characteristic scale of the mesorelief. The integration at each step is carried out for depth values corresponding to the measured mesorelief. Thus, using (4.12) in the case of rough boundaries, one can calculate the relative intensity of the acoustic field in the waveguide $\hat{\delta} = 10\lg(I/I_0) + 10\lg(r/r_0^*)$, characterising the attenuation of the field.

The effect of rough boundaries has been studied in a region of the Barents Sea with clearly expressed roughnesses of the mesorelief. For this region $\hat{\delta}$ was measured and calculated for a radiation frequency of 230 Hz and the value of the coefficient α was estimated. Figure 6.12 shows the experimentally measured sound speed profile in the water layer $c(r, z)$, and the relief of the bottom including both the general relief $H(r)$ and the mesorelief $h(r)$. Measurement of the mesorelief was performed using a narrow-beam echo sounder, capable of recording roughnesses with horizontal dimensions ≥ 10 m. The spatial density of the spectrum of roughnesses was satisfactorily described by a Gaussian function, where the mean-square value of the mesoscale roughnesses of the bottom was around 1.4 m. To calculate the effect of

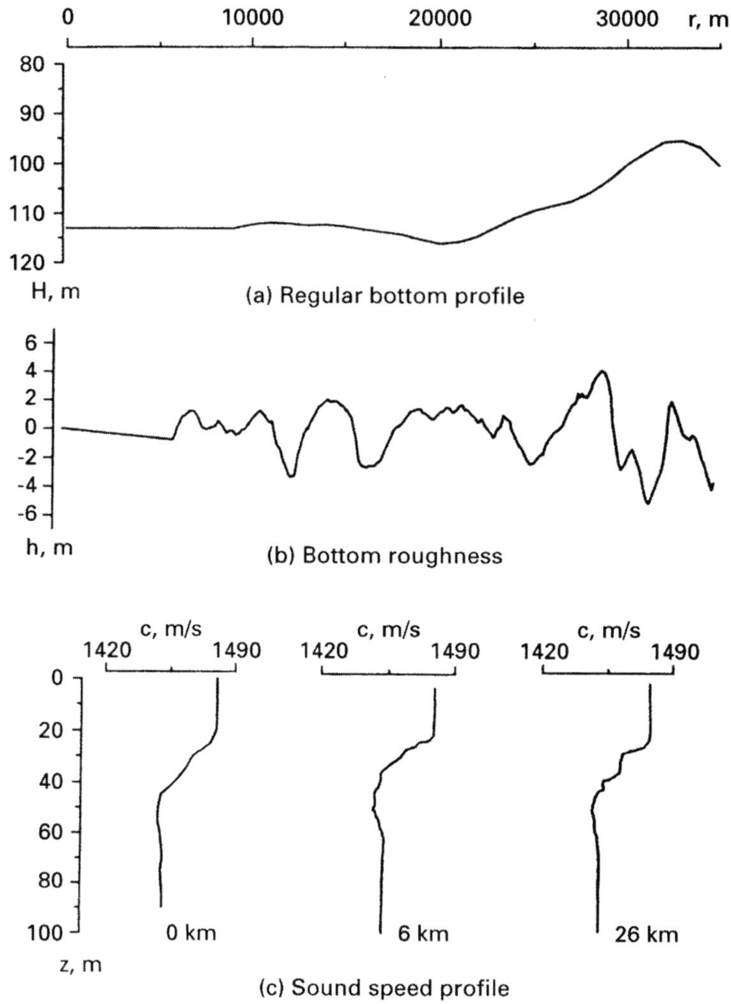

Figure 6.12. Bottom relief (*a*) general (*b*) mesoscale and sound speed profiles (*c*) on the studied track.

surface disturbance, the wind speed was measured during the experiment; its value was $10 \pm 2\,\mathrm{m/s}$, and its direction relative to the acoustic track was $\approx 30°$. The deviations of sea surface are modelled by the Gauss random value with zeroth mean and Pierson–Moskovitz spectrum, which depends on sound speed (Pierson and Moskovitz, 1964).

Figures 6.13 and 6.14 show the results of calculations and experiments for depths of the sound source and receiver $z_0 = 54\,\mathrm{m}$ and $z = 75\,\mathrm{m}$, respectively. All the calculated curves correspond to the best possible agreement (maximum K (6.6)) between experiment and theory for different computational schemes. The first thing

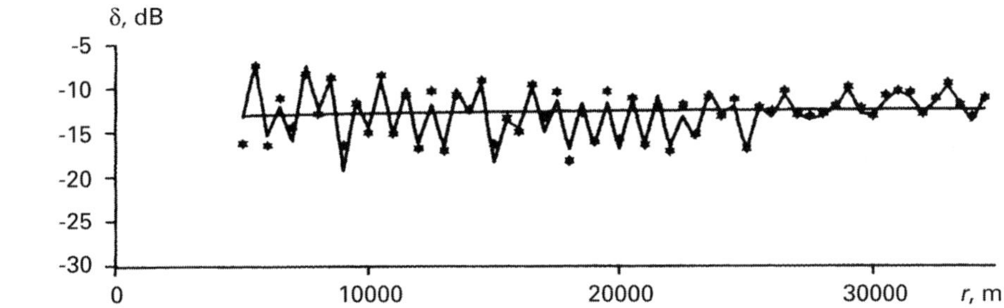

Figure 6.13. Calculated intensity of sound as a function of distance, taking into account scattering on surface waves (asterisks), without taking into account scattering (broken curve), and dependence averaged over the track (smooth curve). Here $h(r) = 0$.

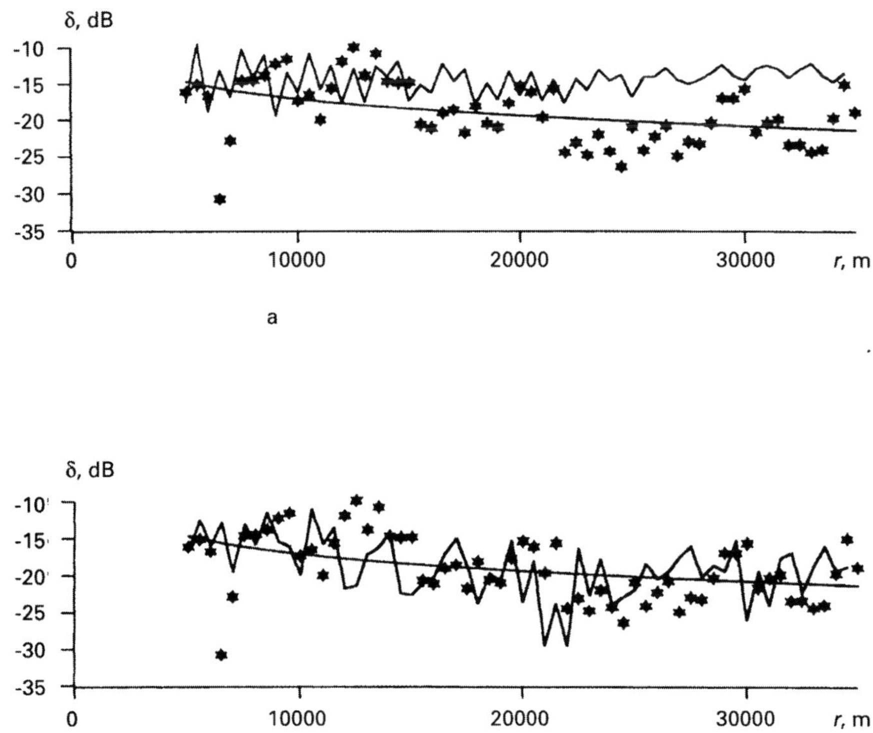

Figure 6.14. Intensity of sound as a function of distance. Asterisks denote experimental values, the broken line denotes the calculated dependence, and the smooth curve denotes the averaged dependence along the track. (*a*) Calculation without taking account of roughnesses of the bottom; (*b*) taking into account roughnesses.

that can be said, based on the calculations, is that the surface waves do not have a significant effect on the attenuation of the sound field. This can be seen in Figure 6.13, which shows the results of calculations of the relative intensity, without taking into account scattering on surface waves (asterisks) and taking into account scattering (broken curve), where these dependences are practically identical.

At the same time, it is important to take account of the scattering of sound waves on roughnesses of the bottom, since this plays an important role in the sound attenuation mechanism. Figures 6.14(*a*) and (*b*) show the results of the calculation of the relative intensity (broken curve) without taking into account scattering on roughnesses of the bottom (*a*), and taking scattering into account (*b*); the asterisks on both figures denote the same experimental dependence $\hat{\delta}_e$, a smooth curve, where $\hat{\delta}_e$ represents averaging over the scale of the intermode beats. In the first case, the set of parameters giving the best agreement for $K = 0.6$ is $\alpha = 0.1$, $c_1 = 1750\,\text{m/s}$, $\rho_1 = 1.9\,\text{g/cm}^3$; in the second case, $\alpha = 0.06$, $c_1 = 1750\,\text{m/s}$, $\rho_1 = 1.9\,\text{g/cm}^3$ for $K = 0.8$. It is clear that by taking scattering on roughnesses of the bottom into account (Figure 6.14(*b*)) we can obtain a better agreement of the theoretical calculations and the experimental data (the proximity coefficient K is greater in the second case). We note that in the second case the parameter α has the physical sense of the loss due precisely to absorption in the bottom; losses due to scattering are considered separately in this approach. In the first case, the α effectively includes both loss mechanisms, due to absorption and scattering; the second model is understandably more appropriate, and in the second case the value of α is close to the values recorded in the experiments described earlier in regions with a flattened mesorelief.

6.2 FLUCTUATIONS OF THE SOUND FIELD IN SHALLOW WATER, INDUCED BY THE LARGE-SCALE VARIABILITY OF THE MEDIUM

Shelf zones of the World Ocean have a significant hydrodynamic variability, which also has clearly expressed geographical traits. First and foremost, this relates to the fluctuating frontal zones, to sea tides, whose amplitude may be greater than 10 m, and to internal waves induced by the tides. The latter, as a rule, are formed in areas with sharp roughnesses of the bottom – for example, in shelf-break areas – and consist of trains of nonlinear soliton-like waves of large amplitude (~3–20 m) and relative weak linear background internal waves. Unlike in the deep ocean, the field of internal waves on the oceanic shelf has a considerable spatial anisotropy with a maximum in the spatial spectrum for waves propagating towards the coast (in the direction of decreasing depth). We also note that in the modal distribution of the amplitude of the displacement of a surface of uniform density, the first mode of an internal wave dominates. These differences and the clearly expressed geographical specifics, mean that in shallow water we cannot use the Garrett–Munk model of the spectrum, which provides a good description of the field of internal waves in the deep ocean.

The hydrodynamic variability gives rise to considerable fluctuations of the low-frequency sound field in shallow water. For acoustic sound sources this reduces to fluctuations of the amplitude $|P(r,z,t)|$ and phase $\varphi(r,z,t)$ of the sound field, as functions of the space–time coordinates, determined by the equation:[6]

$$P(r,z,t) = |P(r,z,t)|e^{i[\varphi(r,z,t)-\omega t]} \tag{6.12}$$

If there are no hydrodynamic perturbations, the parameters of the waveguide do not depend on time, and the sound field of a pure-tone source is given by expression (2.41). If we consider this field at a fixed point in space as a function only of time t, then it is a finite sum (because of the dying out of modes) of terms which are harmonic in time with a different phase shift, i.e. ultimately, an expression of the form $P(r,z,t) = |P(r,z)|e^{i[\varphi(r,z)-\omega t]}$, whose amplitude and phase do not vary with time.

In the presence of time-dependent hydrodynamic perturbations this picture changes, and comparatively slow fluctuations of amplitude and phase appear.

Naturally, both the variability itself and the fluctuation of the sound field depend on the region studied. Figure 6.15 shows the fluctuations of the amplitude and phase of the sound field, recorded on stationary acoustic tracks in different shallow-water regions by various authors (Essen *et al.* (1987), Shmelerv (1990), Rutenko (1997)). As an example, we chose 12-hour fragments of recordings of amplitude and phase in experiments with pure-tone signals with values as close as possible to the carrier frequency ω_0. All the experiments we chose were carried out in near-bottom sound channels, where the vertical sound speed profiles $c(z)$ had a significant negative gradient. The corresponding dependences $c(z)$ are shown in Figure 6.16. As one can see from Figure 6.15, the fluctuations measured in different experiments are very different in both amplitude and frequency. Moreover, the variations in the nature of the fluctuations in different seasons of the year within the same region are very significant. For example, on the shelf of the Sea of Japan, the maximum fluctuations of the phase of signals with frequency 315 Hz, recorded in winter, spring and autumn experiments are $1,500^\circ$, $2,600^\circ$ and $5,000^\circ$, respectively (Bondar). In the spectrum of these fluctuations $S_\varphi(\omega)$, defined for the function $\varphi(t)$ according to (6.12), in spring and autumn there are marked peaks corresponding to the periods of typical internal waves for the season. There are no such peaks in winter. These variations are caused by seasonal variations of the stratification in the Sea of Japan.

Naturally, such fluctuations of amplitude and phase lead to a broadening of the spectrum of the pure-tone signal, transmitted on stationary acoustic tracks. An example of the modulus of such a spectrum $|S(r, 2\pi f)|$, measured in the Barents

[6] In measurements – i.e. in work with a real signal – the amplitude and phase are determined somewhat differently: $|P(r,z,t)| = \sqrt{p^2_{\cos}(r,z,t) + p^2_{\sin}(r,z,t)}$, $\varphi(r,z,t) = \arctan(p_{\sin}(r,z,t)/p_{\cos}(r,z,t))$ where $p_{\sin}(r,z,t) = \langle p(r,z,t)\cdot\sin\omega t\rangle$ and $p_{\cos}(r,z,t) = \langle p(r,z,t)\cdot\cos\omega t\rangle$ are the sine and cosine quadrature components of the real (measured) field $p(r,z,t)$, related to the field $P(r,z,t)$ by equation (1.11). (The pointed brackets denote averaging over the period). Easy transformations show that these two definitions are equivalent.

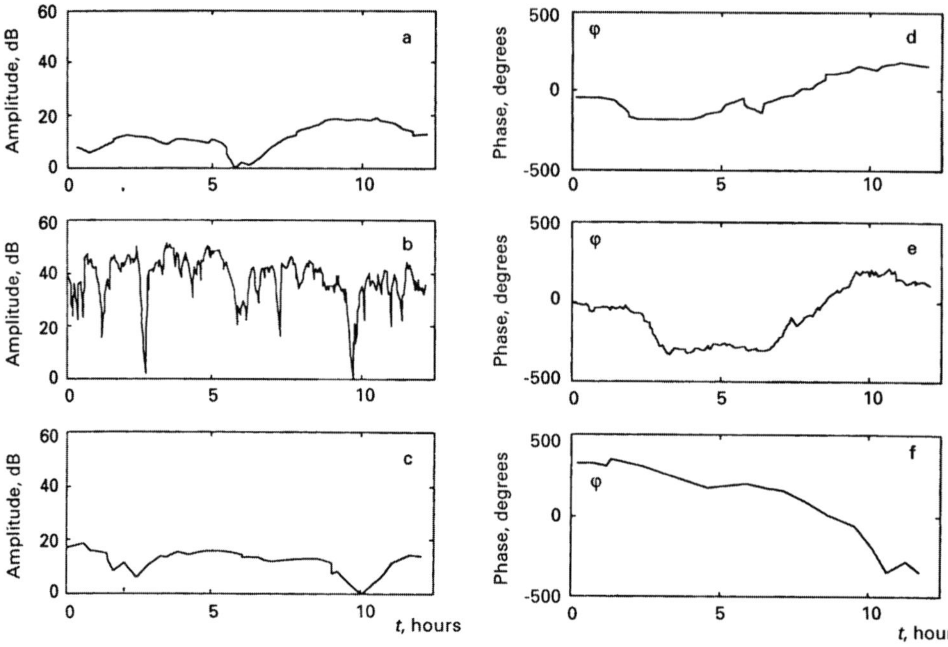

Figure 6.15. Fluctuations of the sound field amplitude and phase, recorded on stationary acoustic tracks. (a), (d) North Sea, $f = 295\,\text{Hz}$, distance between sound source and receiver $r \cong 1.1\,\text{km}$; (b), (e) Barents Sea, $f = 300\,\text{Hz}$, $r \cong 70\,\text{km}$; (c), (f) Sea of Japan, $f = 315\,\text{Hz}$, $r \cong 16\,\text{km}$.

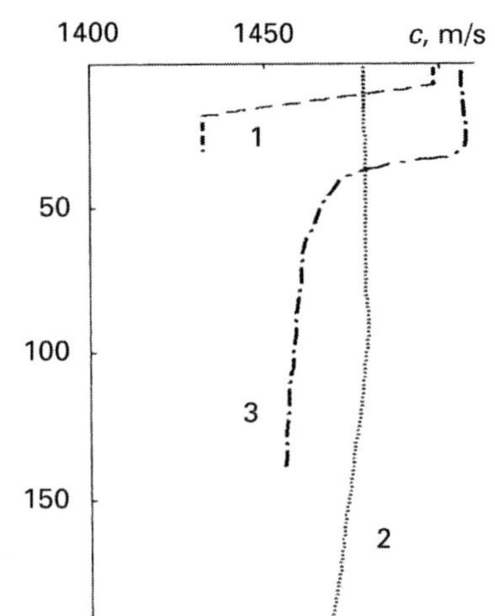

Figure 6.16. Vertical sound speed profiles. 1 – North Sea, 2 – Barents Sea, 3 – Sea of Japan.

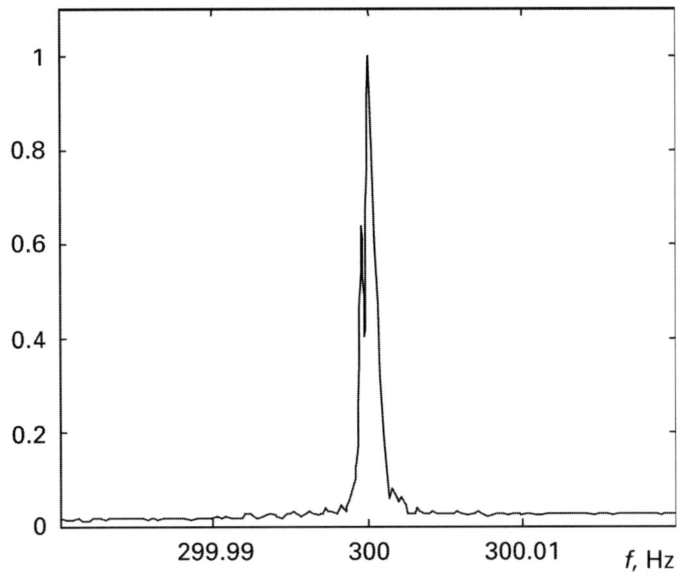

Figure 6.17. Spectrum of a pure-tone signal travelling a distance of 70 km.

Sea for a distance between a static source and sound receiver of $\cong$ 70 km, is shown in Figure 6.17. The width of the spectrum at the 0.5 level is $\delta f \cong 10^{-3}$ Hz. (We note that the width of the spectrum of the radiated signal in this case was $\Delta f = \Delta\omega/2\pi \cong 3 \times 10^{-5}$ Hz).

The geographic specifics of the large-scale hydrodynamic variability in shallow water make it impossible to create universal models describing the fluctuation of the sound field. One can only point to a number of characteristic properties, inherent in these fluctuations, and outline general approaches to constructing models oriented towards specific regions of the ocean shelf. One such property is the significant effect on and specific manifestation of the tide in these fluctuations. As noted in a number of experiments (Zhuravlev *et al.* (1991), Essen *et al.* (1983)) the fluctuations of the phase of the sound field on stationary tracks bear a marked correlation to the tidal variations of the sea depth, although in an analysis of the fluctuations of amplitudes such a correlation was not noted. This is illustrated in Figure 6.18, which shows, for the same moment in time, the variations of the depth of the sea and amplitudes and phases of the sound field with frequency 100 Hz, measured in the Barents Sea. (A corresponding profile of the sound speed is shown by the dashed curve in Figure 6.16. The distance between the sound source and receiver was $r \cong 70$ km, and the average local depth was $H \cong 200$ m).

Very simple estimates also support this conclusion. Indeed, we conjecture that the main cause of phase fluctuations is a tidal wave of height $h(t)$, which is much less than the depth of the waveguide H: $|h(t)| \ll H$. According to (2.41), the field at a considerable distance from the acoustic sound source in an 'unperturbed' waveguide

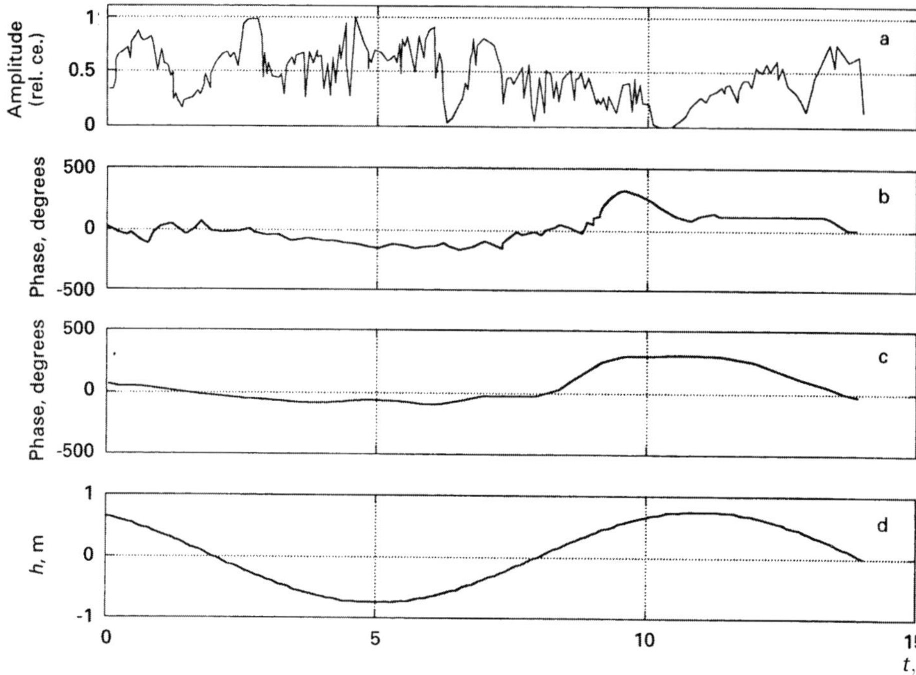

Figure 6.18. Experimental graphs of the variations of the amplitude of a pure-tone signal (*a*) and of the phase of the acoustic field (*b*) against time; calculated graph of the variation of the phase of the pure-tone signal (*c*) against time, variations of the sea level under the effect of the tide (*d*).

can be represented in the form of a sum of normal waves:

$$P(r,z,t) = \sum_l P_l(r,z)e^{i(\varphi_l - \omega t)} \tag{6.13}$$

where $P_l(r,z) = \sqrt{\rho c W_0 / q_l r}\psi_l(z_0)\psi_l(z)e^{-\gamma_l r/2}$ is the amplitude of a normal wave, and $\varphi_l = q_l r + \varphi_0$ is its phase (φ_0 is the initial phase. We neglect the phase of the complex eigenvalue, i.e. in the denominator of the expression within the square root we set $|\xi_l| \approx q_l$).

In this case, the field amplitude and phase can be expressed in terms of the mode amplitudes and phases

$$|P(r,z)| = \sqrt{\left(\sum_l P_l \sin\varphi_l\right)^2 + \left(\sum_l P_l \cos\varphi_l\right)^2} \tag{6.14.1}$$

$$\varphi(r,z) = \arctan\left(\frac{\sum_l P_l \sin\varphi_l}{\sum_l P_l \cos\varphi_l}\right) \tag{6.14.2}$$

Clearly, the values[7] of φ and $|P|$ are sensitive in the first place to small relative changes in the phase of normal waves, associated with variations in the propagation constants q_l, and are practically independent of the amplitudes P_l, which can be assumed to be time invariant. We shall take depth variations into account in the framework of perturbation theory, assuming, moreover, that the variation of $h(t)$ is sufficiently smooth that the temporal dependence of the field is given in terms of the dependence of the waveguide parameters (adiabatic). Setting $H(t) = H + h(t)$, $q_l(t) = q_l + (\partial q_l/\partial H)h(t)$, we obtain the following expression for the variation of the phase $\delta\varphi(t)$:

$$\varphi + \delta\varphi(t) = \arctan\left(\frac{\sum_l P_l \sin(\varphi_l + \delta\varphi_l)}{\sum_l P_l \cos(\varphi_l + \delta\varphi_l)}\right) \tag{6.15}$$

where $\delta\varphi_l = q_l{}'rh(t)$, where $q_l{}' = \partial q_l/\partial H$. Assuming that $\delta\varphi_l$ is small, decomposing the right-hand side of (6.15) in a series up to terms of first order in $\delta\varphi_l$, and taking the average, we obtain

$$\delta\varphi(t) \approx \sum_l W_l \delta\varphi_l = rh(t) \sum_l W_l q_l' \tag{6.16}$$

where $W_l = P_l^2/|P|^2$ is the specific weight of the mode with index l. The phase fluctuations of the field are approximately equal to the 'mean' modal fluctuation and proportional both to the amplitude of the tide and to the horizontal distance from the source; however, the coefficients of proportionality depend heavily on the type of the waveguide and, in particular, on the sound speed profile.

However, the phase $\varphi(t)$, as a function of depth, only in the first very coarse approximation depends linearly on depth. This linear nature of the dependence provides correlation between the variation of signal phase and of the sea level. In reality, as can be seen from equation (6.15), the dependence $\varphi(t)$ as a function of depth is more complex. Analysis of this nonlinear dependence enables us to give a qualitive explanation of the characteristic properties of temporal variations of phase, such as asymptotic behaviour of phase at high and low tides, the strong distortions of phase as a function of time in comparison with the sinusoidal nature of the tide, and, in particular, the flattening near maxima and the appearance of the second harmonic with amplitude comparable with the first. These properties are a consequence of the fact that the dependence $\delta\varphi(h(t))$ is not at all symmetric under a change of sign of $h(t)$ and has an oscillating structure with a characteristic period of oscillations $\Delta_0 \sim 2\pi/r|q_m' - q_l'|$, as a result of which these oscillations are manifest in the temporal dependence, provided the maximum height of the tide is greater than Δ_0.

Here, however, we note that the sound field in a waveguide is characterised by the presence of special points – so-called phase dislocations – where, as a result of the interference of normal waves the amplitude of the field is zero and the phase is indeterminate (Zhuravlev *et al.* (1989)). Applying the theory of dislocations of the

[7] In what follows, we shall consider the behaviour of the amplitude and phase at a fixed point, and the arguments r, z will be omitted.

phase front, one can show that if there is a dislocation near a point of receipt, then the above distortions occur only when $|h(t)| < \Delta_0$ (Zhuravlev *et al.* (1989)).

As an example, we use equation (6.15) to calculate the phase change of an acoustic field under the action of a tide of height 1.5 m for a pure-tone signal with frequency 100 Hz. We used a simplified model of a waveguide with a linear dependence of the sound speed profile $\varepsilon = 2 \times 10^{-4}\,\mathrm{m}^{-1}$ with a rigid absorbing bottom. As in the experiment described earlier, the depth was taken to be $H = 200$ m, and the distance $r = 70$ km. The results of the calculation are shown in Figure 6.18 (*b*), from which it is clear that both the order of magnitude of $\delta\varphi(t)$ and the characteristic distortions of the temporal variation of the phase (flattening near maxima, appearance of the second harmonics) are in qualitative agreement with the experimental data, shown in the same figure. The differences between the experimental and calculated data are primarily associated with the assumption that the tidal wave is the only cause of phase fluctuations. In reality, internal waves had a considerable effect on the phase fluctuations in this experiment.

The dependence of the amplitude of the sound field $|P(t)|$ on time under tidal perturbations of depth is even more complicated. Under the same hypotheses, the amplitude fluctuations are represented by the expression:

$$\delta|P(t)| = |P + \delta P(t)| - |P|$$

where

$$|P + \delta P(t)| = \sqrt{\left(\sum_l P_l \sin(\varphi_l + \delta\varphi_l)\right)^2 + \left(\sum_l P_l \cos(\varphi_l + \delta\varphi_l)\right)^2} \tag{6.17}$$

If perturbation theory is used, then for the corrections $\delta|P(t)|$, one can obtain the expression

$$\delta|P(t)| = \sum_{l,m} \frac{P_l P_m}{|P|} \sin(\varphi_m - \varphi_l)\delta\varphi_m \tag{6.18}$$

It is clear that the sum on the right-hand side of equation (6.18) due to the interference of normal waves, depending on the relations between the parameters, which are in turn determined by the value of $h(t)$, can be taken to have positive and negative values and also to be zero. This means, first, that there is no synchronicity in the variations of the amplitude of the field and of the depth of the waveguide and, second, that nonlinear corrections in the variations of the waveguide depth to the amplitude of the field may play a role.

Ultimately, this means that the variations of amplitude are completely dissimilar to the perturbations of level associated with a tidal wave (see Figure 6.18).

Here, we note that the absence, as a rule, of a well marked correlation between the intense hydrodynamic perturbations and variations of the amplitude of sound signals in shallow water is also a characteristic property of these variations. As an example, we present the results of calculation of variations of the amplitude of the sound field due to a soliton-like internal wave (IW), propagating perpendicularly to a stationary acoustic track (Petnikov (1977)). In the modelling of the soliton, the results of contact measurements of the oscillations of the isodensity surfaces in the thermocline layer (10–35 m), obtained using a towed submerged CTD (conductivity, temperature, depth) sensor, were used. (The measurements were carried out as part of the joint US–Russian experiment (JUSREX), in 1992, in New York Bay (Gasparovic *et al.* (1994))). In these measurements, oscillations with amplitude 10.5 m at depth 25 m, associated with a soliton-like internal wave, moving with speed $\nu_s = 0.6$ m/s, were recorded.

Let us construct a theoretical model of the soliton, reflecting the main experimentally established properties.

- We introduce the horizontal coordinate y-axis, perpendicular to the r-axis, (i.e. the direction of the stationary track) with origin at the point of intersection of the r and z axes. Suppose that the y-axis coincides with the direction of movement of the soliton.
- The transverse dimensions of the soliton will be assumed to be larger than the length of the stationary track; moreover, we shall suppose that the curvature of the IW wavefront is such that a wave may be assumed to be planar on the length of the stationary track. This means that the vertical movements of water particles from an equilibrium position on the stationary track will not depend on r: $\zeta = \zeta(y, z, t)$.
- The movements of water particles with depth take place synchronously $\zeta(y, z, t) = \tilde{\Phi}_{1\tilde{\kappa}}(z)\zeta(y, t)$, where $\tilde{\Phi}_{1\tilde{\kappa}}(z)$ is the first (dominant) mode of the internal wave for the stratification in the New York Bay. (The function $\tilde{\Phi}_{1\tilde{\kappa}}(z)$ has a constant sign and is such that below and above the thermocline layer the IW amplitude decreases almost as a linear function of depth z to a zero value of y on the bottom and on the surface).
- The amplitude of the vertical movement of fluid particles (movement at the depth of the thermocline layer) as a function of y, is described by the expression $\zeta(y, t) = \zeta_0 \mathrm{sech}^2[(y - \nu_s t)/b_s]$; i.e. the wave is a Korteweg de Vries soliton (according to experimental data the maximum value $\zeta_0 \approx 10.5$ m and $b_s \approx 75$ m).
- The IW leads to synchronous perturbations of the vertical sound speed profile, calculated from the formula $c(y, z, t) = c_0(z - \zeta(y, z, t))$, where $c_0(z)$ is the background (unperturbed) sound speed profile.
- The change in the sound speed due to the movement of the soliton, as a result of the slowness of that movement, can be calculated using an adiabatic (in time) approach; i.e. in terms of the temporal dependence of the waveguide parameters.

As a result of the modelling, Figure 6.19 shows seven consecutive sound speed profiles, which occur over 150 s during the passage of the leading soliton front. The bold curve shows the background profile.

In the calculations of the variations of the amplitude of the sound field due to IW, for each profile at different times t_j the spectrum of the eigenfunctions ψ_l and the eigenvalues ξ_l for the acoustic waveguide, which depend on t_j as a parameter, was calculated. Then, a linear interpolation method was applied to find their values at intermediate points, and equation (6.14.1) was used to compute the amplitude of the sound field. The calculations were performed for a hypothetical stationary track of length $r = 15$ km and depth $H = 65$ m, with a homogeneous fluid bottom, whose

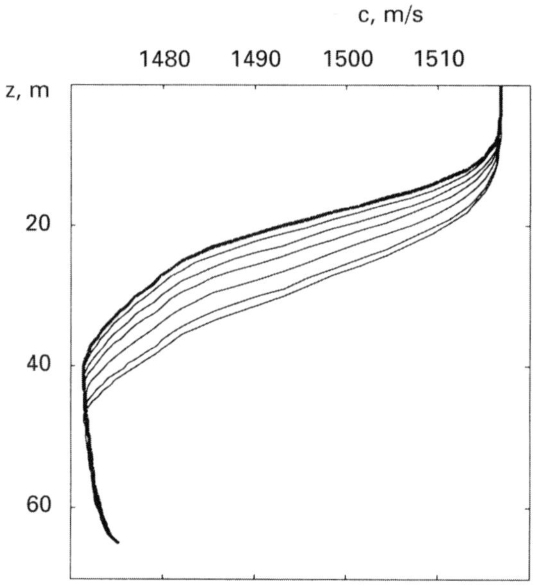

Figure 6.19. Perturbations of the sound speed profile due to a soliton.

characteristics were assumed to be $c_1 = 1,800$ m/s, $\rho_1 = 1.8$ g/cm^3 m $\alpha = 0.0225$. The parameters of the source were $f = 240$ Hz, $W_0 = 130$ W. The results of the calculation are shown in Figure 6.20. This figure shows the corresponding variations of the sound speed at depths 17 m, 24 m and 31 m, due to the passage of the soliton. As can be seen from the figure, the soliton-like internal wave induces intense variations of the sound field amplitude. However, these variations are of a higher frequency and are not at all similar to the variations of $c(t)$. Zhou *et al.* (1996) came to the same conclusion, when they modelled the effect of a soliton, recorded in Massachusets Bay, on a sound field with frequency $f = 200$ Hz.

Furthermore, we note that an absence of a well marked correlation between the dependence $c(t)$ and the amplitude variations always arises only for sufficiently intense hydrodynamic variations, and, importantly, for a significant sound speed gradient. This result is very graphically demonstrated in Figure 6.22, taken from Belov (1997), where the amplitude variations are calculated for a single train of internal waves crossing the same stationary track at right angles, but with different sound speed profiles. (Corresponding characteristic experimental sound speed profiles for the same region in summer and in autumn are shown in Figure 6.22. The calculations were performed for frequency $f = 470$ Hz.)

One further characteristic property, inherent in fluctuations of the sound field in shallow water, is the considerable dependence of these fluctuations on the direction of propagation of the sound waves, and, more precisely, on the angle α_0 between the directions of propagation of sound and internal waves (Petnikov *et al.* (1997), Zhou *et al.* (1996)). (As mentioned earlier, internal wave fields on the shelf are charac-

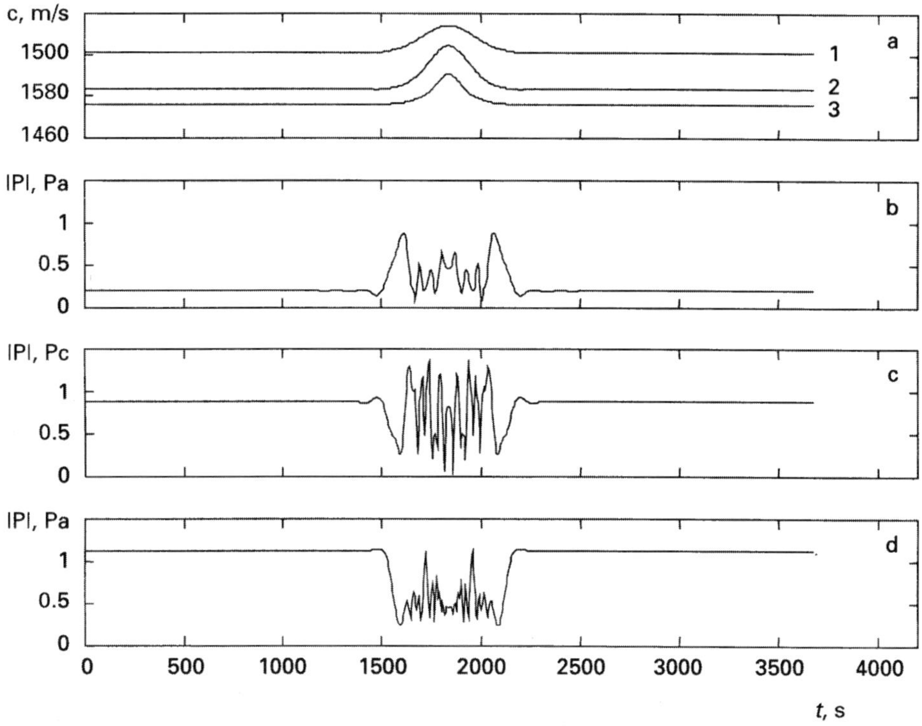

Figure 6.20. Changes in the sound speed at depths 17 m (1), 24 m (2) and 31 m (3) with time, due to a soliton (*a*), and corresponding variations of the sound wave amplitude (*b*)–(*d*) at depths 6.6 m (*b*), 28.5 m (*c*) and 65 m (*d*).

terised by a preferential direction, determined by the relief of the bottom). To illustrate this dependence, let us look in more detail at the results of modelling of the fluctuations of the sound field due to background linear internal waves of small amplitude (Petnikov *et al.* (1997)). (Such waves in shallow water are, as a rule, generated by solitons and, consequently, they have the same inherent properties: a dominance of the first mode and a sharply expressed anisotropy). Let us suppose that, in the presence of IW perturbing the vertical sound speed profile, the square of the refractive index is given by equation (4.8). For a realistic model, we shall use the results of recordings of temperature fluctuations recorded in the JUSREX experiment using moored buoy stations. (These fluctuations were associated with background internal waves). Let us also suppose that we are dealing with plane IW, propagated at angle α_0 to the direction of the track. This can be represented as the existence of a set of perturbations of the sound speed profile, which are propagated along the acoustic track with phase velocity $u\cos(\alpha_0)$, where u_0 is the frequency-independent phase velocity of an internal wave. (Further analysis shows that for the stratification of New York Bay, dispersions can be neglected for practically all energy-bearing areas of the internal wave spectrum and $u = 0.5\,\text{m/s}$). In

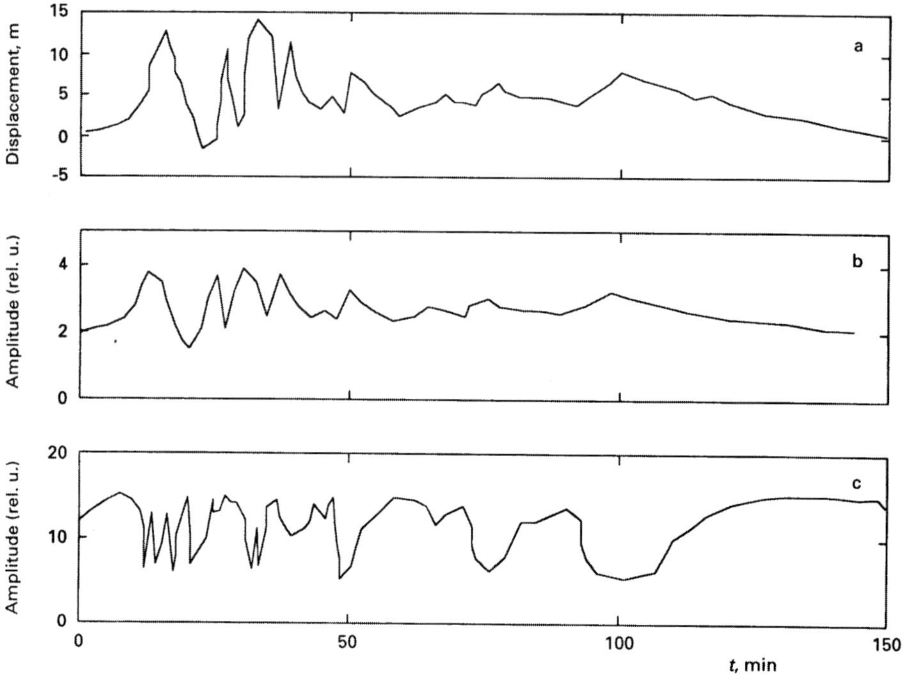

Figure 6.21. Displacement of fluid particles due to an internal wave (a), and corresponding model variations of the sound field amplitude for autumn (b) and summer (c) sound speed profiles.

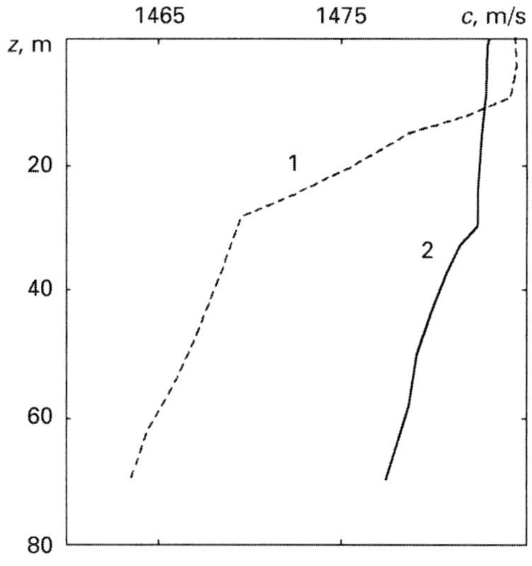

Figure 6.22. Sound speed profiles recorded at the same position in the Barents Sea in summer (1) and autumn (2) (Belov *et al.* (1997)).

this case, the following algorithm can be used to determine the random component of the square of the refractive index $\mu(r,z,t)$.

- Based on experimental recordings of temperature variations $\delta\tilde{T}(z_1,t)$, define the dependence of the vertical displacement of fluid particles on time at depth z_1, as

$$\zeta(z_1,t) = \delta\tilde{T}(z_1,t)\left(\left.\frac{\partial\tilde{T}(z)}{\partial z}\right|_{z=z_1}\right)^{-1}$$

  Here, z_1 is the depth at which the temperature measurements were performed, and $\tilde{T}(z)$ is the mean unperturbed vertical temperature profile, which is also determined based on experimental data.
- Calculate the field of vertical movements of fluid particles, using the following equation:

$$\zeta(r,z,t) = \zeta\left(r - \frac{ut}{\cos\alpha_0}, z_1\right)\cdot\frac{\tilde{\Phi}_{1\tilde{\kappa}}(z)}{\tilde{\Phi}_{1\tilde{\kappa}}(z_1)}$$

  where $\tilde{\Phi}_{1\tilde{\kappa}}(z)$ is a numerically calculated eigenfunction of the first mode of the internal wave.
- Compute the random component $\mu(r,z,t)$ from the following formula (Flatte (*ed.*) 1979):

$$\mu(r,z,t) = -2\tilde{Q}\tilde{N}^2(z)\zeta(r,z,t) \tag{6.19}$$

  Here, $\tilde{Q}$ is a coefficient depending on the physical parameters of the water. (For water of the World Ocean the value of $\tilde{Q}$ is practically invariant and equal to $\tilde{Q} \cong 2.4\,\mathrm{s^2/m}$.) $\tilde{N}(z)$ is the buoyancy frequency, computed from the mean unperturbed profile $c(z)$ ($\tilde{N}^2(z) \cong (1/\tilde{Q}c)(dc/dz)$).

Having determined $\mu(r,z,t)$ in this way, one can then use the diffusion approach described in Section 4.4 to calculate the fluctuations of the sound field.

The results of such a calculation are shown in Figure 6.23(*b*)–(*d*). (The parameters of the stationary track were chosen to be the same as in modelling of the soliton.) The ordinate axis in the figures shows $\delta P = |P_{\mathrm{iw}}| - |P_0|$ (Here $|P_{\mathrm{iw}}|$ and $|P_0|$ are the amplitudes of the sound pressure at the point of receipt at depth 28.5 m, in the presence and in the absence of internal waves, respectively. Figure 6.23(*a*) also shows the variations of the sound field due to background internal waves. As can be clearly seen in the figure, the amplitude fluctuations grow markedly as the angle α_0 increases. The maximum value is observed at $\alpha_0 = 90^\circ$, when there are equiphase oscillations of the sound speed along the length of the track.

Comparison of Figures 6.20 and 6.23 again shows that intense perturbations of the sound field lead not only to an increase in the amplitude fluctuations of the sound field, but also to a significant broadening of the spectrum of these fluctuations.

The complex highly nonlinear dependence of the amplitude of the sound field $|P(t)|$ on the hydrodynamics, and, in particular, on random perturbations of the parameters of the medium in which the waves propagate, enables us to consider $|P(t)|$ as a stochastic quantity. From that point of view, we can construct a statistical

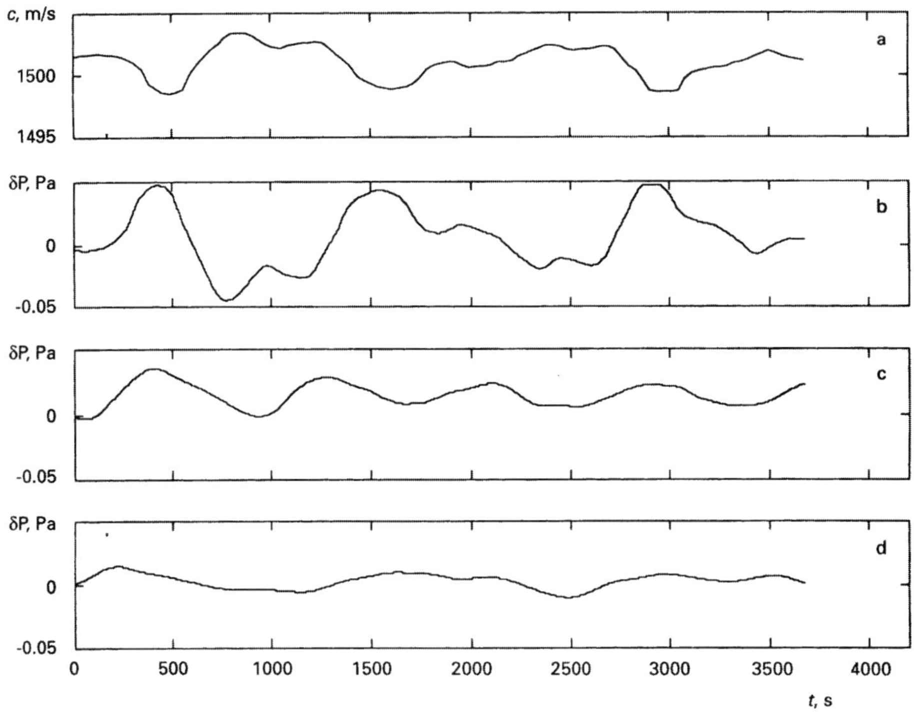

Figure 6.23. Variations of the sound speed at depth 17 m with time, due to background internal waves (*a*), and corresponding fluctuations of the sound speed amplitude (*b*)–(*d*) for angles between the directions of propagation of the internal waves and the sound waves 90° (*b*) 45° (*c*) and 0° (*d*).

model of the amplitude fluctuations, which may be important for practical applications. Lukin *et al.* (1993) attempted to construct such a model for two experiments on a stationary track of length 70 km in the Barents Sea, carried out at the different times of the year. The sound speed profiles in these two experiments are shown in Figure 6.24. Amplitude fluctuations of pure-tone signals at frequencies 100 Hz and 300 Hz were studied. According to Box *et al.* (1974), the construction of the model involved the application of nonlinear and linear transformations to the observed data, as a result of which the data were transformed to white Gaussian noise. The nonlinear transformations were chosen based on a comparison of the one-dimensional probability density of an observed instance, with a Gaussian probability density. This choice was based on the invertibility of the model in the presence of a linear inertial transformation, which, according to the central limit theorem, forms a Gaussian random process from white noise. To simplify the nonlinear transformation, the one-dimensional distribution of the amplitude of harmonic oscillation was approximated by a lognormal distribution and the quantity $\Gamma = \ln|P(t)|$ was used

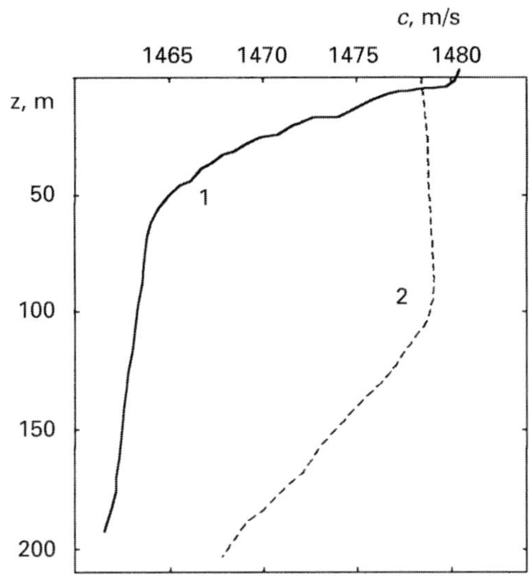

Figure 6.24. Sound speed profiles in experiments on a stationary track in the Barents Sea in summer (1) and in autumn (2) (Lukin *et al.* (1993)).

for the linear analysis. In constructing the statistical model from a single instance, it is naturally to assume that the ergodicity assumption is satisfied, a sufficient evidence for which is the condition

$$B_\Gamma(\tau) \to 0 \text{ as } \tau \to 0 \tag{6.20}$$

where $B_\Gamma(\tau)$ is the sample autocorrelation function. Processing of the experimental data showed that in both experiments and for both frequencies, condition (6.18) was satisfied with probability 95% in the observed time interval of 13.65 hours. (An example of the dependence $B_\Gamma(\tau)$ for frequency 100 Hz in the second experiment is shown in Figure 6.25). This forms the basis for extrapolation of $B_\Gamma(\tau)$ as a function decaying at infinity and for concluding that the amplitude fluctuations of the pure-tone signal form a stationary process at least over several hours.

Using the method for constructing autoregression models (Box *et al.* (1970)) to verify the hypothesis on the order of the autoregression process, partial correlations between the readings Γ_i and Γ_{i+p+1} for fixed values of $\Gamma_{i+1}, \ldots, \Gamma_{i+p}$ were computed. When the latter are non-zero, the partial correlation determines the order of the autoregression process. The analysis showed that with probability 95% the first four values of the partial correlation function can be assumed to be non-zero. Hence, the stochastic difference equation describing the amplitude fluctuations has the form

$$\hat{\Gamma}_i - \hat{k}_1\hat{\Gamma}_{i-1} - \hat{k}_2\hat{\Gamma}_{i-2} - \hat{k}_3\hat{\Gamma}_{i-3} - \hat{k}_4\hat{\Gamma}_{i-4} = \hat{a}_i \tag{6.21}$$

$\hat{\Gamma}_i$ are centralised to the mean value of the readings, Γ_i, $\hat{k}_1$, $\hat{k}_2$, $\hat{k}_3$, $\hat{k}_4$ are model coefficients, and $\hat{a}_i$ are independent Gaussian readings with zero mean and variance $\sigma^2_{\hat{a}}$. The model coefficients were estimated using the maximum likelihood method

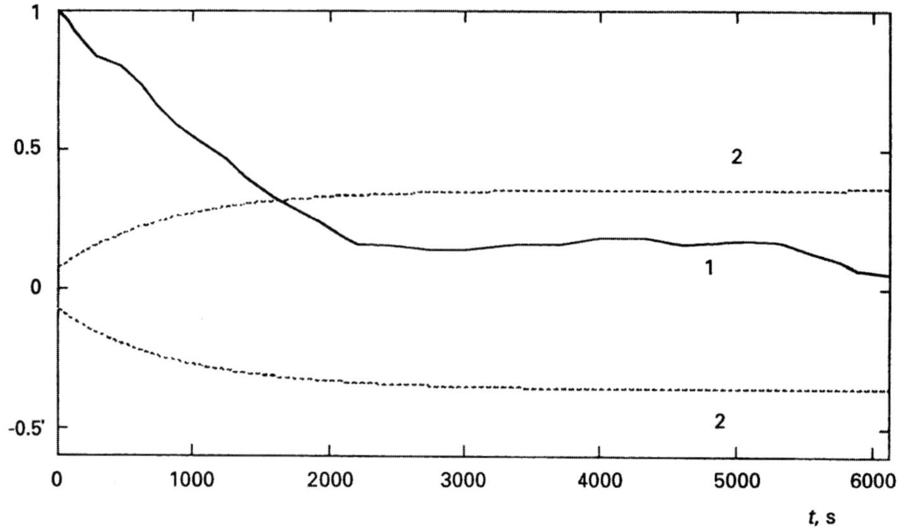

Figure 6.25. Autocorrelation function of the fluctuations of the sound field amplitude (1) for confidence limits of zero values, constructed using Bartlett's formula (2).

Table 6.1. Model regression coefficients for experiments I and II

Experiment, frequency	$\hat{k}_1$	$\hat{k}_2$	$\hat{k}_3$	$\hat{k}_4$	$\sigma_{\hat{a}}^2$
I, 100 Hz	0.881	0.309	−0.008	−0.19	0.035
I, 300 Hz	0.917	0.300	0.015	−0.245	0.024
II, 100 Hz	0.736	0.318	0.094	−0.159	0.015
II, 300 Hz	0.877	0.238	0.145	−0.273	0.020

(Box *et al.* (1970)) and are given in Table 6.1 for the corresponding experiments and frequencies.

The variations of $\hat{k}_1$, $\hat{k}_2$, $\hat{k}_4$ are at most 26%, and $\hat{k}_3$ and the variance of the residual errors $\sigma_{\hat{a}}^2$ are less stable.

This statistical model, of course, requires a verification of its stability on a large number of tracks, at different times of the year and over a longer period of observations, in terms of both the order parameter of the model and the values of its coefficients. However, the results obtained enable us to hypothesise that over a period of 10 hours, the variation of the amplitude of a pure-tone signal in shallow water can be assumed to be a stationary process, and can be described as a fourth-order autoregressive process.

In addition to amplitude and phase fluctuations in shallow water, there may also be significant fluctuations of the phase front of the sound waves, due to the considerable anisotropy of hydrodynamic perturbations. These fluctuations have been studied in less detail and not for all seas, which complicates their comparative

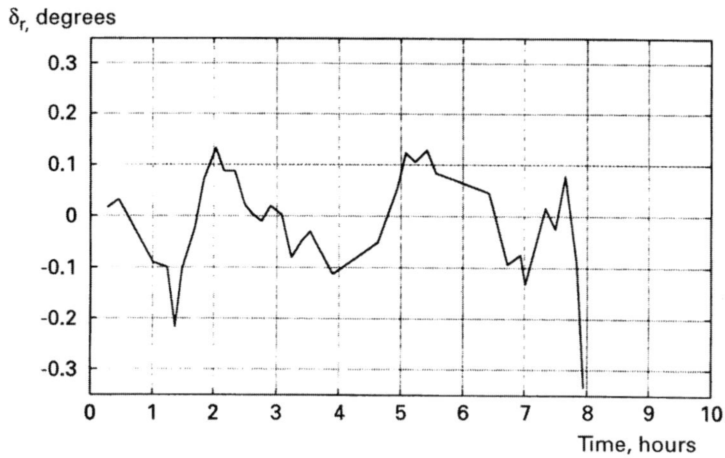

Figure 6.26. Graph of the variation of the horizontal angle of refraction against time.

analysis and generalisation. In this connection, let us consider a number of individual examples of such fluctuations. In the Barents Sea, fluctuations of the phase front were recorded at frequency 100 Hz, on a stationary acoustic track of length $\cong 70$ km in experiments with a horizontal linear array of length $L = 420$ m lying on the sea bed (Shmelerv (1992)). To describe the experiment, we introduce a coordinate system (x, y, z) analogously to that in Section 3.3, with x-axis directed along the track. We shall assume that the sound source is located at the point with coordinates $x = y = 0$, $z = z_0$. The receiving array comprised 48 receivers and was rotated practically perpendicularly to the stationary track, crossing it at the point $x = x_1$, so that the coordinates of the receivers (in metres) were determined by the expression $[x_1, y_j = (j-1)\Delta L]$, $j = 1, \ldots, 48$, $\Delta L = 8.94$ m. In the measurements, phase changes were recorded at each receiver, for an averaging time of 20 min. In order to determine the angle of rotation of the phase front of the acoustic waves, as a whole, all the simultaneous phase readings were plotted on a graph of $\varphi(y_j)$, and the function

$$\varphi = \delta_r k y_j \tag{6.22}$$

was constructed using the least squares method, and hence the angle of rotation of the phase front δ_r was determined. The phase values in the individual hydrophones were time dependent, and for the angle of refraction we had the following set of values, obtained at different time instants. The mean square value of the angle δ_r (in the literature this is called the angle of horizontal refraction) was $0^\circ.15$. Figure 6.26 shows the changes of the angle of refraction with time. It is clear that the characteristic time of variation δ_r is $T_i \sim 3$–3.5 hours.

Since the measurements of the phase-front fluctuations were not accompanied by detailed experimental studies of the hydrodynamic perturbations we shall use the results of Section 3.3 for simple estimates of the angle of horizontal refraction, which enable us to surmise that these fluctuations are of a physical nature. The stationary acoustic track in the Barents Sea, where the angle of refraction was measured, for the most part passed along the coastal slope. In this connection, it is reasonable to suppose that the fluctuations were due to low-frequency ($\Omega_i = 2\pi/T_i \sim 5\text{–}6 \times 10^{-4}$ rad/s) internal waves, crossing the stationary track at right angles. The width of the wavefront of the internal waves $\tilde{L}$ was considerably less than the length of the track. (The above hypothesis is reinforced by the results of measurements of the temperature variations at certain depths, taken in approximately the same region at the same time of year, in Belov *et al.* (1997). The spectrum of the temperature variations had a maximum at frequencies $\sim 5 \times 10^{-4}$ rad/s). The sound speed profile in this experiment is shown in Figure 6.24 (dashed line) and is modelled by a bilinear equilibrium (unperturbed) distribution of the sound speed with depth:

$$c(z) = \begin{cases} c_S & 0 \le z \le h_0 \\ c_S(1 - \varepsilon(z - h_0)) & h_0 \le z \le H \end{cases} \tag{6.23}$$

where $c_S = 1{,}478$ m/s, $\varepsilon = 6.5 \times 10^{-5}\,\text{m}^{-1}$, $h_0 = 100$ m, $H = 200$ m. For these parameters of the sound speed profiles according to (2.90) at frequency $f = 100$ Hz, there exists approximately one near-bottom normal mode which, however, for a track length of $x_1 = 70$ km, carries most of the energy of the sound field. Then, in perturbation theory for horizontal rays (Section 3.3) the expression for the angle of horizontal refraction (3.50) of a sound wave at the point with horizontal coordinates $(x_1, 0)$ in a first approximation has the form (Kuzkin (1996)):

$$\delta_r \approx \delta_1 = \frac{1}{x_1}\int_0^{x_1} (x_1 - x)\nu_1\,dx \tag{6.24}$$

where the value of $\nu_1 = (q_1)'_y/q_1$ can be found using the invariant

$$\int_{\bar{z}_1}^{H} \sigma_1(z, y)\,dz = \int_{\bar{z}_1}^{H} \sqrt{k^2(z, y) - q_1^2}\,dz = \text{constant}$$

which gives

$$\nu_1 = \frac{k^2}{2q_1^2}\int_{\bar{z}_1}^{H} \frac{1}{\sigma_1}\frac{\partial n'_w}{\partial y}\,dz\left(\int_{\bar{z}_1}^{H}\frac{dz}{\sigma_1}\right)^{-1} \tag{6.25}$$

Here $n'_w \approx r\varepsilon\tilde{\Phi}_{1\kappa}(z)\sin(\tilde{k}_1 y - \Omega_i t)$ is the perturbation of the equilibrium vertical profile of the square of the refractive index ($n_w^2(z) = c_s^2/c^2(z)$) induced by the internal wave with dominant first mode, $\bar{z}_1$ is the turning point of the first acoustic mode, and $\tilde{\phi}_{1\kappa}(z)$ is the eigenfunction of the first mode of the internal wave, equal to (Brekhovskikh *et al.* (1994)):

$$\tilde{\Phi}_{1\kappa}(z) = \begin{cases} b_1 \sinh(\tilde{k}_1 z) & 0 \le z \le h_0 \\ a_1 \sin[\tilde{k}_1\tilde{N}(z + H)/\Omega_i]] & h_0 \le z \le H \end{cases} \tag{6.26}$$

$\tilde{k}_1$ is the propagation constant of the first mode of the internal wave, determined by the dispersion relation

$$\tan(\tilde{k}_1\tilde{N}(H - h_0)/\Omega_i) = -\tilde{k}_1\tilde{N}h_0/\Omega_i \tag{6.27}$$

(for the sound speed profile (6.23), frequency $\Omega_i = 5 \times 10^{-4}$ rad/s, buoyancy frequency $\tilde{N} \approx 3 \times 10^{-2}$ rad/s, we have $\tilde{k}_1 l = 1.9 \times 10^{-3}\,\text{m}^{-1})$[8], and q_1 is the unperturbed value of the longitudinal propagation constant of the first mode of an acoustic wave in the approximation of a rigid bottom with a pressure-release surface (in subsequent estimates, we shall assume that $q_1 \approx k$).

Using all the assumptions listed, and substituting equations (6.25)–(6.27) in (6.27), we determine that the maximum angle of refraction is equal to:

$$\delta_r \approx \tilde{k}_1 \varepsilon \tilde{L} a_1 \operatorname{Im}\left[\Phi\left(1, \frac{3}{2}; i\mu_1\right)\right] \tag{6.28}$$

where Φ is a degenerate hypergeometric function, $\mu_1 = (\tilde{k}_1 \tilde{N}/2\Omega_i)(3\pi/4k\sqrt{\varepsilon/2})^{2/3}$. We note that in computing the integral (6.28) we have essentially only considered the section of the track where the function ν_1 is non-zero, defined in (6.25), i.e. the interval of length $\tilde{L}$ corresponding to the width of the wavefront of the internal waves.

We assume that the length of this section is $\tilde{L} = 15$ km, and the amplitude of the internal wave is $a_1 = 5$ m then, according to (6.24), the angle of refraction at frequency 100 Hz is equal to $\delta_r = 0^\circ.13$, which is close to the measured value.

6.3 WAVEGUIDE DISPERSION OF SOUND SIGNALS IN SHALLOW WATER

As mentioned earlier, when considering the propagation of sound in water we neglect the dispersion of sound, i.e. the dependence of the sound speed c on frequency ν. This is inherently associated with the possibility of using the propagation equation in the form (1.9). However, waveguide propagation of sound is characterised by a special form of dispersion of the sound signals, due to the dispersion properties of the propagation channel itself. The essence of the matter is that each normal wave in the waveguide propagates with its group velocity (equation (2.38)) which, in turn, depends on frequency. Thus, for propagation in a sound channel, one can separately identify the intermode dispersion (distortion of the signal shape due to the different velocities of the individual modes) and the intramode dispersion (due to the frequency dependence of the group velocities of the individual modes). Clearly, the dispersion distortions of the sound signals become marked only from certain distances r_b when the value of τ_1 (see equation (5.2)) characterising the spreading of a wavepacket due to intermode (or intramode) dispersion becomes comparable with the signal duration $\Delta\tau \simeq 2\pi/\Delta\omega$. At small distances ($r \ll r_b$) the signal passes

[8] Here, we note that the corresponding length of an internal wave $\Lambda = 3.3 \times 10^3$ m is much greater than the array length $L = 420$ m, used in this experiment, which explains the linear variations of the phase front.

through the waveguide without distortions. From equation (2.63) it follows that as the mode index grows the group velocity decreases, which implies that the effect of the dispersion delay becomes manifest in the higher modes, capable of reaching the point of observation. Assuming that at this distance there are M energy-bearing modes, for the Pekeris waveguide with a rigid bottom ($n_1 \to 0$, $m_1 \to 0$), r_b is equal to:[9]

$$r_b \cong \frac{16\pi c H^2}{\Delta\omega\lambda^2 M^2} \tag{6.29}$$

As mentioned in Chapter 5, the intermode dispersion is used, for example, in acoustic tomography, to identify signals propagating in the individual modes. For every sound channel, there exists some distance r_{bm} ($r_{bm} \gg r_b$) beyond which sound signals corresponding to two adjacent modes are distinguished by their propagation time in the channel. The value of r_{bm} depends on the signal bandwidth, the sound speed profile $c(z)$, the depth of the waveguide and the acoustic properties of the bottom. For a Pekeris waveguide the distance r_{bm} is determined by expression (5.3). However, it has to be said that for shallow water, for strong attenuation of sound, the condition $r > r_{bm}$ is necessary, but not sufficient for identification of an individual mode from the signal propagation time. It is a further requirement that the signal at the output of an optimal receiver should exceed the noise level. In practice, for a limited power of radiation, this appears possible only for the lowest weakly decaying modes.

The intramode dispersion leads to signal distortions, corresponding to the individual modes. However, for narrow-band ($\Delta\omega/\omega_0 \ll 1$) signals, these distortions are small and occur at greater distances than the analogous distances for the intermode dispersion. If r_{im} is defined to be the minimum distance beyond which intramode distortions are marked for one of the lower modes with index m, then for a Pekeris waveguide with a rigid bottom, r_{im} is equal to:

$$r_{im} \cong \frac{16\pi^2 c^2 H^2}{\Delta\omega^2 m^2 \lambda^3} \tag{6.30}$$

When the following inequality is satisfied:

$$\frac{\omega_0}{2\Delta\omega m} > 1 \tag{6.31}$$

r_{im} is greater than r_{bm} and, consequently, there exists an interval of distances, where signals corresponding to individual modes are easily distinguished according to their propagation time, and also contain no marked dispersion distortions.

Let us consider the manifestation of the effect of the intermode dispersion, for the propagation of a narrow-band signal with carrier frequency 230 Hz in a waveguide. The results of numerical and full-scale experiments are described in Vavilin *et al.* (1987), Experimental studies were performed in the Barents Sea for the conditions of the existence of a near-bottom channel, whose characteristics, the dependence of the depth H on distance r and the sound speed profile $c(z)$, are shown

[9] Here, we assume that M is small and $\sigma_M \ll k$.

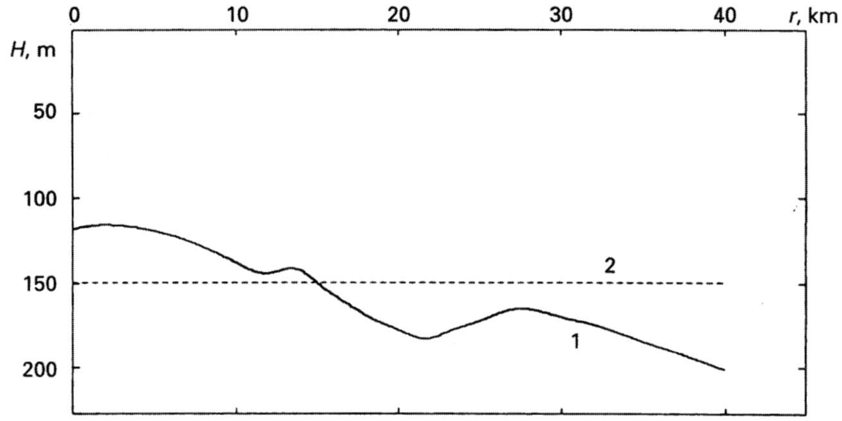

Figure 6.27. Graph of the waveguide depth against distance between sound source and receiver (1). For the calculations, the depth of the waveguide was taken to be 149 m (2).

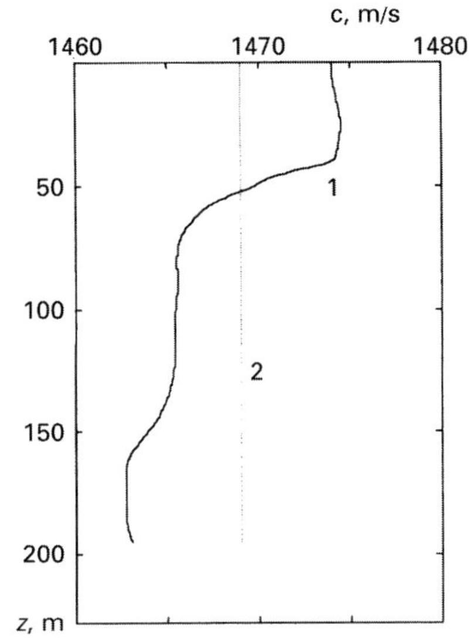

Figure 6.28. Sound speed profile in the studied waveguide (1). For the calculations, the sound speed in the waveguide was taken to be constant: $c = 1,469$ m/s (2).

in Figures 6.27 and 6.28, respectively. In these figures the dashed lines show the depth and the sound speed, used in the calculations. Thus, in order to represent the results in a more graphic form, the Pekeris waveguide was used as the model waveguide. (Calculations of the precise spectrum of the eigenvalues here, demonstrate the applicability of this approach, in the context of the estimates given below). The ratio of the effective waveguide depth to the sound wavelength was taken to be equal to $H/\lambda \cong 23$ ($\lambda = 6.5\,\mathrm{m}$) . The refractive index of sound in the bottom was taken to be $n_1 = 0.8$, and the sound absorption coefficient was $\alpha = 0.015$. The ratio of the densities of the bottom and the water was assumed to be $m_1 = \rho_1/\rho = 1.7$. The sound receiver was located on the bottom, and the source was at a depth of $z_0 = 54\,\mathrm{m}$. The radiated signal $f(t)$ was characterised by a relatively narrow spectrum $S(\omega)$, so that the relative width of the signal band was $\Delta\omega/\omega = 0.03$, and the duration was $T \approx 25.6\,\mathrm{s}$.

For the quantity to be studied, we shall consider the envelope of the cross correlation function (CCF) between a signal measured at some distance from the source and the reference (radiated) signal:

$$B(\tau) = \left| \int_{-\infty}^{\infty} p(\vec{R}, t) f(t - \tau)\, dt \right| \tag{6.32}$$

Taking into account (1.10)–(1.12) and (1.16), and bearing in mind that the modulus of the signal spectrum is an even function of frequency, we obtain the following for $B(\tau)$:

$$B(\tau) = 4\pi \left| \int_{0}^{\infty} \Psi(\vec{R}, \vec{R}_0; \omega) |S(\omega)|^2 \mathrm{e}^{-i\omega\tau}\, d\omega \right| \tag{6.33}$$

In fact, the value of $B(\tau)$ is the response of an optimal receiver, constructed to receive the known signal. The Green function (transfer function of the waveguide) $\Psi(\vec{R}, \vec{R}_0; \omega)$ in the Pekeris model is given by expression (2.51), where we limit ourselves to summation over only those modes of the discrete spectrum which reach the observation point. We shall transform this expression somewhat using a number of approximations. First, at large distances in comparison with the wavelength, instead of the Hankel functions, we substitute their asymptotics (2.39); second, for the eigenvalues and functions given by equations (2.44) and (2.52), we obtain approximate expressions using the fact that under our conditions, the parameters $kH = 2\pi H/\lambda \approx 10^2$ and $M = \frac{H}{\pi}\sqrt{k^2 - k_1^2} \approx 28$ (number of propagating modes) are sufficiently large (the so-called thick layer (Brekhovskikh (1980))). In addition, the number of modes reaching a distance of the order of several tens of kilometres is not too large (in practice, not more than 10) and for these modes the vertical wavenumber satisfies the inequality $\sigma_l \ll k, \frac{M\pi}{H}$. The real part of the eigenvalues obtained under this condition from (2.44) has the form $q_l = \sqrt{k^2 - \sigma_l^2}$, and $\sigma_l = l\pi/H_1$, where $H_1 = H + m_1/k\sqrt{1 - n_1^2}$. The eigenfunctions are accordingly equal to $\psi_l(z) = \sqrt{2/H}\sin((\pi l/H_1)z)$, and the mode absorption coefficient is determined by equation (2.62). Taking this into account, for the Green function we have:

$$\Psi(r, z, 0; z_0; \omega) \cong \frac{1}{\sqrt{8\pi k r}} \sum_l \psi_l(z)\psi_l(z_0) \exp\left(\left(iq_l - \frac{\gamma_l}{2}\right) r + i\frac{\pi}{4}\right) \tag{6.34}$$

As part of the computations according to (6.33), the normalised envelope of the cross correlation function was determined and compared with the analogous

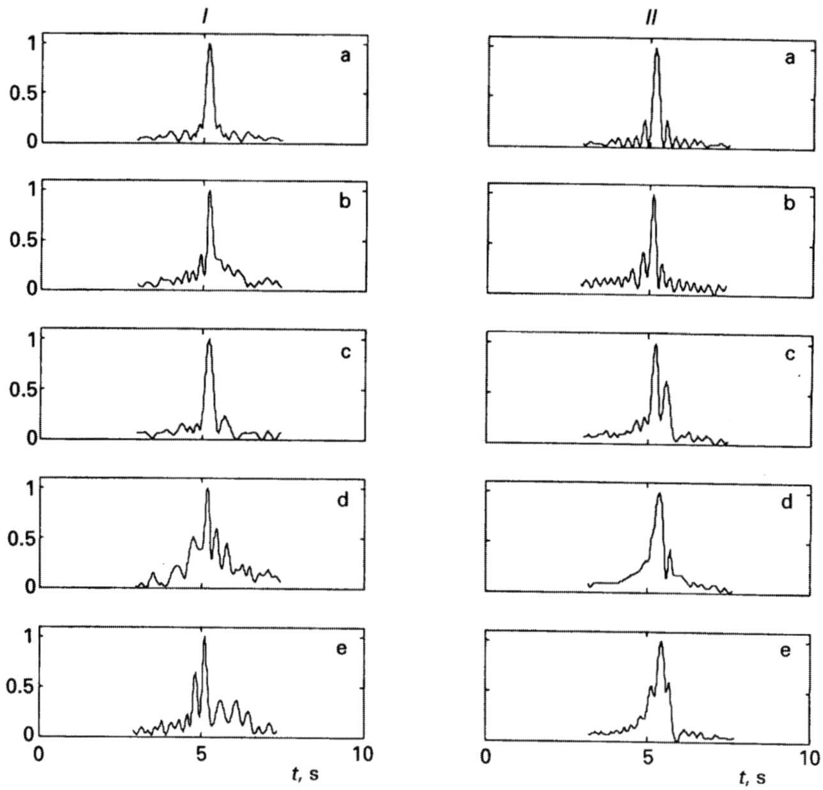

Figure 6.29. Evolution of the CCF envelope between the reference signal and the signal passing through the waveguide, as the distance r between source and receiver increases. I – Experiment; II – calculation; (a) – $r = 3$ km, (b) – $r = 32$ km, (c) – $r = 36$ km, (d) – $r = 38$ km, (e) – $r = 40$ km.

quantity obtained based on experimental data. (The normalisation was carried out so that the maximum value of the envelope is equal to unity).

The method of the experiment involved the receipt of signals with hyperbolic frequency modulation, radiated by a drifting sound source. In the data processing, the signal received at distance $r_0^* = 3\,\text{km} < r_b$ was taken as a reference. (Formally, this means that instead of the radiated signal $S(\omega)$ the product $S(\omega) \cdot \Psi(r_0^*, z, 0, z_0; \omega)$ was used in equation (6.33). However, in view of the smallness of r_0^*, $\Psi(r_0^*, z, 0, z_0; \omega)$ does not depend on frequency in the band $\Delta\omega$ and, consequently, by virtue of the normalisation, it may be set to be unity).

The results of numerical and full-scale experiment – namely, the normalized cross correlation function $B(\tau)$ for different values of the distance between sound source and receiver – are shown in Figure 6.29(a) and (b). (The differences in the signal propagation times from different distances are artificially compensated, since all the signals are ostensibly received at the same time.) Comparison of the results of calculation and experiment in an interval of distances shows their qualitative

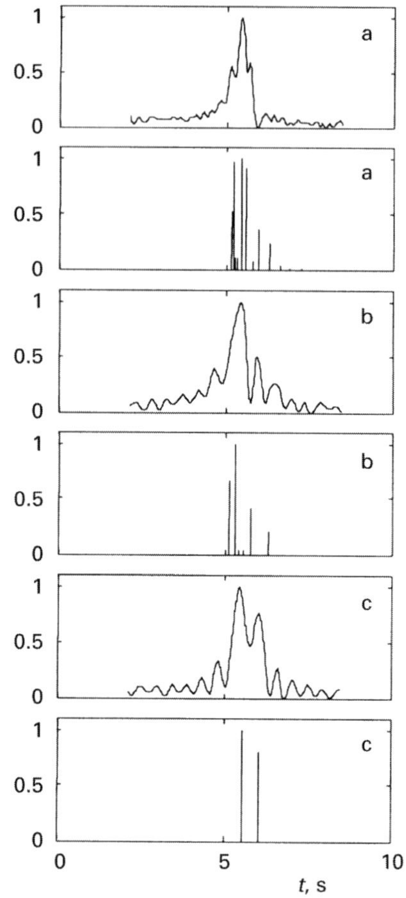

Figure 6.30. CCF envelope (smooth curves) and corresponding bar graphs of the energy-bearing modes as the distance r between sound source and receiver increases (calculations): (a) – $r = 40$ km, (b) – $r = 100$ km, (c) – $r = 300$ km.

correspondence on the interval 32–40 km, and demonstrates the intermode dispersion distortions of signals at the output of the optimal correlation receiver at these distances. Indeed, on average, the signal duration at the output of such a receiver is the same in both theory and experiment, being greater than the duration of a signal at small distances ($\sim$3 km) where the dispersion is not yet marked. The detailed differences in the shape of the CCF envelope, which are noticeable when one compares calculated and experimental curves, are evidently associated with the coarseness of the chosen model, which does not take into account variation of the waveguide depth with distance or the dependence of the sound speed on depth. These differences may also be associated with the experimental positioning error ($\sim$500 m) in the determination of the distance between the sound source and receiver.

In Figure 6.30, calculated curves for the CCF envelope for distances 40 km,

100 km and 300 km are shown by smooth curves. Maximum values of the CCF envelope of the reference signal and of the signal passing through the model medium are shown as bar graphs, for each mode separately. The position of the vertical lines on the time axis shows the signal propagation time for each mode. Figure 6.30 again demonstrates that in the shallow-water waveguide studied, the observed distortions of the received signals at distances up to 100 km are the result of the addition of signals corresponding to individual modes. It is important to note that these signals cannot be identified from the propagation time. As mentioned earlier, because of the strong absorption – and this is a specific property of shallow water – signals corresponding to the highest modes 'die out', while signals of the lower modes cannot be distinguished by time. Such a distinction can be made only at great distances. In the studied waveguide, this was possible at 300˜km, where signals corresponding to the first and second mode could be successfully distinguished (see Figure 6.30). The difference in their arrival times for this distance was approximately 0.2 s, which does not contradict the calculated difference between the group velocities of the first two modes, which, according to equation (2.63), is equal to 1.5 m/s. At a distance of 100 km, the analogous time interval is 0.07 s.

These results on the whole correspond to the estimate (5.3), according to which, r_{b1} is equal to $\cong$ 450 km. We note that the distance of 300 km is considerably less than $r_{i1} \cong 15,000$ km, and, as the figure shows, signals corresponding to the first and second modes are not subject to dispersion distortions.

The difficulty of identifying pulse signals corresponding to individual modes, from the propagation time in shallow water, requires the application of special algorithms, differing from those used in deep-ocean acoustic tomography. These include the algorithms mentioned earlier based on the selection of normal waves using a vertical array (see Chapter 5) and algorithms based on matched field processing of pure-tone signals (see Section 6.4) for shallow water, to which one can also add the acoustic tomography technique, based on matched field processing of pulsed signals with marked dispersion distortions. The essence of this method can be easily understood in the example studied in Bunkin *et al.* (1989), where this technique was used to determine the distance between sound source and receiver in a shallow-water waveguide at a practically constant depth $H \cong 230$ m (depth variations were at most $\sim$10–15 m). The sound speed profile measured at the beginning (a) and the end (b) of the waveguide is shown in Figure 6.31. (The profile used in the calculations is shown by dashes.) There were in fact two sound sources, rigidly attached to a single underwater carrier, towed away from the sound receivers at a depth of 35–40 m. Signals with hyperbolic frequency modulation with carrier frequencies $f_1 \cong 100$ and $f_2 \cong 230$ Hz, were radiated simultaneously. The signal duration was $T = 25.6$ s, and the relative bandwidth was $\Delta f_1/f_1 = 0.018$ and $\Delta\tau \cong \Delta f_2/f_2 = 0.03$. The signals were received by a vertical chain of six equidistant hydrophones, lowered from on board a drifting ship. Depth sensors were located near the first and last hydrophones.

In order to estimate the distance between sound sources and receivers, normalised (calculated and experimental) envelopes of the CCF of a reference signal and the signal passing through the waveguide were also determined. (For the experimental data, as the reference signal the signal received at distance $r_0^* \cong 4\,\text{km}$ was used. As before, the normalisation was carried out at the maximum value). In the computations, the CCF envelope was determined for both frequencies and for all six depths of receipt. The calculation was performed for a model waveguide with a fluid absorbing bottom and a linear dependence on depth of the sound speed in the water layer: $c(z) = c_S(1 - \varepsilon z)$, $\varepsilon \approx 1.8 \times 10^{-5}\,\text{m}^{-1}$. The coefficient determining the absorption of sound in the bottom was taken to equal $\alpha = 0.026$. (This value of the effective coefficient α in the given waveguide was obtained using the method studied in Section 6.1.) The real part of the refractive index of sound in the bottom was assumed to equal $n_1 = 0.92$, and the ratio of the densities of the bottom and the water was $m_1 = 2$. (These parameters of the bottom correspond to sedimentary rocks, sands and gravels, which are characteristic of the study region.) The Green function of the waveguide was taken in the form (6.34) with eigenfunctions and values corresponding to the conditions indicated here.

For the value of the vertical gradient of the sound speed ε, existing in the waveguide, its effect on sound propagation is small. One can estimate somewhat conceptually the extent to which this gradient changes the pattern of sound propagation in comparison with a channel with a constant velocity. It follows from equation (2.90) that at frequency 230 Hz there are at most four modes, and at frequency 100 Hz, at most one mode, not grazing the surface, which differs quite markedly from the corresponding Pekeris mode. The remaining modes propagating are essentially little different from the modes of a Pekeris waveguide, and we therefore use perturbation theory to calculate the eigenfunctions and eigenvalues for equation (6.34). For the unperturbed case we use the Pekeris waveguide with a value $c = c_S$ for the sound speed in the water layer. There is another side to the use of perturbation theory in this instance. The fact is that because of the narrowness of the spectrum of the studied signal, we also obtain the eigenfunctions and values for equation (6.34) at a frequency near the carrier frequency $\omega = \omega_0 + \delta\omega$, from perturbation theory, which enables us to obtain a vivid picture of the dependence on frequency. Thus, we represent the value of the wavevector $k^2(z)$ in the equation in the eigenvalues in the form of an unperturbed part and a perturbation in accordance with Section 3.1 (equation (3.7)):

$$[k^0(z)]^2 = \begin{cases} \omega_0^2 c^{-2} & 0 < z < H \\ \omega_0^2 c^{-2} n_1^2 (1 + i\alpha) & z > H \end{cases} \tag{6.35}$$

$$\delta k^2(z) = \begin{cases} 2\omega_0^2 c^{-2}((\delta\omega/\omega_0) + \varepsilon z) & 0 < z < H \\ 2\omega_0 \delta\omega c^{-2} n_1^2 (1 + i\alpha) & z > H \end{cases} \tag{6.36}$$

A correction to the complex eigenvalue ξ_l^0 (which is also complex), is obtained from equation (3.10):

$$\Delta\xi_l = \frac{k_0^2}{\xi_l^0} \int_0^H \left(\frac{\delta\omega}{\omega_0} + \varepsilon z\right) [\psi_l^0]^2 \, dz + \frac{k_1^2}{m_1 \xi_l^0} \frac{\delta\omega}{\omega_0} \int_H^\infty [\psi_l^0]^2 \, dz \tag{6.37}$$

where the eigenvalues and functions of the unperturbed problem ξ_l^0 and $\psi_l^0(z)$ are determined, for example, by equations (2.44), (2.52), (2.53). Eigenfunctions of the waveguide of the same order in perturbation theory are obtained from equation (3.8.2), although, for calculations using equation (6.34) it is more convenient to represent the expression for an eigenfunction using a rather different approach to perturbation theory (Burov and Sergeev (1992), Turbiner (1984)), where corrections to the functions are sought, for example, in the form

$$\psi_l(z) = \psi_l^0(z)\exp\left\{\int_0^z Z_l(z')dz'\right\} \tag{6.38}$$

where for the function $Z_l(z)$, a nonlinear equation is derived which is solved using a decomposition in a small parameter of the theory. The distinctive property of this approach (nonlinearisation method) is that the corrections to the lth eigenfunction are expressed in terms of the eigenvalues and functions of the same index. In our case the function $Z_l(z)$ is given by the relation:

$$Z_l(z) = [\psi_l^{(0)}(z)]^{-2}\int_0^z [\psi_l^0(z')]^2\left[2\xi_l^0\Delta\xi_l - 2k_0^2\varepsilon z' - k_0^2\frac{\delta\omega}{\omega_0}\right]dz' \tag{6.39}$$

The calculation of the Green function of the model waveguide takes into account the contributions from only M weakly decaying modes. For frequency 100 Hz there are seven such modes, and there are 17 for frequency 230 Hz.

To determine the distance between sound source and receiver, the experimental value of the envelope of the cross correlation function $B_e(\tau)$ was compared with the calculated value $B_{th}(\tau)$, computed for various distances. The true distance was taken to be the distance corresponding to the best agreement of $B_e(\tau)$ and $B_{th}(\tau)$. Quantitative agreement was assessed using a criterion analogous to (6.6):

$$K = \frac{\tilde{K}(r) - K_{min}}{K_{max} - K_{min}} \qquad \tilde{K}(r) = 2\frac{\int_0^\infty B_{th}(\tau)B_e(\tau)d\tau}{\int_0^\infty B_{th}^2(\tau)d\tau + \int_0^\infty B_e^2(\tau)d\tau} \tag{6.40}$$

when the true distance was taken to be the distance which maximises the value of K at all receiver depths. (Here K_{max} and K_{min} are, respectively, the maximum and minimum value of $K(r)$ in the chosen range of variation of r.)

Figure 6.32 shows the results of calculation of the value of K on the distance interval $r = 26$–28 km with step 20 m, for the different receiver levels and carrier frequencies (The range of variation of distances of 2 km, chosen for the calculations, was approximately four times greater than the accuracy with which the distance between the sound source and receiver was determined, in the experiment using monitoring devices.) Clearly, the dependence $\tilde{K}(r)$ is of an irregular nature with a large number of local extrema. From a single such dependence, even over the range of the navigation, it is difficult to determine the distance $\hat{r}$ at which the best agreement between theory and experiment takes place. However, the set of all such curves for the different receiver depths and carrier frequencies enables one to determine this distance. Figure 6.33 shows the dependence $K_{\Pi}(r)$ normalised to the maximum,

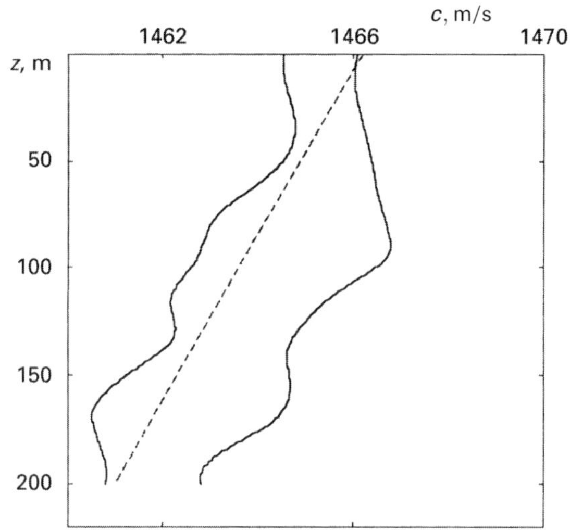

Figure 6.31. Dependence of the sound speed on depth. 1 – At the point where the sound source is; 2 – at the point of the receivers; 3 – dependence used in the calculations.

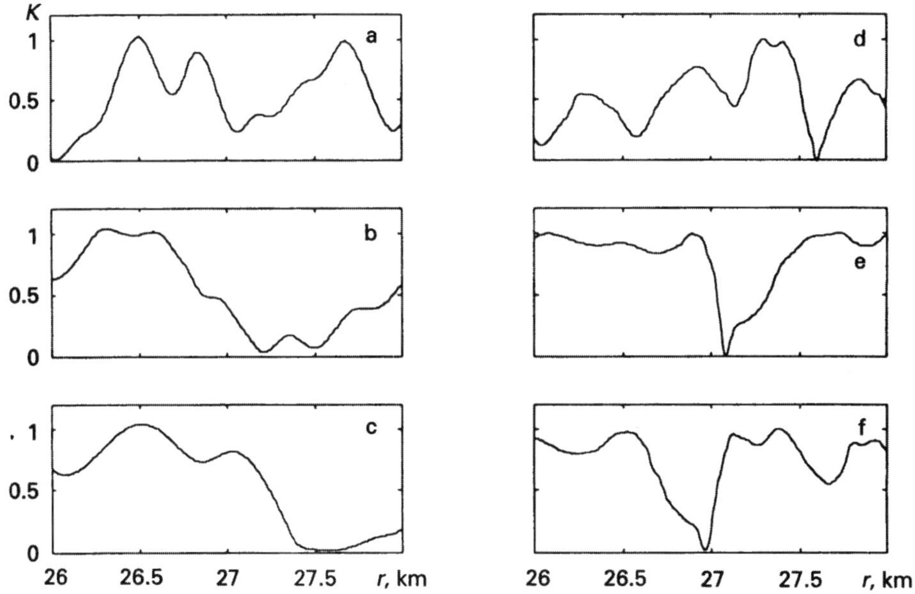

Figure 6.32. Dependence of the coefficient K on distance for various depths of receipt and carrier frequencies: (*a*) – $f = 230$ Hz, $z = 125$ m; (*b*) – $f = 230$ Hz, $z = 69$ m; (*c*) – $f = 100$ Hz, $z = 125$ m; (*d*) – $f = 100$ Hz, $z = 97$ m; (*e*) – $f = 100$ Hz, $z = 83$ m; (*f*) – $f = 100$ Hz, $z = 69$ m.

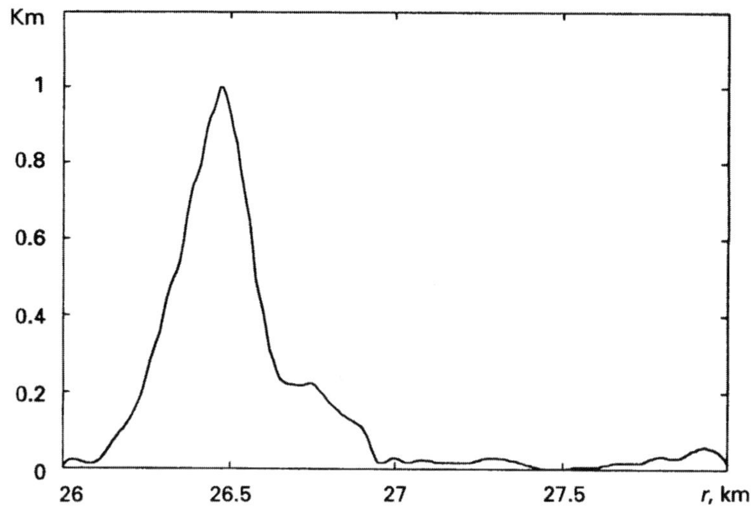

Figure 6.33. Graph of the coefficient $K_{\prod}$ against distance.

which was calculated on the same interval of distances $r = 26$–$28\,\mathrm{km}$ using the following formula:

$$K_{\prod}(r) = \prod_{i,j} K_{i,j}(r) \tag{6.41}$$

where i is the index of the signal carrier frequency ($i = 1, 2$) and j is the index of the receiving hydrophone (index of the receiver level). The curve $K_{\prod}(r)$ has a sharp maximum at $r = \hat{r} = 26.5\,\mathrm{km}$, where the best agreement of theory and experiment takes place. One may conjecture that $\hat{r}$ is close to the true distance between sound source and receiver. (From the navigation data the distance between ships was equal to $r \cong 27 \pm 0.5\,\mathrm{km}$.) Analogously, one can determine not only the distance between sound source and receiver, but also the waveguide parameters; i.e. one can solve the tomographic problem. In fact, this method is a version of a technique of acoustic tomography, based on matched field signal processing.

In conclusion, we note that the very high degree of correlation between the experimental and theoretical CCF envelopes enables one, if necessary, to correct the dispersion distortions of the received pulse signals. Figure 6.34 shows an example of such correction using a filter with transfer characteristic equal to $\Psi^{-1}(\hat{r}, z, 0, z_0; \omega)$, where the receiver depth is $z = 125\,\mathrm{m}$ and the frequency is $f_2 = 230\,\mathrm{Hz}$. However, the discrepancy between the model Green function and the real function does not allow one to carry out such a correction in full.

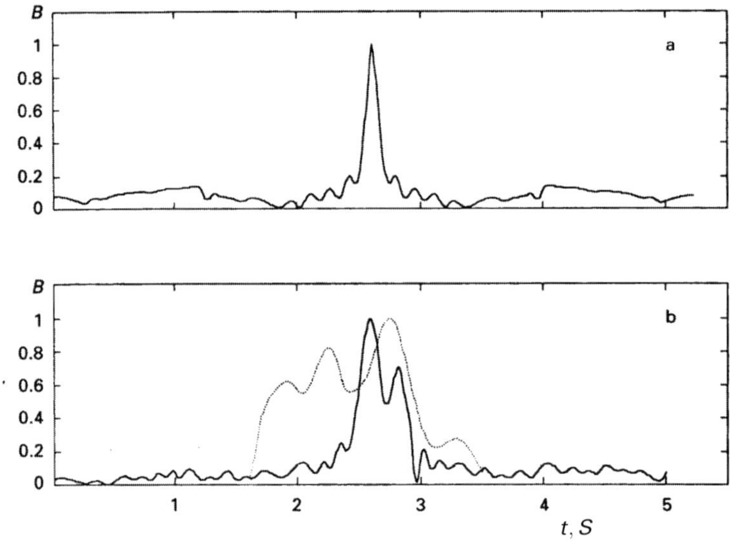

Figure 6.34. Results from the processing of experimental data: (*a*) – enevelope of the auto-correlation function of the reference signal; (*b*) – CCF envelope of the corrected signal (the dotted line is the CCF envelope of the signal without correction).

6.4 MATCHED FIELD SIGNAL PROCESSING IN ACOUSTIC MONITORING OF SHALLOW WATER

When considering the acoustic monitoring of shallow water, one must in the first place mention the idea, proposed by Munk (1979), of acoustic tomography of large-scale inhomogeneities of the ocean, based on measurement of the propagation time for pulse-tone signals passing through the studied waters on different trajectories (rays). As far as shallow water is concerned, this involves measuring the propagation times of signals corresponding to different modes of the acoustic waveguide, which traverse the same inhomogeneity in different ways. The difficulties discussed in the earlier sections of the book, associated with the identification of signals corresponding to individual rays or modes in shallow water, required the application of special algorithms for large-scale acoustic monitoring. These algorithms are based on the use of certain *a priori* information about the medium and on so-called 'matched field processing' (MFP) (Baggeroer *et al.* (1992)). We must at the same time mention that the possibility of using these algorithms in such a complex medium as shallow water is associated, on the one hand, with the sharp increase in the level of computing technology which is currently taking place and, on the other hand, with the preservation of a high degree of coherence of the pure-tone signals travelling a considerable distance in shallow water. (See, for example Section 6.2, Figure 6.17). This latter fact shows that it is in principle possible to provide an adequate description of the sound field in shallow water in the framework of a particular model of the medium, and consequently, the inverse problem of determining the parameters of the medium from acoustic sensing data can be solved.

At the present time, there are a number of approaches to the acoustic monitoring of shallow water based on matched field signal processing. Some of these are very close to Munk's idea, and are based on the identification of individual modes using vertical linear arrays. A suitable algorithm for these purposes, using *a priori* information about the eigenfunctions of the waveguide, was described in detail in Chapter 5. Using such an approach, Parsons *et al.* (1996) succeeded in realising a tomographic scheme for recovering the dynamics of the sound speed field (and, consequently, a temperature chart) in the shelf zone of the Barents Sea in the presence of a moving polar front. The basis for the reconstruction at each moment in time (with a certain step) was a set of time delays $\Delta\tau_l$ for all modes propagating in the waveguide (there were 30 such). The modes were selected using a vertical array of 16 hydrophones, attached to the bottom (see Figure 5.3).

It is interesting to note that in the conditions of a polar front with very abrupt longitudinal changes (large gradients) of temperature, and also in the conditions of a slope ($dH/dr \geq 5 \times 10^{-3}$), with the polar front moving perpendicularly to it, one must take into account mode coupling (the non-adiabaticity parameter for the sound frequency 224 Hz used in the experiment and depth $H \approx 350$, has value $\tilde{\kappa} \approx k|H'H|/2| \approx 0.8$, and an adiabatic approximation is not appropriate).

One development of the above approach, proposed in Nechaev *et al.* (1987), is the differential diagnostics method, where a knowledge of the medium is used to identify modes not only on the receipt of signals, but also on their radiation using a vertical chain of sound sources. This method is based on the excitation of one waveguide mode and receipt of a signal corresponding to another mode. The measured coefficient of conversion from mode to mode makes it possible in the framework of specific models to diagnose inhomogeneities in shallow water. It is important to note that in differential diagnostics, unlike in classical tomography, one can localise (determine the position of) inhomogeneities from just one radiated signal (signal corresponding to a single mode). Indeed, let us suppose that the radiating system excites a mode with index l and the receiving system identifies a mode with index m, where the group velocities of the modes ν_l^{gr} and ν_m^{gr} are very different. Suppose that at distance r_{in} from the source there is an inhomogeneity, which induces a 'transfer' of energy from mode l to mode m. Then the propagation time of the signal along the track will be

$$t = \frac{r}{\nu_m^{gr}} + r_{in}\left(\frac{1}{\nu_l^{gr}} - \frac{1}{\nu_m^{gr}}\right) \tag{6.42}$$

As can be seen from (6.42) the propagation time depends on the position of the inhomogeneity on the track and, hence, it can be localised. (The length of the track r is assumed known, and the group velocities belong to the set of *a priori* known parameters of the medium.) It is important to note that for differential diagnostics, in some situations, one can get by without using long arrays, performing the mode selection. For example, in the case when in shallow water, there exist natural zones of 'mode shadow' due to inhomogeneities of the relief.

In another approach different from classical tomography, measurements of the values of the sound field characteristics (for example, of amplitude and phase when acoustic sound sources are used) are compared with theoretical values based on a quantitative criterion. In this case, by varying the parameters of the theoretical model, the best agreement between theory and experiment is obtained and the parameters corresponding to this agreement are assumed to be the desired parameters. Here, we note that the parameters of the theoretical model may vary; they may be parameters of the bottom layer or layers, the sound speed profile, the depth of the sea, the position of the source, etc. In addition, the set of experimental data used in the search for the model parameters will be distinguished by a great diversity, being related to the vertical and horizontal interference pattern of the sound field at fixed frequency and various forms of frequency dependence of the field characteristics etc.

Let us illustrate this approach with a number of examples. In particular, let us consider the method for estimating the parameters of a shallow-water waveguide proposed in Andreev *et al.* (1996) and Katsnelson *et al.* (1992). Sensing was carried out using pure-tone sound signals of quartz frequency standard on a stationary acoustic track in the central area of the Barents Sea. The sound source and the vertical receiver chain of ten hydrophones were installed statically on the sea bed. The length of the chain was 66 m. The lowest hydrophone was at a distance of 18.5 m from the bottom. The installation of the acoustic systems on the bottom was carried out using two scientific research ships, each of which was attached to two anchored mooring buoys (see Figure 5.2). The distance between the ships (length of the stationary track) was determined using range-finder radar, and was equal to 13.82 km. The sea depth varied along the track and was $\cong 170$ m. The sound speed profiles $c(z)$, measured at the point of radiation with an interval of 2 hours, are shown in Figure 6.35. The marked changes in $c(z)$ with time were evidently associated with tides. This conclusion is indirectly confirmed by the results of measurements of the current velocities $\vec{v}_c$ of the main water volume, at depth 10–150 m, carried out using an Acoustical Doppler Current Profiler (ADCP) for measuring current velocity. The vector $\vec{v}_c$ changed direction and value with period 12 hours (see Figure 6.36), and these changes were practically synchronous with the variations of the local depth of ~ 1.5 m. It is important to note that the tidal current was directed perpendicularly to the stationary track. In this situation, a tidal wave, on the one hand, perturbed the whole track simultaneously and, on the other hand, tilted the receiver chain in a plane also perpendicular to the stationary track.[10] This means that the distance r between the sound source and receivers did not vary with time. The experiment lasted just over four hours, during which time a pure-tone signal with frequency 100 Hz was continuously radiated. As mentioned earlier in Section 6.1, seismography of the upper layer of the bottom sediments failed to identify a stable

[10] The maximum angle of inclination was $\Delta\varphi_0 \cong 9°$.

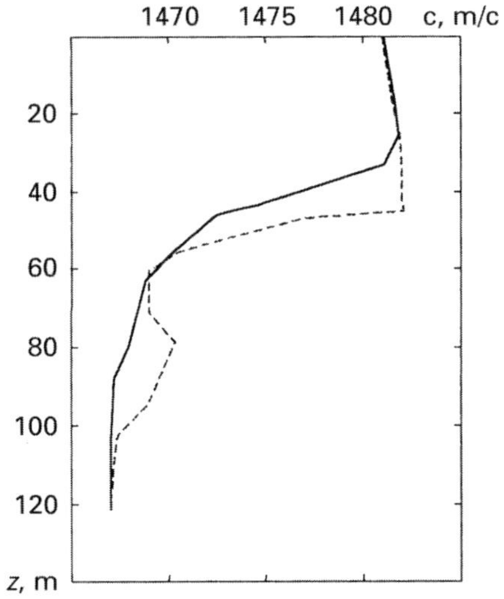

Figure 6.35. Sound speed profile at times $t_1 = 16$ hrs 07 min (1) and $t_2 = 18$ hrs 15 min (2).

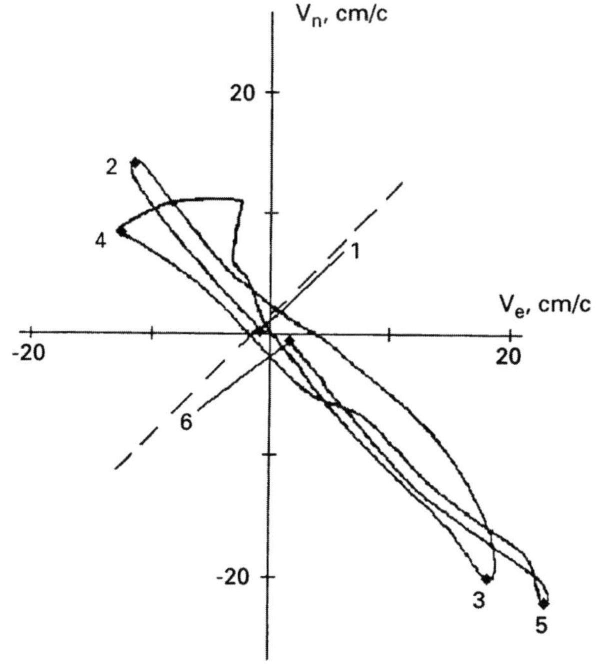

Figure 6.36. Variations of direction and current velocity over a day. The points denote time instants: 11 hrs 00 min (1), 14 hrs 00 min (2), 20 hrs 00 min (3), 02 hrs 00 min (4), 8 hrs 00 min (5), 11 hrs 30 min (6). V_N is the current velocity in a northerly direction; V_E is the flow rate in an easterly direction. The dotted line shows the direction of the stationary track.

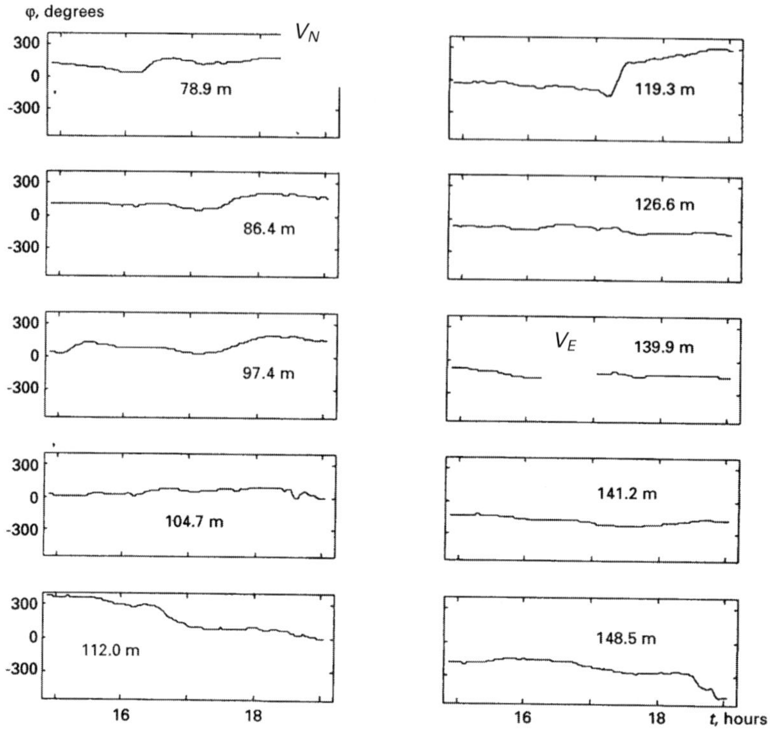

Figure 6.37. Time dependence of the signal phase at the individual hydrophones. The numbers in figures denote the depths in waters at which the hydrophones were located.

layer structure in the study region; thus here, a model of a homogeneous bottom could be used.

The processing of the received signals involved identifying the amplitudes $|P(z_j, t)|$ and the phases $\varphi(z_j, t)$ (see the first footnote to Section 6.2) for their subsequent comparison with the theoretical values. One result of the processing is the time dependence of the signal phase at the depths z_j of the receiver hydrophones, as shown in Figure 6.37 (see colour plate section). The dependences obtained were used to construct a time diagram of $\cos(\varphi(z, t))$, in the plane of time t and depth z (Figure 6.38 in the colour plate section). The whole of the dynamic range of variations of $\cos(\varphi(z, t0)$ $(-1, \ldots, +1)$ was divided into eight gradations of colour, which in Figure 6.38(a) depends on the value of $\cos(\varphi(z, t))$ (see the scale on Figure 6.38(a)). The phase of the field in the intervals between the space–time coordinates of the hydrophones, was determined by linear interpolation. In an analogous manner, a time diagram for the variations of the amplitude of the sound field (variations of the interference pattern) was constructed. This is shown in Figure 6.39(a) in the colour plate section where the yellow areas correspond to a minimum of the amplitude and the dark-blue areas to a maximum.

For the calculations of the amplitude and phase of the sound field according to equations (6.14.1) and (6.14.2), equation (6.4) was solved numerically for a set of parameters corresponding to experimental values. (First and foremost for the calculations, measured dependences $c(z)$ (see Figure 6.35) were used.) The method of the calculation was studied in detail in Section 6.1. The bottom was assumed to be an homogeneous fluid half space. Comparison of theoretical and experimental values was performed based on the following quantitative criteria:

$$\tilde{\sigma}_\varphi = \frac{1}{J_h}\sqrt{\sum_j^{J_h}\left(\hat{F}_{\mathrm{e}}(z_j,t) - \hat{F}_{\mathrm{th}}(z_j,t)\right)^2}$$

$$\tilde{\sigma}_P = \frac{1}{J_h}\sqrt{\sum_j^{J_h}\left(|P_{\mathrm{e}}(z_j,t)|^2 - |P_{\mathrm{th}}(z_j,t)|^2\right)^2} \qquad (6.43)$$

where $\hat{F}(z_j,t) \equiv \cos(\varphi(z_j,t))$ and the subscripts 'e' and 'th' refer respectively to measured (experimental) and calculated (theoretical) values. The minimum of $\tilde{\sigma}$ (ideally 0) corresponds to the best agreement between theory and experiment. The variable parameters of the model included the coefficient α, the sound speed in the bottom c_1, the bottom density ρ_1, i.e. the parameters which were not measured in the experiment, together with the depth H and the correction to the sound speed Δc in the water layer, being parameters which were not measured with sufficient accuracy. The absolute error in the echo sounding measurements of the depth was associated with an error in the determination of the draught of the ship. The value Δc carries the sense of the absolute error in the determination of the sound speed in the waveguide[11] $c(z)$, Δc was assumed to be independent of depth. The parameters varied around the reference values known from the literature, or measured experimentally. We note that in order to eliminate an arbitrary constant phase difference between theory and experiment a constant phase value was added to the theoretical value of $\varphi_{\mathrm{th}}(z,t)$, ensuring equality of the phases at the first hydrophone at the first measurement time for the sound speed velocity t_1 ($\varphi_{\mathrm{th}}(z_1,t_1) = \varphi_{\mathrm{e}}(z_1,t_1)$).

For this same time instant model parameters indicated above were determined, which provided the best agreement between theory and experiment, The choice of the parameters was made independently for the amplitude and phase distribution of the sound field. The dependences of the coefficient $\tilde{\sigma}_\varphi$ (residual errors) on a number of parameters of the waveguide for the phase distribution are shown in Figure 6.40. As can be seen from the figure, the agreement between theory and experiment depends strongly on the sound speed in the bottom and the depth of the

[11] The sound speed was calculated according to the Chen–Millero formula and the measured vertical profiles of temperature and conductivity. The correction Δc includes both the absolute errors in measurements of the temperature and conductivity and the errors of the given empirical formula.

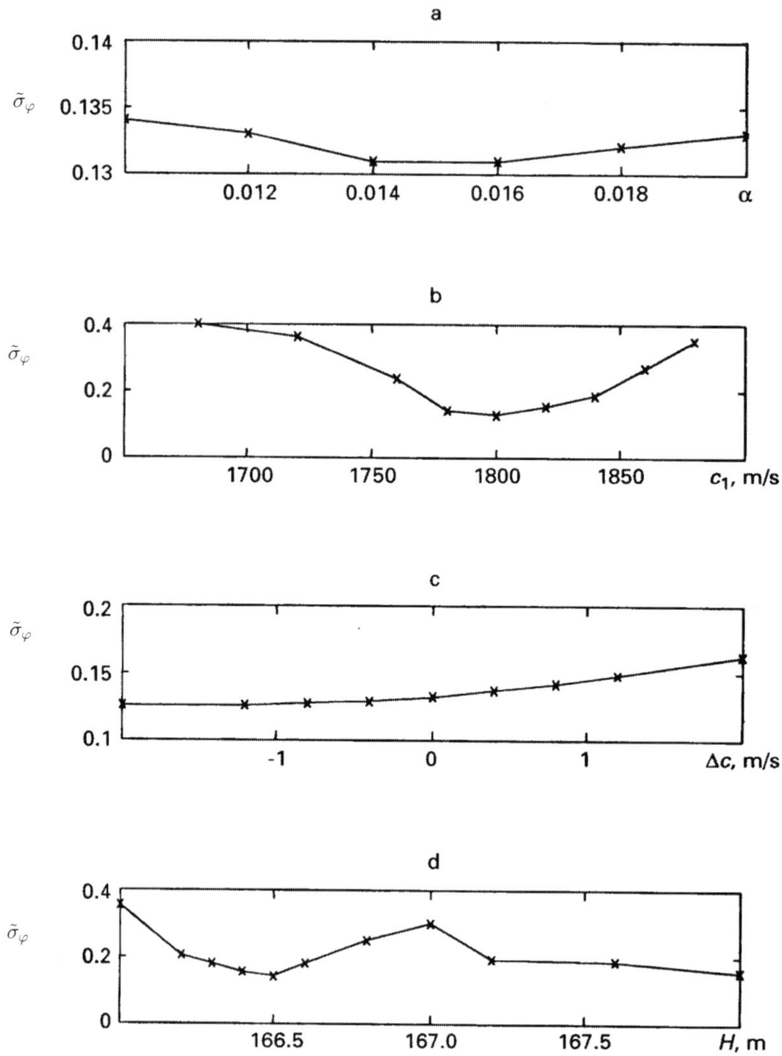

Figure 6.40. Dependence of the coefficient $\tilde{\sigma}_\varphi$ on the attenuation coefficient α (*a*), on the sound speed in the bottom c_1 (*b*), on the error in measurements of the sound speed in the waveguide (*c*), and on the waveguide depth H.

waveguide (Figure 6.40(*b*) and (*d*)). The effect of the attenuation coefficient in the bottom and the sound speed in the waveguide on the relative vertical phase distribution was insignificant, and the value of $\tilde{\sigma}_\varphi$ is weakly dependent upon these quantities (Figure 6.40(*a*) and (*c*)). Calculations showed that the density of the bottom also has an insignificant effect on the distribution of $\cos(\varphi(z))$. The values of the waveguide parameters corresponding to the best agreement between theory and experiment for

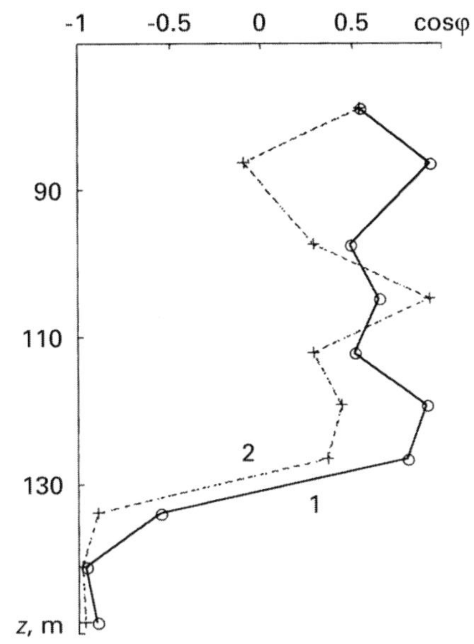

Figure 6.41. Vertical distribution of $\cos\varphi$ at time t_1. 1 – Experiment, 2 – theory.

the phase distribution were as follows: $H = 166.4\,\text{m}$, $c_1 = 1{,}800\,\text{m/s}$, $\alpha = 0.016$, $\rho_1 = 1.8\,\text{g/cm}^3$. The theoretical vertical dependence of $\cos(\varphi(z))$ corresponding to these parameters is shown in Figure 6.41, which also shows the corresponding experimental dependence.

An analogous procedure for the amplitude distribution produced the following set of parameters: $H = 169\,\text{m}$, $c_1 = 1{,}780\,\text{m/s}$, $\alpha = 0.015$, $\rho_1 = 1.8\,\text{g/cm}^3$. The theoretical and the experimental dependences of $|P(z)|$ corresponding to these parameters are shown in Figure 6.42. It follows from the given data that the choice of waveguide parameters for the amplitude and phase distribution produced similar but not coincident values. The observed spread of the values is most probably related to the fact that the parameters of the waveguide vary along the track and their values in the specific ranges can be considered as random quantities. We are concerned with the parameters of the bottom (as mentioned earlier in Chapter 1 and Section 6.1) and with the profile $c(z)$ perturbed by internal waves, which have a very much smaller spatial and temporal scale of variation than a tidal wave. From this point of view, chosen parameter values carry the sense of average (in space) effective values, which may differ from estimates of them from the amplitude and phase distribution.

The derived values of H, c_1, α, ρ_1 were used to compute the eigenvalues and eigenfunctions of the waveguide (6.2) at time t_2 (time of the second measurement of the sound speed profile). Then, according to equation (6.14), the amplitude and phase of the sound field were computed in the interval $[t_1 - t_2]$, where the eigenfunctions and the eigenvalues at intermediate times were computed according to the formula:

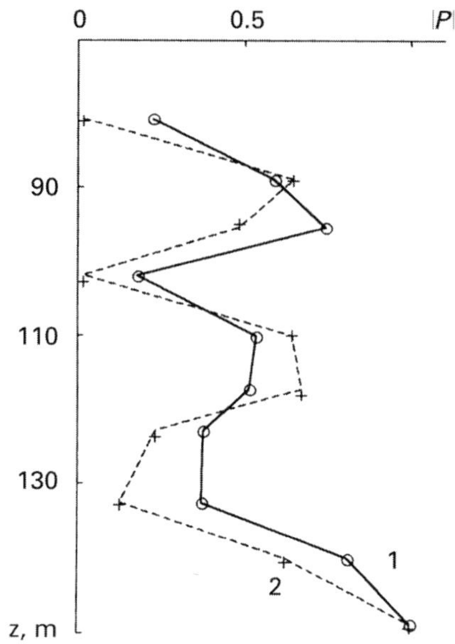

Figure 6.42. Vertical distribution of the amplitude $|P|$ at time t_1. 1 – experiment, 2 – theory.

$$\xi_l(t) = \xi_l(t_1) + (t - t_1)\frac{\xi_l(t_2) - \xi_l(t_1)}{t_2 - t_1}$$

$$\psi_l(t) = \psi_l(t_1) + (t - t_1)\frac{\psi_l(t_2) - \psi_l(t_1)}{t_2 - t_1} \tag{6.44}$$

The variations of amplitude and phase computed in this way made it possible to calculate time diagrams for the changes in the values of $\cos(\varphi(z,t))$ and $|P(z,t)|$ in the plane of the time t and the waveguide depth z (see Figures 6.38(b) and 6.40(b)). Comparison of the theoretical and experimental (Figures 6.38(a) and 6.40(a)) diagrams demonstrates a good correspondence between the results of the computations and the experimental data at least over the course of an hour. The same conclusion follows from Figure 6.43, which shows the dependence of the coefficient $\tilde{\sigma}_\varphi$ on time. The increase in $\tilde{\sigma}_\varphi$ with time indicates that in order to determine the effective parameters of the bottom with greater accuracy using the present method, temporal averaging must be used; i.e. they must be chosen at different time instants (for different profiles $c(z)$), with subsequent averaging of the results obtained.

An approach similar to the studied approach to determining the parameters of the bottom was used in Ageeva *et al.* (1980), where the possibility of determining the parameters of the bottom based on a comparison of the measured and calculated vertical interference pattern of the sound field was also considered. However, here a more complex model of the waveguide was used, with a sound speed profile in the water layer obtained from experiment and a bottom in the form of a fluid layer of thickness h_1, with parameters c_1, ρ_1, α_1, lying on an homogeneous fluid half space ρ_2, c_2. The main characteristics of the interference pattern (position and value of the

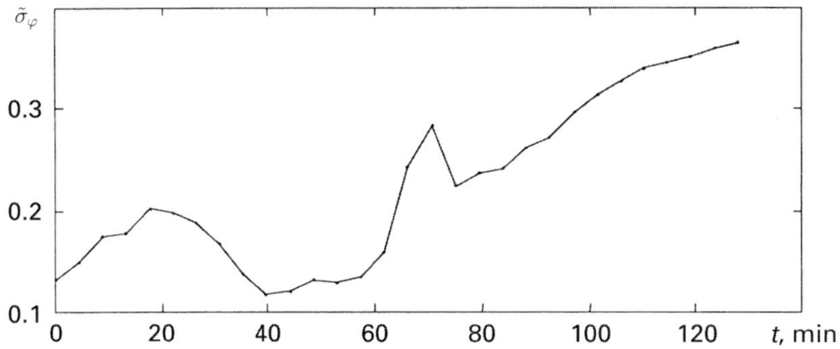

Figure 6.43. Dependence of the coefficient $\tilde{\sigma}_\varphi$ on time for waveguide parameters H, c_1, α, ρ_1, corresponding to the best correspondence between experiment and theory at time t_1.

interference minima) were chosen by varying the listed parameters of the layer at several distances from the source. The sound field of a point source, located at depth $z_1 = 30$ m was considered. The thickness of the water layer was taken to be 65 m, and the sound frequency was 63 Hz. The sound field (vertical interference pattern) was considered at three distances from the source: 0.5 km, 0.95 km and 2.4 km. Calculations of the interference pattern showed that in the model with an homogeneous fluid (or solid) layer, there is no minimum in the interference pattern in the experiment. To explain the occurrence of a minimum, another bottom layer was introduced, whose thickness, together with the characteristics of the bottom listed above, was a tunable parameter. The best agreement of experiment and theory was observed for the values $h_1 = 60$ m, $c_1 = 1,512$ m/s, $\alpha_1 = 3.6 \times 10^{-3}$. However, the cited work did not indicate the quantitative criterion upon which the comparison of experiment and theory was based.

In the general outline the given method of acoustic monitoring involves the following main points.

- Construction of a geo-acoustic model of the channel (with various sound speed profiles $c(z)$, with an homogeneous bed, with a layered bed, etc.), including the set of parameters to be determined. We note that these parameters may be very different in their physical nature: characteristics of the bottom $(h_n, c_n, \rho_n, \alpha_n)$, sound speed profiles, coordinates of the source, etc.
- Choice of the method for calculating the sound field within the chosen model (mode method, parabolic equation method, etc.)
- Choice of criterion for comparison of experimental and calculated data.
- Determination of the method for searching for unknown parameters.

It remains to discuss the last two points in more detail. Presently, in the literature, a very broad set of all possible comparison criteria of type (6.6), (6.9), (6.43) has been described (see also Voronovich *et al.* (1992), Daugherty *et al.* (1990)) for the determination of various parameters. Without giving a comparative assessment of these criteria, we simply note that the problem of their choice depends upon the conditions of the task and is determined by the stability and the convergence to optimal model parameters in the process of matching experiment and theory. Mathematically, the process of searching for tunable parameters, based on minimisation of some function corresponds to a search for an extremum of a function of many variables. It is known that in that search, the problem of 'false' (local) maxima does not have an easy solution, which leads to certain difficulties (discussed in relation to the localization of the source in Daugherty *et al.* (1990)). In connection with this problem, some authors use the so-called simulated annealing (SA) method (Dosso *et al.* (1993)). The essence of the SA method is the following. The matching process is split into individual stages (statistical tests). At each stage, the tunable parameters undergo random perturbations, which affect the value of the residual error. Here, each set of parameters (we shall speak of configurations) obtained as a result of perturbations is attributed a certain probability; the greater this is, the less the residual error.

Suppose we have a set of tunable parameters (configuration) $x_1, x_2, \ldots, x_N$, each of which varies in the range $x^i_{\min} < x_i < x^i_{\max}$. At each iteration (with index $k+1$) the value of the parameters is chosen according to the formula $x_i = x^i_{\min} + \zeta(x^i_{\max} - x^i_{\min})$, where ζ is a random number between 0 and 1. Then, this parameter (set of parameters) is used to compute the residual error $\tilde{\sigma}_{k+1}$ between the experimental and theoretical values, determined according to a given criterion. If the residual error decreases – that is $\tilde{\sigma}_{k+1} < \tilde{\sigma}_k$ – then the new configuration of parameters x is accepted unconditionally. If the residual error increases, then the set x is assigned some probability; for example, $p(\Delta\tilde{\sigma}) = [1 + \exp(-\Delta\tilde{\sigma}/T)]^{-1}$, ($\Delta\tilde{\sigma} = \tilde{\sigma}_{k+1} - \tilde{\sigma}_k$ is a T-'controlling parameter'). In other words, this is the probability of accepting the new 'perturbed' configuration (set of parameters). As the number of iterations increases, the decrease in the residual error almost ceases and the parameters of the model emerge at certain almost constant values.

In Dosso *et al.* (1993) this method is applied to determine a set of parameters of the bottom from the vertical interference pattern of the field. An experiment in acoustic sensing was carried out in the northern Pacific Ocean with a sound source of frequency 15 Hz, submerged at depth $z_0 = 100$ m. The depth was $H = 1,000$ m, along the track of length 10 km. Despite the great depth, the relationship between the frequency f and H (1.1) is such that the conditions for sound propagation are typical for shallow water. The channel is characterised by a sound speed profile with, on average, a small negative gradient; that is, it is a near-bottom channel. The number of propagating modes is around 20. The signal is received by a vertical array, lowered from a ship, comprising 39 hydrophones, arranged with a step of 25 m. The model of the bottom (Dosso *et al.* (1993)) taken for the calculation of the vertical interference structure, is shown in Figure 6.44, where the tunable parameters were chosen to be the characteristics of the fluid layer and the homogeneous fluid half space. If we add the unknown thickness of the layer h to the parameters of the bottom, then we will have seven unknown tunable parameters, forming a vector $\vec{m} = (c_1, \rho_1, \alpha_1, c_2, \rho_2, \alpha_2, h)$. The proximity coefficient (residual error) between the measured and calculated fields was determined by the expression:

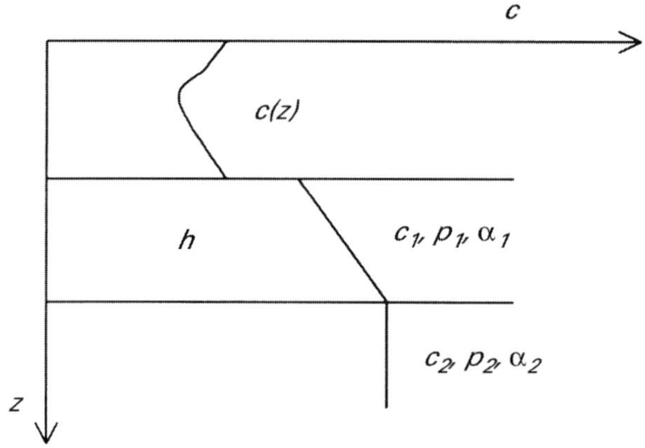

Figure 6.44. Waveguide model.

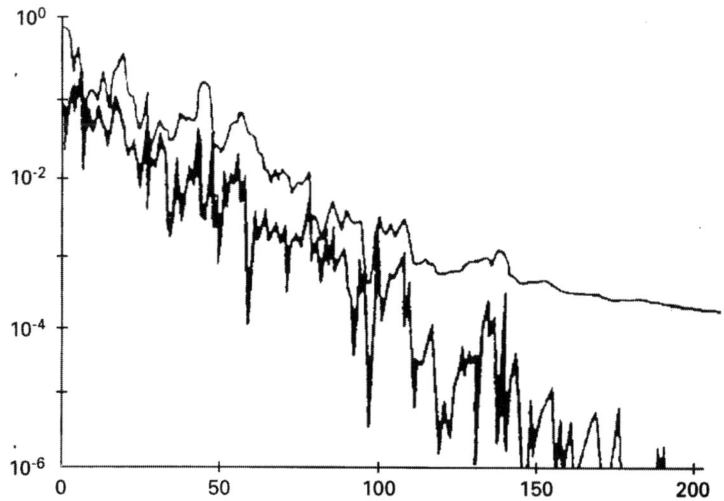

Figure 6.45. Dependence of the coefficient $\tilde{\sigma}$ on the number of iterations.

$$\tilde{\sigma}(\hat{m}) = 1 - \frac{|\sum_{j=1}^{J_h} P_{\text{th}}(\hat{m}, z_j, r_j) P_{\text{e}}^*(z_j, r_j)|^2}{\sum_{j=1}^{J_h} |P_{\text{th}}(\hat{m}, z_j, r_j)|^2 \sum_{j=1}^{J_h} |P_{\text{e}}(z_j, r_j)|^2} \tag{6.45}$$

(here j is the number of the hydrophone), depending on the position of a point in seven-dimensional space. Using the SA method, parameters of the bottom were found, which minimise (with some probability) the coefficient $\tilde{\sigma}$ (6.45). Figure 6.45 shows the convergence of the residual error for successive statistical trials. The values of the parameters of the bottom, $c_1 = 1{,}474\,\text{m/s}$, $\rho_1 = 1.33\,\text{g/cm}^3$, $c_2 = 1{,}724\,\text{m/s}$,

$\rho_2 = 2.11\ \mathrm{g/cm^3}$, $h = 249\ \mathrm{m}$, obtained using this method, differ from the values chosen for the first iteration by just a small percentage. The exception is the absorption coefficients, whose values ($\alpha_1 \approx 10^{-3}$, $\alpha_2 \approx 0.5 \times 10^{-3}$) are significantly different from the original values.

To conclude this section, we note that at the present time, the term tomography, based on matched field processing ('matched field tomography', MFT, (Baggeroer *et al.* (1992))) is taken to combine all the approaches to acoustic monitoring listed above, based on the use of *a priori* information about the medium in the signal processing.

6.5 FREQUENCY SHIFT OF THE SOUND FIELD INTERFERENCE PATTERN AND REMOTE MONITORING OF PERTURBATIONS OF THE MEDIUM

It is well known that acoustic fields in the ocean have a clearly expressed interference pattern, which is manifest in oscillations of the amplitude of the sound pressure as the distance r between the sound source and receiver varies for a fixed frequency, or as the frequency varies for a fixed observation point. Indeed, equation (4.14.1) for the amplitude of the sound field can be rewritten in the following form:

$$|P(r,z)| = \sqrt{\sum_l P_l^2 + 2\sum_{l<m}\sum P_l P_m \cos[(q_l - q_m)r]} \tag{6.46}$$

where the first term under the square root corresponds to the averaged intensity of the sound field (see Section 3.5), and the second term determines its spatial and frequency interference variations, taking into account the fact that the real parts of the eigenvalues $q_{lm} = q_l - q_m$ in shallow water depend to a considerable extent upon frequency (most often $|q_{lm}|$ decreases as the frequency grows; in particular, for the Pekeris model, $|q_{lm}| \sim \omega^{-1}$).

The characteristics of the space–frequency interference pattern of the field in oceanic waveguides with an homogeneous track have been studied in detail in Zverev and Orlou (*ed.*) (1984) and Chuprov (1982). The opinion has been voiced that when the distance between source and receiver (and their depths) is constant, as a result of variations of amplitude and phase of the arriving signals, due to temporal variations of the stratification of the medium along the track, the extrema (maxima and minima) of the interference pattern remain at practically the same frequency, and their values only change gradually. However, frequency shifts of the oscillations of intensity are determined by the temporal variability of the dispersion characteristic of the sound channel, and in shallow water they are manifest in different ways for different groups of received modes. As a result, marked shifts in frequency of the interference pattern may occur and may be very considerable.

Such shifts were noted in the experiment with a wideband sound source on a stationary track, described in Section 6.1 and in Kuzkin *et al.* (1998). Without detailing the conditions of the experiment in full here, we note only the important

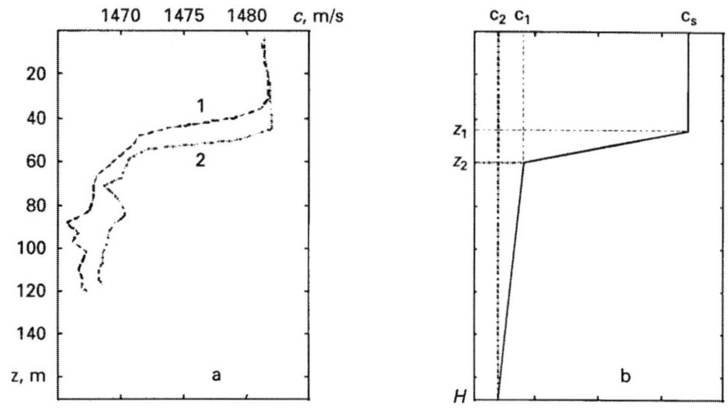

Figure 6.46. Sound speed profile: (*a*) – measured at the start t_0 (dotted line) and the end (dashed line) of the experiment; (*b*) – used in the calculations.

aspects as far as we are concerned. The mean square deviation of the distance between range finder marker beacons, on the 'receiving' and the 'emitting' ships over the measurement period, was $\sigma_r \approx 3\,\text{m}$, and we may hypothesise that the horizontal variations of the position of the source were approximately the same. The horizontal variations of the lowest hydrophone of the receiver chain located on the sea bed were less, since the inclination of the chain under the action of the current took place in the plane perpendicular to the stationary track. The sound speed profile measured before the start and at the end of the experiment with LFM signals, is shown in Figure 6.46(*a*).

Figure 6.47 shows the moduli of the spectra $|S_e(r, z; \omega)|$ in the frequency band 60–90 Hz for five signals received consecutively in a 25-minute interval at the lowest hydrophone of the chain (r and z are the coordinates of this hydrophone). This limited frequency band, used for the analysis, was associated with the presence of strong discrete interference in the spectrum of signals outside this frequency band, due to noise from the receiving ship. We note that unlike the spectra used to estimate the frequency dependence of absorption in the bottom (Figure 6.9), the curves in Figure 6.47 were not subjected to any averaging over frequency, and the frequency resolution for them is 0.025 Hz.

As one can see from the figure, the interference maxima and minima are well marked in the frequency dependence $|S_e(r, z; \omega)|$. One's attention is drawn to the constant presence of deep minima at frequency 67 Hz, where the signal amplitude is close to zero. From the wave point of view, these minima correspond to phase dislocations of the wavefront (Zhuravlev *et al.* (1989)) at whose intersections phase jumps of the sound wave phase are observed. These jumps can be clearly

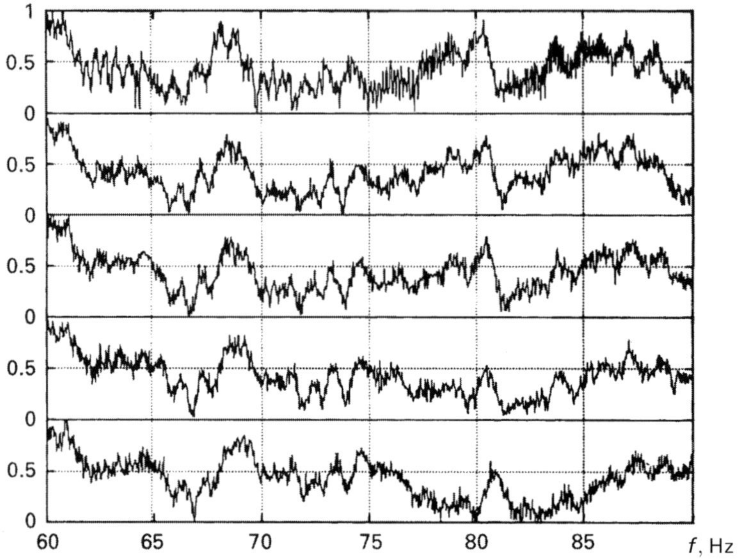

Figure 6.47. Signal spectra normalized to the maximum value. The lower curve corresponds to time t_0.

seen in the graph of the dependence of the derivative of the phase difference $\Delta\dot{\varphi}(r,z,\omega) = d(\Delta\varphi)/d\omega$ on frequency (Figure 6.48). Here,

$$\Delta\varphi(r,z,\omega) = \varphi(r,z,\omega) - \varphi_0(\omega) \tag{6.47}$$

is the difference between the phases of the received and the radiated signal. We note that the phase dislocations, according to the theory of Zhuravlev *et al.* (1989), may disappear or emerge from local interference minima. Such an effect was observed, for example, at frequency 72 Hz.

An analysis of the dependences $|S_e(r,z;\omega)|$ and $\Delta\dot{\varphi}(r,z,\omega)$ for all received signals revealed marked frequency shifts of the extrema of the interference pattern over time, associated with changes in the propagation medium. Against the background of random displacements, apparently due to the stochastic field of internal waves, there was also a marked regular drift. For the frequencies 67 Hz, 74 Hz and 89 Hz, over a measurement time of 100 min, the drift amounted to ~ 0.40 Hz, 0.44 Hz and 0.52 Hz, respectively. As one can see, the frequency shifts of the oscillations of intensity are proportional to the frequency. An analogous result was also observed for the other nine hydrophones, attesting to the independence of the shift in frequency of the signal spectrum from the depth of receipt.

We stress that the recording of the frequency shifts of the phase dislocations, as was done for the dependences $\Delta(\dot{\varphi}(r,z,\omega)$ at frequency 67 Hz, was a more convenient means of measuring the frequency shifts of the interference pattern.

To explain the observed shifts, in the first place, we must mention the following. The conditions of sound propagation in shallow water are such that the field is effectively formed by a small number of modes of the same type (for example,

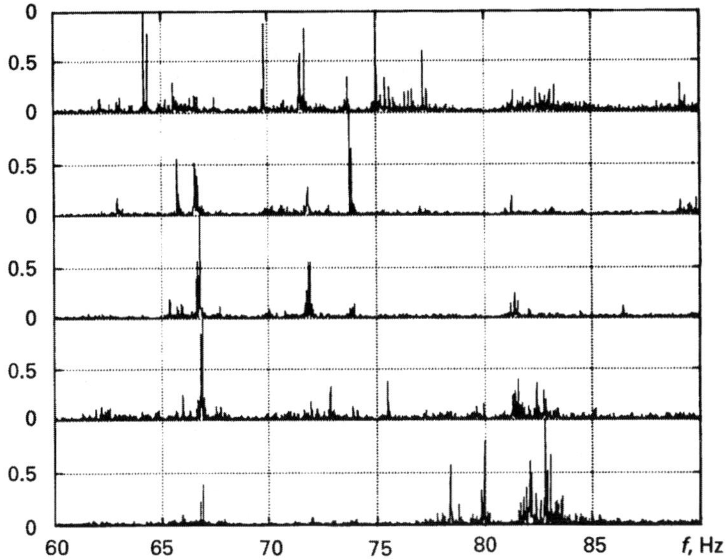

Figure 6.48. Dependence of the derivative of the phase difference on frequency. The lower curve corresponds to time t_0.

bottom–surface) with similar mode indices. The frequency shifts of the oscillations of intensity for such groups of modes with similar phases and group velocities (as we shall see later) to a first approximation are independent of their indices, and of the depth of radiation and receipt. In this connection, we analyse the shift of the interference pattern, formed by just a pair of propagating modes.

Suppose the phase difference of modes l and m at the point (r, z) is given by the expression:

$$\Delta\varphi_{lm}(r, z, \omega) = \int_0^r q_{lm}(r', z, \omega)dr' \tag{6.48}$$

Suppose further, that some parameter of the waveguide (let us denote it by h) experiences a variation $h + \Delta h$, which leads to a variation of the eigenvalues and, correspondingly, to a reconstruction of the field interference pattern. If, for example, the phase difference (6.48) corresponds to a maximum of the amplitude, then the new position of the maximum in the (r, ω)-plane will be given by the equation

$$d[\Delta\varphi(r, z, \omega)] = \frac{\partial}{\partial h}\int_0^r q_{lm}(r', z, \omega)dr' \cdot \Delta h + \frac{\partial}{\partial \omega}\int_0^r q_{lm}(r', z, \omega)dr' \cdot \Delta\omega = 0 \tag{6.49}$$

Hence, we obtain the expression for the frequency shift of the maximum, corresponding to the variation of the waveguide parameter (Kuzkin *et al.* (1998)):

$$\Delta\omega = \frac{-\displaystyle\int_0^r \frac{\partial q_{lm}(r', z, \omega, t)}{\partial h}\,dr'}{\displaystyle\int_0^r \frac{\partial q_{lm}(r', z, \omega, t)}{\partial \omega}\,dr'}\Delta h \tag{6.50}$$

Let us now consider the frequency shift of the interference pattern, occurring in our experiment. Under the conditions of the experiment, the characteristic times for a pulse (pulse duration 1 min, pulse propagation time from source to receiver 9.2 s) are far less than the characteristic times of changes in the medium (several hours), so that the medium may be assumed to be 'frozen' at the time of propagation of each pulse. For simplicity, we shall assume a sound speed profile and depth along the track independent of r, but varying from pulse to pulse. In this case, expression (6.50) can be written in the following simpler form:

$$\Delta\omega = -\left(\frac{\partial q_{ml}(t_0)}{\partial \omega}\right)^{-1}\Delta q_{ml} \tag{6.51}$$

where $\Delta q_{lm} = q_{lm}(t) - q_{lm}(t_0) = (\partial q_{lm}/\partial h)\Delta h$ is the variation of the difference of the mode propagation constants over the time $\Delta t = t - t_0$ between pulses, due to the change in the parameter h: $\Delta h = h(t) - h(t_0)$.

In order to interpret the results of the experiment, as was done in Kuzkin *et al.* (1998), let us turn to an analytic description of the propagation constants. The sound speed profiles measured in the experiment are very well approximated by the piecewise-linear dependence $c(z)$ shown in Figure 6.46(b). For this dependence, the perturbations of the water medium may lead to changes in the boundaries of the seasonal thermocline and the depth of the bottom; i.e. to variations in the parameters c, t, z_1, z_2, H. The values of the sound speed $c(H) - c_H$ on the bottom and c_S in the variable upper layer are constant and independent of time. Based on an estimate of the number of near-bottom modes (which do not graze the surface) (equation (2.90)), it is clear that at the frequencies used (< 90 Hz) there are practically no such modes, and the extrema of the sound field observed in the experiment may be due to interference of modes, reflected from the bottom and from the upper boundary. This means that in order to obtain an explicit expression for the propagation constant q_l, there are grounds for using perturbation theory (Section 3.1), taking the Pekeris model for the unperturbed waveguide, with a constant depth H along the track, and a sound speed and density of $c = c_S =$ constant, $\rho =$ constant, respectively in the water layer, and c_1, ρ_1 in the bottom. With a view to the further analysis of the frequency dependence, we express the perturbation $\delta\tilde{k}^2(z)$ in terms of the perturbation of the sound field profile

$$\delta\tilde{k}^2(z) \approx -2\frac{\omega^2}{c_S^3}\delta c(z) \tag{6.52}$$

Using equations (2.52), (2.53) and (2.50), we derive an expression for the eigenfunctions and values of the unperturbed Sturm–Liouville problem, bearing in mind that because of absorption a small number of modes are present at the observation point,

$$\psi_l^0(z) \approx \sqrt{\frac{2}{H}}\sin\sigma_l^0 z \quad q_l^0 = \sqrt{\frac{\omega^2}{c_S^2} - (\sigma_l^0)^2} \approx \frac{\omega}{c_S} - \frac{\pi^2 c_S (l^*)^2}{2\omega H^2} \tag{6.53}$$

Using these expressions, for the first-order correction Δq_l to the real part of the eigenvalue for mode l we obtain

$$\Delta q_l \approx -\frac{\omega^2}{q_l^0 c_S^3 H}\int_0^H \delta c(z)dz = \frac{\omega^2 \hat{\alpha}}{q_l^0 c_S^3 H}$$

$$\hat{\alpha} = -\int_0^H \delta c(z)dz \tag{6.54}$$

where $\delta c(z)$ is the perturbation of the sound speed profile corresponding to Figure 6.46(b):

$$\delta c(z) = c(z) - c_0 = \begin{cases} 0 & 0 \le z \le z_1 \\ \dfrac{\Delta c_1}{\Delta z_1}(z_1 - z) & z_1 \le z \le z_2 \\ -\Delta c_1 + \dfrac{\Delta c_2}{\Delta z_2}(z_2 - z) & z_2 \le z \le H \end{cases} \tag{6.55}$$

As a result of integration we obtain:

$$\hat{\alpha} = \Delta z_2 \Delta c_1 + 0.5(\Delta z_2 \Delta c_2 + \Delta z_1 \Delta c_1) \tag{6.56}$$

the integral characteristic of the sound speed profile,

$$\Delta c_1 = c_S - c_1 \quad \Delta c_2 = c_1 - c_H \quad \Delta z_1 = z_2 - z_1 \quad \Delta z_2 = H - z_2$$

We shall assume that the depth $H(t)$, the sound speed at the thermocline boundary $c_1(t)$ and the position of this boundary, $z_1(t)$, $z_2(t)$ are time-dependent parameters. When substituted in (6.54) and (6.56) these parameters give the function of time $q_l(t) = q_l^0 + \Delta q_l(t)$ and, in turn, the characteristics of the interference pattern as a function of time.

Namely, for the propagation constant at frequencies far from the critical frequency we have:

$$q_l(t) = \frac{\omega}{c_S}\left[1 + \frac{\hat{\alpha}(t)}{c_S H(t)}\right] - \frac{\pi^2 (l^*)^2 c_S}{2\omega H^2(t)}\left[1 - \frac{\hat{\alpha}(t)}{c_S H(t)}\right] \tag{6.57}$$

One possible reason for the frequency drift of the extrema of the field over the period of the experiment may be the sea tide. In our case, corresponding to that we have a perturbation, of depth associated with diurnal tidal oscillations, which is assumed to be periodic: $H(t) = H + h(t)$

$$h(t) \approx h \sin\left[\frac{\pi}{6}(t - t_1)\right] \tag{6.58}$$

(For the experiment described above, $H = 170\,\text{m}$, $h = 0.75\,\text{m}$, the time t_1 is determined by the phase of the tide, in our case $t_1 \approx t_0$, and in equation (6.58) the time is measured in hours). The corresponding changes of the thermocline boundary of and the sound speed at this boundary vary with the same period. Let us consider the frequency shift of the interference pattern which occurs between pulses, separated by $\Delta t \approx 1$ hour and 40 minutes; i.e. corresponding to the lower and upper plot on Figures 6.47 and 6.48. Then, substituting (6.57) and (6.58) in equation (6.51), it is not difficult to obtain:

$$\Delta\omega = \frac{-\omega}{c_S H[1 - (\hat{\alpha}(t_0)/c_S H)]}\left\{2c_S \Delta H + \Delta\hat{\alpha} - \frac{3\Delta H \hat{\alpha}(t_0)}{H}\right\} \qquad (6.59)$$

Here, $\Delta H = h(t) - h(t_0)$, $\Delta\hat{\alpha} = \hat{\alpha}(t) - \hat{\alpha}(t_0)$ are the variations of $\hat{\alpha}$ and H over the time Δt. Thus, the shift of the frequency spectrum is proportional to the frequency, which agrees with the results of experimental data. The first, second and third terms in curly brackets in equation (6.59) describe the shifts of the oscillations of the frequency spectrum, due, respectively, to the change in depth transformations of the sound speed profile and the joint effect of changes in the sound speed and the depth.

According to measurements at times t_0 and $t = t_0 + \Delta t$, $\Delta t = 1$ hour and 40 minutes (Figure 6.46(a)), the parameters approximating the sound speed profile are respectively estimated as $c_1 = 1{,}468.9\,\text{m/s}$ and 1,471.1 m/s; $z_1 = 35.2\,\text{m}$ and 48.3 m; $z_2 = 54.2\,\text{m}$ and 55.3 m for unchanged $c_0 = 1{,}482.1\,\text{m/s}$ and $c_2 = 1{,}466.5\,\text{m/s}$. The value of $\hat{\alpha}$ is, correspondingly equal to 1,782.9 m^2/s and 1,564.0 m^2/s. The estimated values of the frequencies shifts at frequencies 67 Hz, 74 Hz and 89 Hz were equal to 0.38 Hz, 0.40 Hz and 0.50 Hz, respectively. The results obtained lead to the conclusion that there is a good agreement between theory and experiment, not only with respect to the functional dependence, but also in the absolute values of the frequency shifts of the interference pattern.

The estimates showed that for our conditions the second and third terms on the right-hand side of equation (6.59) can be neglected, and the expression for $\Delta\omega$ has the form

$$\Delta\omega \approx -\frac{2\omega\Delta H}{H} \qquad (6.60)$$

Then, according to (6.60) the estimated values of the frequency shifts are, respectively, equal to 0.44 Hz, 0.49 Hz and 0.59 Hz. This means that the frequency shifts of the oscillations of the spectrum, due to interference of lower modes, reflecting from the boundary of the waveguide system, are weakly dependent on the variability of the sound speed in the vertical direction and are primarily determined by changes in depth. It is interesting to note that this fact has been studied in a number of other works (see, for example, Zverev (1984) and references cited therein). However this has been in connection with the study of different questions, namely in connection with the study of the space–frequency invariant of the interference pattern of the sound field in a layered ocean. The physical meaning of this is that for the given group of modes, forming the field, the quasi-random component associated with internal waves has practically no effect on the frequency shifts of the intensity oscillations. The effect due to the tidal component is very significant, despite the very short duration of the measurement interval.

In the case of dominance of modes of another type, so-called 'bottom modes', which propagate in an underwater sound channel between the thermocline layer and the sea bed, the situation may be reversed. Modes of this type predominate for similar sound speed profiles at higher frequencies and greater distances between sound source and receiver.

We further note that changes in the distance r between source and receiver due to instability of the position of the anchored ships also cause a frequency shift of the interference pattern. This shift can be estimated using equation (6.50) if one takes the distance r as the variable parameter. Then

$$q_{lm} \cdot \Delta r + \frac{\partial q_{lm}}{\partial \omega} r \cdot \Delta\omega = 0$$

whence, using equation (6.53), for q_l^0 we obtain

$$\Delta\omega \approx \frac{\Delta r}{r} \cdot \frac{q_{lm}^0}{\partial q_{lm}^0 / \partial \omega} \approx \frac{\Delta r}{r} \omega$$

Setting $|\Delta r| \approx \sigma_r \approx 3\,\text{m}$, $r \approx 13.8\,\text{km}$, $f \approx 100\,\text{Hz}$, we obtain $\Delta f = \Delta\omega/2\pi \approx 0.02\,\text{Hz}$. The computed value of Δf is far less than the frequency shift recorded in experiment. This again indirectly confirms that the frequency shift is caused by a hydrodynamic perturbation in the medium.

We note that based on the general integral equation (6.50), it is in principle possible to solve the inverse problem in the monitoring of the perturbations of waveguide parameters, by measuring the frequency shift of the interference pattern of the sound field. It is important to emphasise that it is possible to monitor regular perturbations associated, for example, with tides, and also random perturbations, due to background internal waves. This is significant for the construction of a statistical model of shallow water. In practice, this monitoring can be performed using long vertical arrays, capable of modes selection of different types.

7

Noise field in shallow water

7.1 DESCRIPTION OF THE FIELD OF A DISTRIBUTED WIDEBAND SOURCE

One of the most important problems of ocean acoustics is the study (measurement, forecasting) of noise; that is, of the sound fields formed by various unpitched (wideband) non-point (distributed) sources. These sources usually include wind-generated waves (surface disturbance), various engineering constructions (ships, drilling platforms, etc.) which generate industrial and mechanical noises, marine organisms (biological noises), earthquakes, volcanic eruptions, etc. (seismic noises), meteorological noises, noises from under ice, etc. The effect of these sources and, correspondingly, the characteristics of the noise field generated, depends on many factors including weather, geographical position, and time of day and season, and is manifest, in various frequency bands. In particular, of course, mechanical noises have a marked maximum at the alternating current mains frequency (50 or 60 Hz).

The noise field has a dual role. On the one hand it represents interference for the operation of various hydroacoustic devices, while on the other hand it provides researchers with an instrument with which to study the propagation media or noise sources based on processing of the signal characteristics.

The more or less systematic study of underwater noise began more than 50 years ago, after the end of the Second World War, and since that time a large number of both experimental and theoretical results have been obtained. There is a vast literature devoted to the study of various aspects of this problem, including monographs, in particular (Urick (1975), Ross (1976), Furduev (1974), Kuryanov and Klyachin (1984)). Historically, the first works were devoted to the study of noise in deep water, and the interest has only recently been carried over to shallow water. We note that in what follows, we shall be interested in the following observable characteristics of the noise field, usually at a fixed frequency (spectral characteristics): space–time variations of the level, depth dependence of the intensity, space–time coherence, and vertical and horizontal directivity.

We note that these and other characteristics can be calculated theoretically if one has known information about the propagation medium and the noise sources. Specifically, one needs to know the level (intensity) and angular distribution of radiation from the sources, generally speaking, as a function of the coordinates and of time, together with parameters of the acoustic waveguide.

Although the theory of sound propagation in shallow water presented earlier enables us in principle to solve this problem, in reality, however, many of the data needed to model the noise are completely unknown or insufficiently well known. In the first place, this relates to the space–time distribution and level of radiation of noise sources. This is associated with the diversity of the noise generation mechanisms in the ocean, such as the movement of the water surface, the interaction of surface waves, the formation and collapse of bubbles in surface waves, biological sources, industrial (mechanical) noises, noises from shipping, etc. Many works consider the different mechanisms for noise generation (Ross (1976), Mendus and Postnov (1993), Furduev (1966), Wilson (1979), Oguz (1994)), but not all of these have been experimentally verified. This is because at the present time, the prediction of precise values of the noise characteristics at a receiving array is a very difficult and ill-defined problem.

However, some properties of ocean noise can be modelled with high accuracy. These are properties which are determined by certain general properties of the propagation medium and by certain characteristics of the sources, which are of a quite general nature for all types of source. For example, as mentioned in Yang and Kwang Yoo (1997), the noise in an azimuthal (horizontal) direction will have a maximum in the direction along a ship's track when one is studying shipping noise, or along the line of a storm if the noise is mainly generated by wind-generated waves.

Furthermore, at low frequencies (below 1 kHz) noise sources are considered to be point sources, which because of their proximity to the surface, radiate as effective dipoles. In this case, the vertical direction of the noise is determined by the properties of the waveguide (the propagation medium), and the absolute intensity of the noise source (source level) is not required (Yang and Kwang Yoo (1997), Hamson (1985)). It has also been noted that in a lossy waveguide, the vertical scale of coherence is mainly determined by the angle of total internal reflection of a ray by the bottom (Carbone *et al.* (1998)), which makes it, in principle, possible to determine the characteristics of the bottom from the observed characteristics of the noise.

In this chapter, as the main noise source we shall consider a wind-generated disturbance, responsible for the formation of the noise field at comparatively low frequencies (up to 1–2 kHz) in large open areas of the sea. We shall not discuss detailed hypotheses about noise generation mechanisms, and will consider effects due to the specific waveguide distribution of the noise field in a shallow-water waveguide. Accordingly, we shall consider a sound source as a distribution of a set of individual sources on the surface, characterised by an intensity, a direction and other statistical parameters. This approach, which was apparently first proposed in Cron and Sherman (1962) for a half-space, was subsequently developed and generalised to the case of shallow-water waveguides (Kuperman, Katsnelson). In this approach various results for the observable characteristics of the sound field have been obtained: intensity depth distribution (Kuperman and Ferla (1985), Derevyagina *et al.* (1994)), vertical and horizontal direction (Yang and Kwang Koo (1997), Hamson (1985), Carbone *et al.* (1998)), spatial correlation functions (Buckingham (1980), Kuperman and Ingenito (1980), Abdullaev *et al.* (1986)), the spectral distribution of the intensity at a fixed depth (Kuperman and Ferla (1985)). The effect of various factors relating to the sea on the sound field characteristics has been considered: an absorbing sea bed (Derevyagina *et al.* (1991), Carbone *et al.* (1998)), the sound speed profile (Lyubchenko and Petnikov (1991)), random homogeneities of various kinds (Derevyagina and Katsnelson (1995)) including, in particular, internal waves (Yang and Kwang Yoo (1997)).

As before, our approach is based on a modal decomposition of the sound field, which is completely justified for the comparatively low frequencies considered.

Up to here, we have assumed that the source is a pure-tone point source, but in working with the noise we must now deal with a wideband signal distributed in some spatial region. Let us briefly consider a description of the sound field for this case. Such a description is based on the notion of a unit impulsive point source for which the sound field (space–time Green function), for example, for a homogeneous space, is a solution of the equation

$$\left[\Delta - \frac{1}{c^2}\frac{\partial^2}{\partial t^2}\right]\Psi(\vec{R}, \vec{R}'; t, t') = -\delta(\vec{R} - \vec{R}')\delta(t - t') \tag{7.1}$$

The explicit expression for the Green function in an homogeneous medium is of the form

$$\Psi(\vec{R}, \vec{R}'; t, t') = \frac{1}{4\pi|\vec{R} - \vec{R}'|}\delta\left(t - t' - \frac{|\vec{R} - \vec{R}'|}{c}\right) \tag{7.2}$$

The solution of the inhomogeneous equation determining the field of a source radiating a complex signal with an arbitrary dependence on coordinates and time $Q(\vec{R}, t)$[1]

$$\left[\Delta - \frac{1}{c^2}\frac{\partial^2}{\partial t^2}\right]P(\vec{R}, t) = -Q(\vec{R}, t) \tag{7.3}$$

[1] According to our definitions, the quantity $Q(\vec{R}, t)$ is called the volume density of the radiated signal.

is of the form

$$P(\vec{R},t) = \int d\vec{R}'dt'\Psi(\vec{R},\vec{R}';t,t')Q(\vec{R}',t') = \int d\vec{R}' \frac{Q\left(\vec{R}', t - \frac{|\vec{R}-\vec{R}'|}{c}^{-1}\right)}{4\pi|\vec{R}-\vec{R}'|} \tag{7.4}$$

In particular, for a monopole located at the point with radius vector $\vec{R}_0$ and oscillating according to an arbitrary law

$$Q(\vec{R},t) = F(t)\delta(\vec{R}-\vec{R}_0) \tag{7.5}$$

the complex field has the form

$$P(\vec{R},t) = \frac{F\left(t - \frac{|\vec{R}-\vec{R}_0|}{c}^{-1}\right)}{4\pi|\vec{R}-\vec{R}_0|} \tag{7.6}$$

If $F(t) = \mathrm{e}^{-i\omega t}$, then $P(\vec{R},t) = \exp[i(k|\vec{R}-\vec{R}_0| - \omega t)]/(4\pi|\vec{R}-\vec{R}_0|)$. For a dipole, with dipole moment $\vec{M}(t)$ varying in time, located at the point with radius vector $\vec{R}_0$, the volume density of the radiated signal is of the form

$$Q(\vec{R},t) = \rho\frac{\partial\vec{M}(t)}{\partial t}\nabla\delta(\vec{R}-\vec{R}_0) \tag{7.7}$$

corresponding to which, the sound field has the form

$$P(\vec{R},t) = \rho\nabla\left\{\frac{\partial}{\partial t}\vec{M}\left(t - \frac{|\vec{R}-\vec{R}_0|}{c}\right)\Big/[4\pi|\vec{R}-\vec{R}_0|]\right\} \tag{7.8}$$

In particular, if the dipole moment is directed along the z-axis and depends harmonically on time: $M(t) = M_z(t) = M\mathrm{e}^{-i\omega t}$, then

$$Q(\vec{R},t) = -i\omega\rho M\mathrm{e}^{-i\omega t}\frac{\partial}{\partial z}\delta(\vec{R}-\vec{R}_0) \tag{7.9}$$

and in free space

$$P(\vec{R},t) = -i\omega\rho M\mathrm{e}^{-i\omega t}\frac{\partial}{\partial z}\frac{\exp[ik|\vec{R}-\vec{R}_0|]}{4\pi|\vec{R}-\vec{R}_0|} \tag{7.10}$$

For a complex sound field depending in an arbitrary manner on spatial coordinates (and for the radiated signal), we introduce the spatial wave vector spectrum or simply spatial spectrum according to the equations

$$P(\vec{R}) = \int S(\vec{k})\mathrm{e}^{i\vec{k}\vec{R}}d\vec{k}$$

$$S(\vec{k}) = \frac{1}{(2\pi)^3}\int P(\vec{R})\mathrm{e}^{-i\vec{k}\vec{R}}d\vec{R} \tag{7.11}$$

We also introduce the spatial spectrum in the horizontal plane

$$P(\vec{R}) = P(\vec{r}, z) = \int S(\vec{\xi}, z) e^{i\vec{\xi}\vec{r}} d\vec{\xi}$$

$$S(\vec{\xi}, z) = \frac{1}{(2\pi)^2} \int P(\vec{r}, z) e^{-i\vec{\xi}\vec{r}} d\vec{r} \tag{7.12}$$

which unlike the former will be denoted by a vector in the horizontal plane $\vec{\xi}$. Finally, we introduce the space–time spectrum

$$P(\vec{R}, t) = 2\int_0^{\infty} d\omega \int d\vec{k} S(\vec{k}, \omega) e^{-i(\vec{k}\vec{R} - \omega t)}$$

$$S(\vec{k}, \omega) = \frac{1}{(2\pi)^4} \int_{-\infty}^{\infty} dt \int d\vec{R}\, \mathrm{Re}\,[P(\vec{R}, t)] e^{-i(\vec{k}\vec{R} - \omega t)} \tag{7.13}$$

7.2 MODEL OF A NOISE SOURCE: GENERAL EQUATIONS

As a model of sources of surface noise, we shall consider a set of randomly distributed monopoles in the horizontal plane, which is found at a small depth z_S below the sea surface. For the characteristics of the surface sources of noise, we introduce the complex function (generally speaking, random) $Q(r, t)$ of the surface density of the radiated signal (complex signal from a unit surface). The surface spectral density of the radiated signal

$$Q_\omega(\vec{r}) = \frac{1}{2\pi} \int_{-\infty}^{\infty} \mathrm{Re}\, Q(\vec{r}, t) e^{i\omega t} dt \tag{7.14}$$

$$Q(\vec{r}, t) = 2\int_0^{\infty} Q_\omega(\vec{r}) e^{-i\omega t} d\omega \tag{7.15}$$

is also a random function of space–time variables. We shall assume that $Q(\vec{r}, t)$ is a process stationary in time and homogeneous in the plane, characterised by the spectral cross correlation function

$$B_q(\vec{r}, \omega; \vec{r}', \omega') = \langle Q_\omega(\vec{r}) Q^*_{\omega'}(\vec{r}') \rangle = B_q(\vec{r} - \vec{r}') \delta(\omega - \omega') \tag{7.16}$$

where $B_q(\vec{r} - \vec{r}')$ is the cross-spectral density and the angular brackets denote ensemble averaging. The quantity $Q_\omega(\vec{r})$ can be interpreted as the number of sources per unit surface area with the same amplitude of the radiated signals in a unit frequency interval, that is the spectral strength of the noise sources. The cross-spectral density of the sources $B_q(\vec{r} - \vec{r}')$, which is generally speaking frequency dependent, and characterises the spectral intensity of radiation, is sometimes written in the form

$B_q(\vec{r} - \vec{r}') = q^2_\omega N(\vec{r} - \vec{r}')$. In this expression, the source spectrum level q_ω (Kuperman and Ingenito (1980)), measured in $\mathrm{dB}/\mu\mathrm{Pa}/\sqrt{\mathrm{Hz}}$, and the spatial density of the distribution of the sources $N(\vec{r}-\vec{r}')$, normalised by the number of sources N_0, $\int N(\vec{r})d\vec{r} = N_0$, are distinguished. If we introduce the spectral component of the volume speed of the pulsating volume, radiating the noise field V_ω, then the source spectrum level is related to this quantity by the equation: $q_\omega = \rho\omega V_\omega$. Accordingly, the radiated power from a unit surface in a unit frequency interval (spectral power density) is $W_\omega = q_\omega^2/8\pi\rho c$.

The complex amplitude of the sound field $P(\vec{R}, t)$, formed by such a random distributed source satisfies the equation:

$$\left(\Delta - \frac{1}{c^2}\frac{\partial^2}{\partial t^2}\right)P(\vec{R}, t) = -Q(\vec{r}, t)\delta(z - z_s) \tag{7.17}$$

The sound field at a fixed frequency (spectral amplitude of the sound field $S(\vec{R}, z_s; \omega)$), taken at the point with cylindrical coordinates (r, z) is given by the equation:

$$\left[\Delta + \frac{\omega^2}{c^2(z)}\right]S(\vec{r}, z, z_s; \omega) = -Q_\omega(\vec{r})\delta(z - z_s) \tag{7.18}$$

with corresponding boundary conditions on the bottom and on the surface. Using the Green function for the waveguide (1.14), the solution of (7.18) can be written in the form

$$S(\vec{r}, z, z_s; \omega) = \int \Psi(\vec{R}, \vec{R}_s; \omega)Q_\omega(\vec{r}_s)d\vec{r}_s \tag{7.19}$$

where $\vec{R}_s = (\vec{r}_s, z_s)$. In particular, in an homogeneous space

$$S(\vec{r}, z, z_s; \omega) = \int \frac{\exp(ik|\vec{R} - \vec{R}_s|)}{4\pi|\vec{R} - \vec{R}_s|}Q_\omega(\vec{r}_s)d\vec{r}_s \tag{7.20}$$

Historically, the first model for study of the sound field due to sources distributed on the sea surface was the model of sources uniformly distributed on the surface of an homogeneous half-space, proposed by Cron and Sherman (1962). To move to this model, as an individual source, we imagine a point monopole of unit amplitude, located at a small depth z_s below the sea surface, radiating a monochromatic signal of frequency ω, where the radius vector of this point in the horizontal plane is r_s (see Figure 7.1).

The complex amplitude of the field at the given frequency, given by the equation $P_s(\vec{R}, t) = P_s(\vec{R})\exp(-i\omega t)$ at the point of observation with coordinates $\vec{R} = (\vec{r}, z)$ at a large distance from the plane of the surface in comparison with the wavelength, is equal to

$$P_s(\vec{R}) = \frac{\exp[ik|\vec{R} - \vec{R}_s^+|]}{4\pi|\vec{R} - \vec{R}_s^+|} + V(\vartheta)\frac{\exp[ik|\vec{R} - \vec{R}_s^-|]}{4\pi|\vec{R} - \vec{R}_s^-|} \tag{7.21}$$

where $\vec{R}_s^\pm = (\vec{r}_s, \pm z_s)$, $V(\vartheta)$ is the reflection coefficient, and ϑ is the angle from the vertical, defined by the condition $z + z_s = |\vec{R} - \vec{R}_s^-|\cos\vartheta$. Assuming the depth of the sources is small, we shall suppose that all the angles from the vertical of the vectors

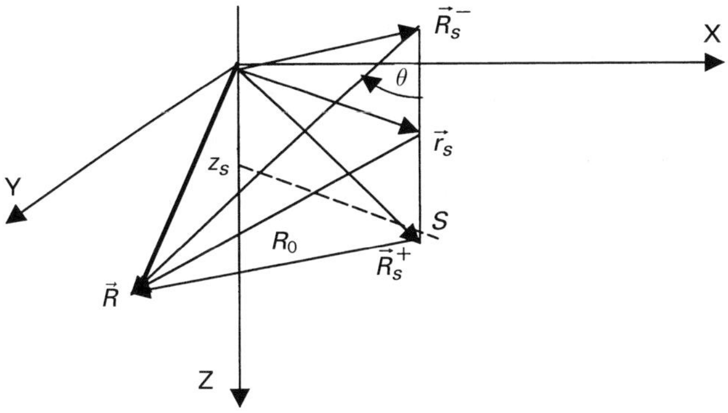

Figure 7.1. Formation of the field of an individual surface source located at the point S near the surface with coordinates $\vec{r}_s, z_s$. The point of observation is given by the radius vector $\vec{R}$. The radius vector from the source to the point of observation is $\vec{R}_s^+$, and the radius vector to the point of observation from the point corresponding to the source image is $\vec{R}_s^-$.

drawn from their source and its representation are equal, and also that $|\vec{R} - \vec{R}_s^{\pm}| \approx R_0(1 \pm z_s \cos\vartheta)$, $R_0 = \sqrt{(\vec{r} - \vec{r}_s)^2 + z^2}$.

Using this approximation, we obtain for the field of an individual effective source formed by the pair comprising the source and its image or by the source field and its signal reflected from the surface,

$$P_s(\vec{R}) = \frac{\exp[ikR_0]}{4\pi R_0}\left[e^{-ikz_s\cos\vartheta} + V(\vartheta)e^{ikz_s\cos\vartheta}\right] = g(\vartheta)\frac{\exp[ikR_0]}{4\pi R_0} \tag{7.22}$$

Equation (7.22) is a subcase of a known property for a field formed by a set of monopoles, occupying a bounded area of space. It allows us to consider the distribution of the sources as the set of all effective sources, located on the surface, each of which has a certain directivity pattern: $g(\vartheta)$ is a function determined by the boundary conditions and by the conditions of radiation, which determines the directivity of an effective point source. In particular, if we have a pressure-release sea surface ($V(\vartheta) = -1$), then an individual effective source is a dipole and its directivity is determined by the function $g(\vartheta) = ikz_s \cos\vartheta$. We note that these expressions (7.21) and (7.22) do not include inhomogeneous waves and are approximately valid only for sufficiently large distances from the surface. The expression (7.22) was used in Cron and Sherman to determine the correlation function. This was done as follows.

Let us suppose that the receiver is located at the point with coordinates $r = 0$, z; that is, on the z-axis. Uncorrelated sources, each of whose directivity patterns is determined by the function $g(\vartheta)$, lie in the XY-plane, and an individual source is located at the point S. The geometry of the problem is shown in Figure 7.2.

The field at the point of observation of the individual source is given by equation (4.22). The intensity of the field from all the noise sources is determined by the total

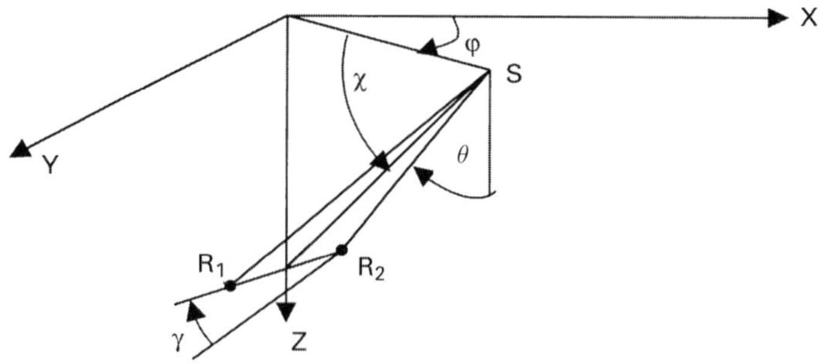

Figure 7.2. Diagram of the coordinates in the calculation of the noise correlation function. The angles χ from the horizontal (or ϑ from the vertical), drawn from the point of observation, are assumed to be identical for all radius vectors.

field from all sources, distributed in the XY-plane:

$$I(z) = 2\pi \int_0^\infty \frac{1}{2\rho c} |P_s(\vec{r})|^2 r dr = \frac{1}{16\pi\rho c} \int_0^{\pi/2} |g(\vartheta)|^2 \tan\vartheta d\vartheta \tag{7.23}$$

which is Cron and Sherman's formula for the noise intensity.

Let us now consider the spatial correlation function of the noise, for which we introduce two different receivers at the points R_1 and R_2, located, for definiteness, in the YZ-plane (Figure 7.2). The distance between the receivers is denoted by d, and the angle of this section with the horizontal axis (Y-axis) is denoted by γ. The spatial correlation function, normalised to unity at the maximum, and measured by summing the sound pressures at the distinct receivers, is determined using the equation:

$$\begin{aligned} B(\vec{R}_1, \vec{R}_2) &= \frac{\int_0^\infty r dr \int_0^{2\pi} d\varphi |P(\vec{R}_1) + P(\vec{R}_2)|^2}{\int_0^\infty r dr \int_0^{2\pi} d\varphi [|P(\vec{R}_1)|^2 + |P(\vec{R}_2)|^2]} - 1 \\ &= \frac{\mathrm{Re} \int_0^\infty r dr \int_0^{2\pi} d\varphi P(\vec{R}_1) P^*(\vec{R}_2)}{I(z)} \end{aligned} \tag{7.24}$$

Here, $\vec{R}_0$, $\vec{R}_1$, $\vec{R}_2$ are the radius vectors from the individual sources, respectively to the mean of the left- or right-hand points of observations. In introducing this formula, we have neglected the difference in the coordinates of the points of observation everywhere, except for the phases of the radiated field. Substituting the expression for the radiated field and integrating, we obtain Cron and Sherman's expression

$$B(\vec{R}_1, \vec{R}_2) = \frac{\int_0^{\pi/2} |g(\vartheta)|^2 \tan\vartheta \cos(kd \sin\gamma \cos\vartheta) J_0(kd \cos\gamma \sin\vartheta) d\vartheta}{\int_0^{\pi/2} |g(\vartheta)|^2 \tan\vartheta d\vartheta} \tag{7.25}$$

It is interesting to note that this expression does not contain z; that is, it does not depend on the depth of the receiver. From it, one can obtain the correlation functions for various subcases of the orientation and directivity of the noise sources. As the general case, it is customary to assume that the directivity of the source corresponds to the radiation of multipoles distributed on the surface; i.e. to the function $g(\vartheta) \sim \cos^m \vartheta$. Moreover, for example, for the horizontal ($\gamma = 0$) and the vertical ($\gamma = \pi/2$) correlation functions, the given expression simplifies to

$$B(d,0) = \frac{\int_0^{\pi/2} \cos^{2m} \vartheta \tan \vartheta J_0(kd \sin \vartheta) d\vartheta}{\int_0^{\pi/2} \cos^{2m} \vartheta \tan \vartheta d\vartheta} = \frac{2^m m! J_m(kd)}{(kd)^m} \tag{7.26}$$

$$B(d,\pi/2) = \frac{\int_0^{\pi/2} \cos^{2m} \vartheta \tan \vartheta \cos(kd \cos \vartheta) d\vartheta}{\int_0^{\pi/2} \cos^{2m} \vartheta \tan \vartheta d\vartheta} = 2m \int_0^1 x^{2m-1} \cos(kdx) dx \tag{7.27}$$

These expressions are easily computed, and numerical results are discussed in Cron and Sherman.

Equation (7.22) which was used to obtain the results, is approximate. It would be more accurate to work with a decomposition of the field in the plane into plane waves, then and we would be dealing with precise solutions of the wave equation, as in Kuryanov (1963). Namely, it is known that a point source in the plane $z_s = 0$ gives a field (complex amplitude at fixed frequency) in the form of a decomposition into plane waves (Brekhovskikh (1980)):

$$\frac{\exp[ikR_0]}{4\pi R_0} = \frac{i}{8\pi^2} \int \frac{1}{\sqrt{k^2 - \xi^2}} \exp[i\vec{\xi}(\vec{r} - \vec{r}_s) + ik_z z] d\vec{\xi}$$

$$R_0 = \sqrt{(\vec{r} - \vec{r}_s)^2 + z^2} \tag{7.28}$$

Accordingly, the complex amplitude of the field from an effective directional point source in the plane can be written as

$$P_s(\vec{r}, z) = \frac{i}{8\pi^2} \int \frac{g(\xi)}{\sqrt{k^2 - \vec{\xi}^2}} \exp[i\vec{\xi}(\vec{r} - \vec{r}_s) + ik_z z] d\xi \tag{7.29}$$

At large distances, this field becomes equation (7.22). Suppose now that sources (7.29), are distributed in the plane $z_s = 0$ with some density $Q_\omega(\vec{r}_s)$, then the total field at frequency ω, at some distance from the plane, is of the form

$$\begin{aligned} S(\vec{r}, z, 0; \omega) &= \int d\vec{r}_s P_s(\vec{r}, z) Q_\omega(\vec{r}_s) \\ &= \frac{i}{8\pi^2} \int\int d\vec{r}_s d\vec{\xi} \frac{Q_\omega(\vec{r}_s)}{\sqrt{k^2 - \xi^2}} g(\xi) \exp[i\vec{\xi}(\vec{r} - \vec{r}_s) + ik_z z] \\ &= \frac{i}{2} \int d\vec{\xi} \frac{Q_\omega(\vec{\xi})}{\sqrt{k^2 - \xi^2}} g(\xi) \exp[i\vec{\xi}\vec{r} + ik_z z] \\ &= \int d\vec{\xi}\, \overline{F}_\omega(\vec{\xi}) \exp[i\vec{\xi}\vec{r} + ik_z z] \end{aligned} \tag{7.30}$$

where

$$Q_\omega(\xi) = \frac{1}{(2\pi)^2}\int d\vec{r}_s Q_\omega(\vec{r}_s)\mathrm{e}^{-i\xi\vec{r}_s}$$

is the spatial spectrum of the source distribution function. Kuryanov (1963) uses a distribution of the field in the plane of the sources and its spatial spectrum. In our case, this is

$$\overline{F}_\omega(\vec{\xi}) = \frac{i}{2}\frac{Q_\omega(\vec{\xi})}{\sqrt{k^2-\xi^2}}g(\xi) \tag{7.31}$$

The correlation functions of these functions are interrelated, and assuming δ-correlation of the spectral distribution function, we have the relation

$$\begin{aligned}\langle \overline{F}_\omega(\vec{\xi})\overline{F}^*_{\omega'}(\vec{\xi}')\rangle &= \frac{1}{4}\frac{\langle Q_\omega(\vec{\xi})Q^*_{\omega'}(\vec{\xi}')\rangle}{\sqrt{(k^2-\xi^2)(k^2-\xi'^2)}}g(\xi)g^*(\xi')\\ &= \frac{|g(\xi)|^2\overline{B}_q(\vec{\xi})}{4(k^2-\xi^2)}\delta(\omega-\omega')\delta(\vec{\xi}-\vec{\xi}')\end{aligned} \tag{7.32}$$

where $\overline{B}_q(\vec{\xi})$ is the spatial spectral density, associated with the correlation function for the sources in the plane

$$B_q(\vec{\rho}_s) = \int d\vec{\xi}\,\overline{B}_q(\vec{\xi})\mathrm{e}^{i\vec{\xi}\vec{\rho}_s} \tag{7.33}$$

Accordingly

$$\overline{B}_q(\vec{\xi}) = \frac{1}{(2\pi)^2}\int d\vec{\rho}_s B_q(\vec{\rho}_s)\mathrm{e}^{-i\vec{\xi}\vec{\rho}_s} \tag{7.34}$$

The function $f(\vec{\xi})$ used in Kuryanov (1963) is related to our notation by

$$\langle \overline{F}_\omega(\vec{\xi})\overline{F}^*_{\omega'}(\vec{\xi}')\rangle = f(\vec{\xi})\delta(\vec{\xi}-\vec{\xi}')\delta(\omega-\omega') \tag{7.35}$$

The correlation function for the field in the plane of the sources is

$$\begin{aligned}\langle S(\vec{r},0,0;\omega)S^*(\vec{r}\,',0,0;\omega')\rangle &= B_\omega(\vec{r}-\vec{r}\,')\delta(\omega-\omega')\\ B_\omega(\vec{\rho}) &= \int d\vec{\xi} f(\vec{\xi})\mathrm{e}^{i\vec{\xi}\vec{\rho}}\end{aligned} \tag{7.36}$$

Let us find the correlation function for the field

$$\langle S(\vec{r},z,0;\omega)S^*(\vec{r}\,',z',0;\omega')\rangle = B_\omega(\vec{r},z;\vec{r}\,',z')\delta(\omega-\omega') = B_\omega(\vec{R},\vec{R}')\delta(\omega-\omega') \tag{7.37}$$

From equation (7.30) for the cross-correlation density of the sound field we obtain

$$B_\omega(\vec{r},z;\vec{r}\,',z') = \iint d\vec{\xi}d\vec{\xi}' f(\vec{\xi})\delta(\vec{\xi}-\vec{\xi}')\mathrm{e}^{i(\vec{\xi}\vec{r}+k_z z)}\left[\mathrm{e}^{i(\vec{\xi}'\vec{r}\,'+k_z{}'z')}\right] \tag{7.38}$$

In the integral, inhomogeneous waves must also be taken into account, which gives, in principle, for the cross-spectral density, a dependence on the coordinates and z and z'. At sufficiently large distances from the radiating planes, inhomogeneous waves can be neglected, and the cross-spectral density has the form

$$B_\omega(\vec{r}-\vec{r}', z-z') = \int d\vec{\xi} f(\vec{\xi}) \exp\left\{i\left[\vec{\xi}(\vec{r}-\vec{r}') + k_z(z-z')\right]\right\} \tag{7.39}$$

Clearly, for $z = z'$ the correlation function at an arbitrary distance from the plane coincides with the correlation function (7.36) for the field in the plane, which is defined in Kuryanov as the 'transport' of correlation function of the field from the plane. For the spectral intensity of the noise field at the point of observation with coordinates $\vec{r}$, z:

$$\begin{aligned} I_\omega(\vec{r}, z) &= \frac{2}{\rho c} B_\omega(0,0) = \frac{2}{\rho c}\int d\vec{\xi} f(\vec{\xi}) \\ &= \frac{2}{\rho c}\iint k^2 f(\vartheta,\varphi)\cos\vartheta \sin\vartheta d\vartheta d\varphi = \int I(\vartheta,\varphi) d\Omega \end{aligned} \tag{7.40}$$

where we have introduced the ray intensity

$$I(\vartheta,\varphi) = \frac{2k^2 f(\vartheta,\varphi)\cos\vartheta}{\rho c} = \frac{|g(\vartheta)|^2 B_q(\vartheta,\varphi)}{2\rho c \cos\vartheta} \tag{7.41}$$

Let us now consider the cross-spectral density of the noise $B_\omega(\vec{R}, \vec{R}')$ of the noise in an arbitrary inhomogeneous medium. With the functions $B_\omega(\vec{R}, \vec{R}')$ one can associate the space–time correlation function using the Fourier transform:

$$B(\vec{R}, \vec{R}', \tau) = \langle p(\vec{R}, t+\tau) p(\vec{R}', t)\rangle = \int_{-\infty}^{\infty} B_\omega(\vec{R}, \vec{R}') \mathrm{e}^{-i\omega\tau} d\omega \tag{7.42}$$

The cross-spectral density function of the complex field is also expressed in terms of the spectral correlation function:

$$\langle P(\vec{R}, t) P^*(\vec{R}', t+\tau)\rangle = 4\int_0^{\infty} B_\omega(\vec{R}, \vec{R}') \mathrm{e}^{-i\omega\tau} d\omega \tag{7.43}$$

Following Kuperman and Ingenito (1980), let us find an expression for the correlation function. For this, we shall use the Hankel transformation of the Green functions $\bar{\Psi}(\xi, z, z_s)$ see equations (2.8) and the footnote on p.14:

$$\Psi(\vec{R}, \vec{R}'; \omega) = \int d\vec{\xi} \bar{\Psi}(\xi, z, z') \mathrm{e}^{i\vec{\xi}(\vec{r}-\vec{r}')} \tag{7.44}$$

where $\vec{\xi}$ is a wave vector in the horizontal plane. Using in addition (7.37), (7.19) and (7.33), we obtain

$$B_\omega(\vec{R},\vec{R}') = \frac{1}{(2\pi)^4}\int d\vec{r}_s \int d\vec{r}_s{}' \int d\vec{\xi} \int d\vec{\xi}' B_q(\vec{r}_s - \vec{r}'_s)$$

$$\times \bar{\Psi}(\xi, z, z_s)\bar{\Psi}^*(\xi', z', z_s)\exp\{i[\vec{\xi}(\vec{r}-\vec{r}_s) - \vec{\xi}'(\vec{r}' - \vec{r}'_s)]\}$$

$$= \frac{1}{(2\pi)^4}\int d\vec{r}_s \int d\vec{\rho} \int d\vec{\xi} \int d\vec{\xi}' B_q(\vec{\rho})$$

$$\times \bar{\Psi}(\xi, z, z_s)\bar{\Psi}^*(\xi', z', z_s)\exp\{i[\vec{\xi}(\vec{r}-\vec{r}_s) - \vec{\xi}'(\vec{r}' - \vec{r}_s - \vec{\rho})]\}$$

$$= \frac{1}{(2\pi)^2}\int d\vec{\rho} \int d\vec{\xi} B_q(\vec{\rho})\bar{\Psi}(\xi, z, z_s)\bar{\Psi}^*(\xi, z', z_s)\exp[i\vec{\xi}(\vec{r}-\vec{r}' - \vec{\rho})]$$

$$= \int d\vec{\xi}\bar{B}_q(\vec{\xi})\bar{\Psi}(\xi, z, z_s)\bar{\Psi}^*(\xi, z', z_s)\exp[i\vec{\xi}(\vec{r}-\vec{r}')]$$

$$= \frac{1}{2\pi}\int d\vec{\rho} B_q(\vec{\rho})\int_0^\infty \xi d\xi J_0(\xi|\vec{r}-\vec{r}'-\vec{\rho}|)\bar{\Psi}(\xi, z, z_s)\bar{\Psi}^*(\xi, z', z_s) \tag{7.45}$$

This formula was obtained in Kuperman and Ingenito (1980), where expressions for the correlation function of the noise for various forms of directivity functions for radiation were obtained. The simplest case corresponds to δ-correlation in the horizontal plane or to dipole sources in the plane.

Using the connection between Bessel functions and Hankel functions, we obtain

$$B_\omega(\vec{R},\vec{R}') = \frac{1}{4\pi}\int d\vec{\rho} B_q(\vec{\rho})\int_{-\infty}^\infty \xi d\xi H_0^{(1)}(\xi|\vec{r}-\vec{r}'-\vec{\rho}|)\bar{\Psi}(\xi, z, z_s)\bar{\Psi}(\xi, z', z_s) \tag{7.46}$$

In particular, for the spectral noise intensity

$$I_\omega(\vec{R}) = \frac{2}{\rho c}B_\omega(\vec{R},\vec{R}) = \frac{1}{2\pi\rho c}\int d\vec{\rho} B_q(\vec{\rho})\int_{-\infty}^\infty \xi d\xi H_0^{(1)}(\xi\rho)|\bar{\Psi}(\xi, z, z_s)|^2$$

$$= \frac{2}{\rho c}\int d\vec{\xi}\bar{B}_q(\vec{\xi})|\bar{\Psi}(\xi, z, z_s)|^2 \tag{7.47}$$

Equations (7.45)–(7.47) include the Cron–Sherman formulae as a subcase if, as the Green function $\bar{\Psi}(\xi, z, z_s)$ one uses the expression for an homogeneous half-space. The following transformation of equations (7.45)–(7.47) is associated with the modal decomposition of the Green function:

$$\bar{\Psi}(\xi, z, z_s) = \sum_l \frac{\psi_l(z)\psi_l(z_s)}{\xi^2 - \xi_l^2} + \int_{k_1}^{i\infty} d\eta \frac{\psi_l(z)\psi_l(z_s)}{\xi^2 - \eta^2} \tag{7.48}$$

which includes both the discrete and the continuous spectra.

Assuming that the main contribution to the noise field comes from the discrete spectrum (the validity of this assumption will be discussed below), that is, taking for the Green function $\bar{\Psi}(\xi, z, z_s)$, only the first sum in (7.48) and substituting in (7.46) we obtain, after integration in the complex plane, the expression for the cross-spectral density function:

$$B_\omega(\vec{R}, \vec{R}') = \frac{1}{4}\sum_{l,m} \int d\vec{\rho} B_q(\vec{\rho}) \frac{\psi_l(z)\psi_l(z_s)\psi_m(z')\psi_m(z_s)}{\xi_l^2 - (\xi_m^*)^2} \times \left[H_0^{(1)}(\xi_l|\vec{r} - \vec{r}' - \vec{\rho}|) - H_0^{(1)}(\xi_m^*|\vec{r} - \vec{r}' - \vec{\rho}|)\right] \quad (7.49)$$

which, as can be easily seen, depends on the difference in coordinates in the horizontal plane $\vec{r} - \vec{r}'$ and on the depths z and z' individually. Retaining only the largest terms in the sum – that is, where the denominator contains a difference of eigenvalues with the same indices $\xi_l^2 - (\xi_l^*)^2 = 2iq_l\gamma_l$ – we obtain the expression

$$B_\omega(\vec{r} - \vec{r}'; z, z') = \frac{1}{4}\sum_l \int d\vec{\rho} B_q(\vec{\rho}) \frac{\psi_l(z)\psi_l(z')\psi_l^2(z_s)}{q_l\gamma_l} J_0(q_l|\vec{r} - \vec{r}' - \vec{\rho}|) \quad (7.50)$$

As a rule, noise sources are assumed to be δ-correlated; that is,

$$B(\vec{\rho}) = q_\omega^2 N_0 \frac{\delta(\rho)}{2\pi\rho} \quad (7.51)$$

Carrying out the integration in (7.50) we obtain

$$B_\omega(\vec{r} - \vec{r}'; z, z') = \frac{q_\omega^2 N_0}{4} \sum_l \frac{\psi_l(z)\psi_l(z')\psi_l^2(z_s)}{q_l\gamma_l} J_0(q_l|\vec{r} - \vec{r}'|) \quad (7.52)$$

The vertical cross-spectral density function corresponding to the case $\vec{r} = \vec{r}'$ is

$$B_\omega(z, z') = \frac{q_\omega^2 N_0}{4} \sum_l \frac{\psi_l(z)\psi_l^2(z_s)\psi_l(z')}{q_l\gamma_l} \quad (7.53)$$

In turn, for the intensity of the noise, we have

$$I_\omega(z) = \frac{q_\omega^2 N_0}{2\rho c} \sum_l \frac{\psi_l^2(z)\psi_l^2(z_s)}{q_l\gamma_l} \quad (7.54)$$

We note that this intensity corresponds to taking into account only the contribution of the discrete spectrum. One can also obtain the intensity corresponding to the contribution of the continuous spectrum.

7.3 RELATIONSHIP BETWEEN THE CONTINUOUS AND THE DISCRETE COMPONENTS OF THE FIELD OF SURFACE NOISE SOURCES IN A WAVEGUIDE

One of the most important things in calculating the noise field is the nature of the role of the continuous spectrum in the formation of the field of surface sources. It is known that for a point source, the role of the continuous spectrum decreases rapidly as the receiver moves away from the source; in analysis, the role of the same field formed by both closely located and distant sources is not at all clear and requires detailed consideration. Clearly, if the sound source is directly above the receiver then the proportion of 'direct' rays is large and, accordingly, the contribution of the continuous spectrum may be large. As the 'sound-suppressing' region increases, the proportion from distant sources – the discrete spectrum – grows. Thus, the question of the role of the continuous spectrum is associated with the study of the field from a bounded region – for example, a sound-suppressing neighbourhood – depending on its radius. Similarly, there is a need to analyse the role of other factors: the frequency of the field, the refractive index in the bottom, and depth.

Of course, the role of the continuous spectrum depends on the specific problem. In some waveguide conditions this contribution is significant, whereas in others in can be neglected. It is intuitively clear that, for example, the contribution of the continuous spectrum must grow with the sound frequency, since this corresponds to an effective increase in the waveguide depth. It is almost impossible to give a general answer to the question of the role of the continuous spectrum, but we shall attempt to consider this question for typical shallow-water conditions and comparatively low frequencies.

According to the results of Kuperman and Ingenito (1980), the relative contribution of the discrete and continuous spectrum is determined by the losses in the bottom. If the losses are large, the 'near field' – the continuous spectrum – predominates, and in the case of a small loss in the bottom, the contribution of the 'far field' – the discrete spectrum – is dominant. Accordingly, in calculations of the noise field at frequencies up to 1 kHz, Buckingham (1980) takes into account only the discrete spectrum of the field. The role of the continuous spectrum was also treated in Hamson (1985), where the response of vertical and horizontal arrays to the noise field was analysed. There, it was noted that the contribution of the components is also determined by the sector of angles in the vertical directional diagram: near the horizontal direction (for small angles from the horizontal) the discrete component dominates, and near the vertical direction the continuous component dominates. It was also shown that there is a dependence on the absorptive properties of the bottom. All the results there were given for a frequency of 480 Hz. The frequency dependence of the relative contribution of the continuous and discrete fields to the noise was considered in Kuperman and Ferala (1985), where for a specific region, the effect of the main cause of noise – wind generated disturbance – was also considered. It was shown that for a comparatively low wind intensity at low frequencies (up to 200 Hz) the contribution of the discrete component dominates, whereas for high frequencies the main contribution comes from the continuous spectrum. A significant role in this analysis was played by the incorporation of the scattering of sound waves on surface waves. See also Schmidt and Kuperman (1986).

Therefore, let us consider the intensity of the noise field, taking into account both the discrete and the continuous spectra. Substituting the expression for the Green function in the form of a decomposition into waveguide modes (2.61), we obtain for the sound field an expression in the form of a sum of two components, corresponding to the discrete and the continuous spectrum:

$$S(r, z, z_s; \omega) = S^d(r, z, z_s; \omega) + S^c(r, z, z_s; \omega) \tag{7.55}$$

where

$$S^d(r, z; \omega) = \frac{i}{4} \sum_l \psi_l(z) \psi_l(z_s) \int Q_\omega(\vec{r}_s) H_0^{(1)}(\xi_l |\vec{r} - \vec{r}_s|) d\vec{r}_s$$

$$S^c(r, z; \omega) = \frac{i}{4} \int_{i\infty}^{k_1} \xi d\xi \psi_\xi(z) \psi_\xi(z_s) \int Q_\omega(\vec{r}_s) H_0^{(1)}(\xi |\vec{r} - \vec{r}_s|) d\vec{r}_s \tag{7.56}$$

In this notation, $|P(r, z, t)|^2/2\rho c$ is the intensity (energy flux density), and $B_\omega(\vec{R}, \vec{R})/2\rho c$ is the spectral density of the noise intensity. This quantity can be found by substituting (7.56) and (7.55) in (7.47), where we assume that the point of observation is at some depth on the z- axis, i.e. $\vec{r} = 0$. Then

$$I_\omega(z) = 2B_\omega(0, z; 0, z)/\rho c = \frac{q_\omega^2 N_0}{8\rho c} \int d\vec{r} \left| \sum_l \psi_l(z) \psi_l(z_s) + \int_{i\infty}^{k_1} \xi d\xi \psi_\xi(z) \psi_\xi(z_s) \right|^2 \tag{7.57}$$

If we neglect cross terms of the discrete and continuous spectrum, then the intensity of the noise field can be approximated in the form of the sum of the intensities of the discrete and continuous components

$$I_\omega(z) = I_\omega^d(z) + I_\omega^c(z) \tag{7.58}$$

where

$$I_\omega^d(z) = \frac{q_\omega^2 N_0}{8\rho c} \int d\vec{r} \left| \sum_l \psi_l(z) \psi_l(z_s) \right|^2$$

$$I_\omega^c(z) = \frac{q_\omega^2 N_0}{8\rho c} \int d\vec{r} \left| \int_{i\infty}^{k_1} \xi d\xi \psi_\xi(z) \psi_\xi(z_s) \right|^2 \tag{7.59}$$

Let us now assess the role of the continuous spectrum for the noise field in a simple waveguide model. We shall suppose that the medium in question is a Pekeris waveguide, where the absorption coefficient of sound in the bottom depends on frequency according to an exponential law (1.2). For the Pekeris mode, functions of the continuous spectrum were given in Section 2.2.

To determine the relative contribution of the continuous spectrum to the second integral in (7.59), the integral is calculated numerically by direct integration, and the intensity of the component given by the discrete spectrum was computed based on expression (7.54). Figure 7.3 shows $\delta_n(f) = I_f^d / I_f^c$ as a function of the sound frequency for waveguide parameters: $m_1 = 1.4$, $n_1 = c/c_1 = 0.83$, $\alpha = 0.025$, $H = 100$ m. Clearly, for low frequencies $f \leq 300$–500 Hz, we have $\delta_n \gg 1$, so that in this frequency region the continuous spectrum can be neglected. These results are in qualitative agreement with the analogous analysis of Kuperman and Ferla (1985), which also concluded that the role of the continuous spectrum grows with the sound frequency, and that naturally, the frequency boundary corresponding to the

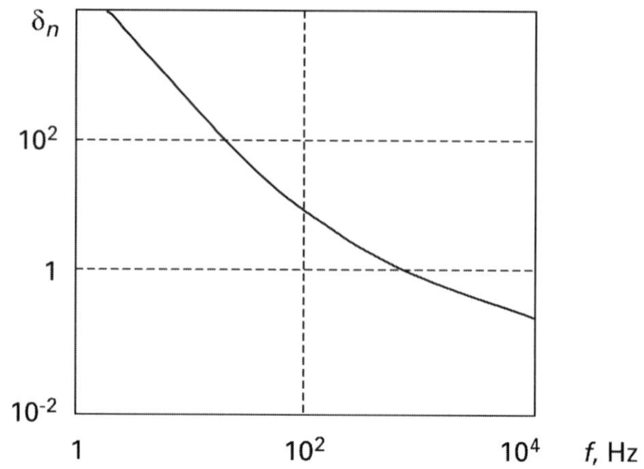

Figure 7.3. Frequency dependence of the relative contribution of the discrete spectrum.

dominance of a particular part of the spectrum depends on the specific conditions. A more precise analysis of the contributions of the continuous and discrete components of the intensity involves an analysis of the dimensions of the radiating region, corresponding to which a dependence on the dimensions and shape of this region is manifest.

It is also interesting to consider the relative contribution of the continuous spectrum as a function of the refractive index of the bottom n_1, which determines the number of trapped modes. It is intuitively clear that the closer n_1 is to unity, the fewer propagating modes there are and the smaller is the proportion of the discrete spectrum in the intensity. Naturally there is a significant frequency dependence, since the frequency also determines the number of trapped modes. This property is illustrated in Figure 7.4, which shows the change in δ_n as a function of the refractive index n_1 for some frequencies, for the above parameters.

7.4 EFFECT OF THE FREQUENCY DEPENDENCE OF THE BOTTOM ABSORPTION COEFFICIENT ON THE NOISE SPECTRUM

One of the most important characteristics of the noise field is its spectral intensity at the point of receipt, measured in decibels, relative to $1\,\mu\text{Pa}/\sqrt{\text{Hz}}$. Many works are devoted to this quantity, in both deep (Knudsen (1948), Ross (1976), Wenz (1962)) and shallow water (Kuperman and Ferla (1985), Pigott (1965)), where (as earlier in other works on this topic) the significant difference of the spectra of deep and shallow water is noted. This difference is manifest in the (usually) higher noise level in shallow water than in deep water in the region of comparatively low frequencies. At high frequencies there is an approximately identical drop-off with frequency of around 5 dB/octave. This apart, the difference is a far greater variability of shallow-water noise as a function of the environmental conditions (weather,

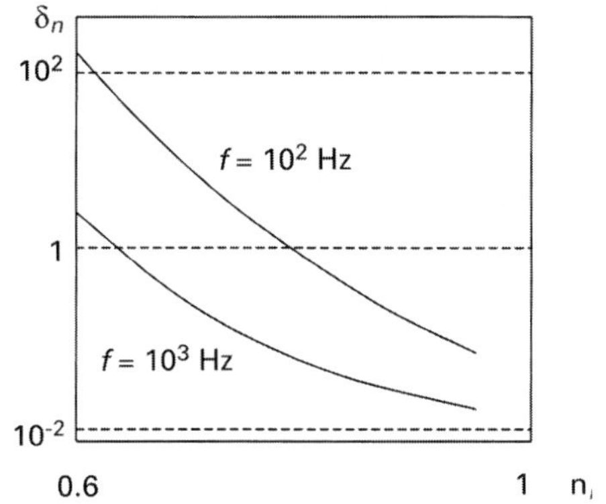

Figure 7.4. Dependence of the relative contribution of the discrete spectrum to the noise field on the refractive index in the bottom.

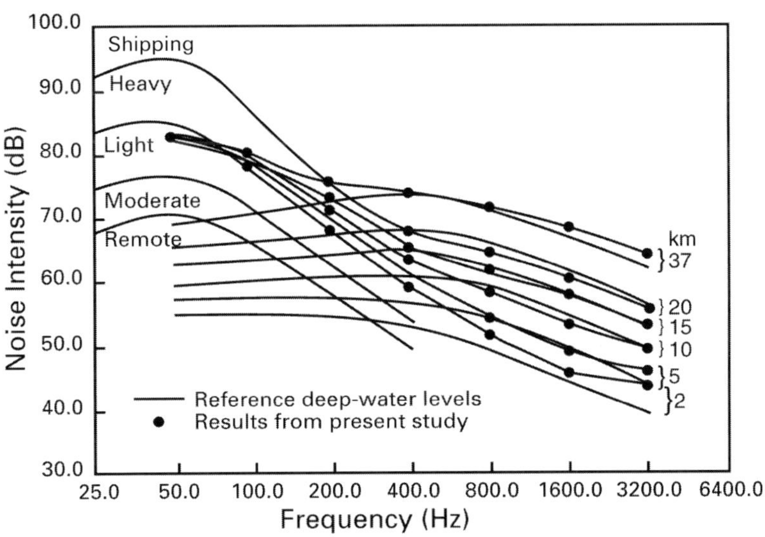

Figure 7.5. Noise spectra in deep (continuous curves) and shallow (shown by dots) water for various wind speeds, in comparison with analogous spectra in deep water (according to the results of Kuperman and Ferla (1985)).

geographical position, etc.). Figure 7.5 shows the results of experimental studies of the spectral intensity of the noise in a region of the Mediterranean Sea, given in Kuperman and Ferla (1985), in comparison with noise spectra in deep water based on the results of Ross (1976). Moreover, Kuperman and Ferla (1985), Crouch (1972) and Perrone (1969) studied the dependence of the spectral characteristics on the wind speed. The physically natural growth of the sound intensity due a wind-generated disturbance, with increasing wind speed v_ω, was demonstrated. Here, the dependence of the spectral intensity on the wind speed is exponential in nature,

$I_\omega = \text{constant} \cdot (v_\omega)^m$ with exponent m varying from 0.4 to 4 as a function of frequency and of the propagation conditions. Of course, the greater variability of the noise in shallow water in comparison with deep water is due to the greater effect of the bottom. Let us analyse, in particular, how the frequency dependence of the absorption coefficient (see, for example, Chapter 6) is manifest in the observed spectrum of the noise field. We shall consider this effect in a simple model of shallow water consisting of a water layer of thickness H with density ρ and sound speed $c(z)$, and a bottom in the form of an homogeneous fluid absorbing half-space with density ρ_1, and sound speed c_1, with absorption coefficient depending on frequency according to an exponential law $\alpha = \alpha_0(f/f_0)^q$, where the parameters α_0, f_0, q are determined by the bottom type. For example, according to Kuperman and Jensen (*ed.*) (1980), for sand $\alpha \approx 0.009 f^{1.78}$ dB/km, and for boulder clay $\alpha \approx 0.0065 f^{1.3}$ dB/km, where the frequency is measured in Hz.

If we assume that the effective sources are distributed at a small depth in comparison with the wavelength z_s, then the spectral intensity of the noise field in this case is determined by equation (7.54). We shall also assume that we are dealing with an homogeneous distribution of sources on the sea surface. The spectral intensity of the noise can be rewritten in the form

$$I_\omega(z) = \frac{q_\omega^2 N_0}{2\rho c} z_s^2 \sum_l \frac{\psi_l^2(z)[\psi_l'(0)]^2}{q_l \gamma_l} \tag{7.60}$$

Instead of the unknown parameter – the depth of the sources – we can introduce another parameter: the moment of the effective dipole source. In our language we shall be dealing with the spectral density of the dipole distribution; in other words, we have N_0 dipoles with moment $M_\omega = 2 z_s V_\omega$ per unit area (V_ω is the spectral component of the volume speed of an individual source). Such a model is often used, and in this case we may suppose that the noise field is formed by dipoles distributed on the sea surface.

$$I_\omega(z) = \frac{\rho \omega^2 N_0 M_\omega^2}{8c} \sum_l \frac{\psi_l^2(z)[\psi_l'(0)]^2}{q_l \gamma_l} \tag{7.61}$$

For the subsequent transformation of equation (7.61), we assume that we can use the WKB approximation for the eigenfunctions and the eigenvalues. We note that in connection with the formation of the field of the sources on the surface, the lower turning point in equations (2.73) will be the point $z = 0$. Thus

$$\psi_l(z) = 2\sqrt{\frac{q_l}{\sigma_l(z) D_l}} \sin \int_0^z \sigma_l(z')\,dz' \qquad \gamma_l \approx \frac{2 s \sigma_l(H)}{k D_l} \tag{7.62}$$

where s is determined by equation (2.87). For values of the eigenfunction at the point z_s, such that $z_s \ll \lambda$, we obtain

$$\psi_l(z_s) = 2\sqrt{\frac{q_l \sigma_l(0)}{D_l}} z_s \tag{7.63}$$

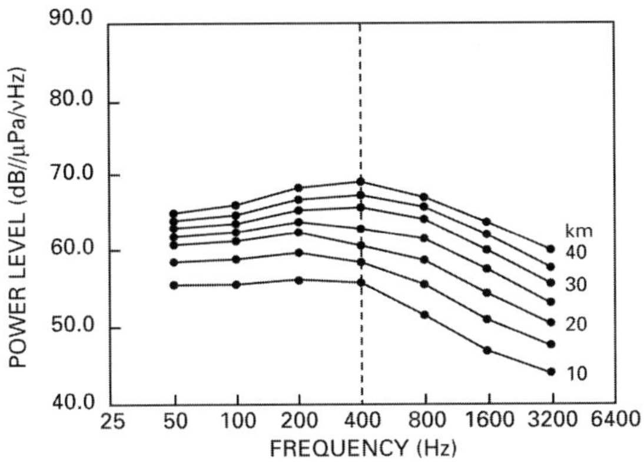

Figure 7.6. Spectral level of radiation in the frequency band from 25 Hz to 3.2 kHz as a function of wind speed (shown at the nodes) (based on the results of Kuperman and Ferla (1985)).

Equation (7.61) thus takes the form

$$I_\omega(z) = \frac{\rho\omega^2 k M_\omega^2 N_0}{2c_s} \sum_l \frac{q_l \sigma_l(0)}{\sigma_l(z)\sigma_l(H) D_l} \sin^2 \int_0^z \sigma_l(z')dz' \tag{7.64}$$

If we consider this expression as a function of frequency, then it is clear that the sum on the right-hand side of the equality depends weakly on frequency, so that the frequency dependence of the intensity is mainly determined by the function

$$G(f) = \frac{4\pi^3 \rho f^3 M_\omega^2 N_0}{c^2 s} \tag{7.65}$$

This factor contains both information about the sources (spectral level of the radiation) and information about the waveguide. As far as the spectral level of radiation is concerned, it is the result of a 'microscopic' mechanism, responsible for the radiation of noise (for example, wind-generated disturbances). This is the topic of separate research, both experimental (for example, Wolf (1982)) and theoretical, based on various model ideas about the radiation mechanism. In particular, Kuperman and Ferla (1985) presented the result from recovery of the spectral level of radiation based on measurement of the spectral intensity of the noise in the Mediterranean Sea. These results for q_ω are shown in Figure 7.6. Clearly, this quantity depends (naturally) on the wind strength and frequency, being very smooth in a broad frequency interval. It has a practically constant value and a small maximum at frequency $\sim$400 Hz, and the subsequent fall in the level in the region up to 1 kHz, amounts to at most 5 dB. If we take a similar dependence for the sources in our case, then carrying out the calculations according to equation (7.64), we obtain results for two types of bottom (sand and boulder clay) shown in Figure 7.7. The frequency is brought to a logarithmic scale with respect to a frequency of 30 Hz. Clearly, the nature of the fall in intensity with frequency

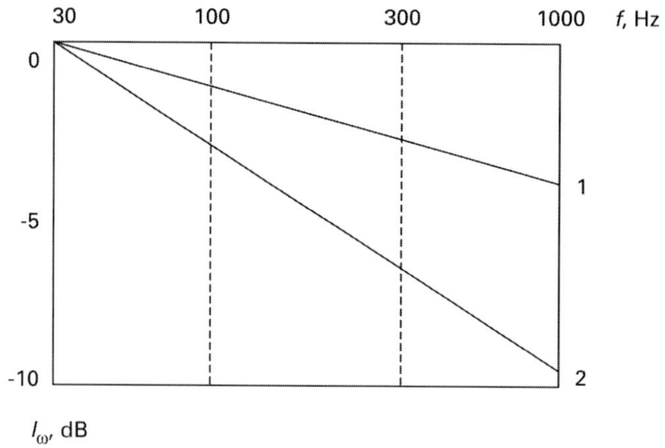

Figure 7.7. Frequency spectrum of the intensity relative to its level at frequency 31 Hz for various types of bottom: 1 – for boulder clay, 2 – for sand.

(slope of the curves) varies; this variation amounts to around 5 dB for frequencies $\sim$400 Hz, which is absolutely the experimentally recorded value.

7.5 VERTICAL DIRECTIVITY OF THE NOISE FIELD: EFFECT OF RANDOM INHOMOGENEITIES

One of the most important measurable characteristics of the noise field is the so-called directivity of the noise; that is, the power (intensity) flux density receivable per unit solid angle. In normalized form this quantity gives the directivity pattern of the noise field. In our formulation of the problem, when we assume anisotropic symmetry, the directivity pattern in the vertical plane is of interest. For the simplest Cron–Sherman model, this quantity can be obtained directly from equation (7.23). Assuming a solid angle element $d\Omega = 2\pi \sin\vartheta d\vartheta$, equation (7.23) can be rewritten in the form:

$$I(z) = \frac{1}{32\pi^2 \rho c} \int_0^{\pi/2} \frac{|g(\vartheta)|^2}{\cos\vartheta} d\Omega = \int I(\vartheta) d\Omega \tag{7.66}$$

where

$$I(\vartheta) = \frac{1}{32\pi^2 \rho c} \frac{|g(\vartheta)|^2}{\cos\vartheta} \tag{7.67}$$

is the intensity transported in a unit solid angle. Often, when it is required only to determine the form of this dependence, a normalised quantity is used (called the directivity pattern), where the nature of the normalisation is chosen in different ways, for example, at the maximum value or with unity taken to be at $\vartheta = 0$ or at $\vartheta = \pi/2$. From equation (7.67) it is clear that the directivity pattern of the noise in the

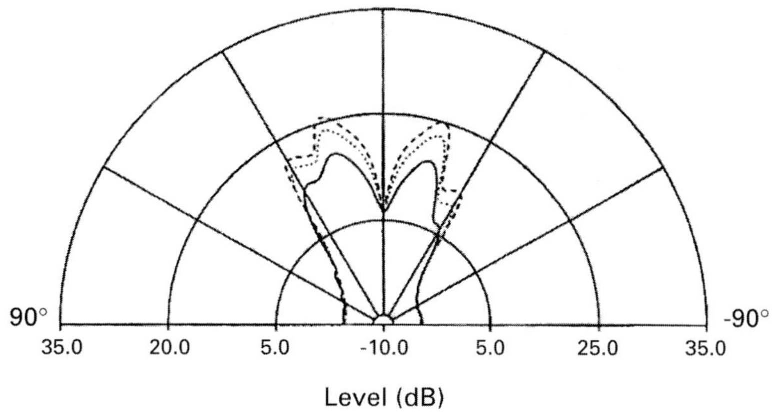

Figure 7.8. Angular dependence of the noise field intensity measured by a vertical array with aperture 8, 16 and 24 wavelengths, respectively, corresponding to the following curves: continuous – 16 hydrophones; dots – 32 hydrophones; dashes – 48 hydrophones (based on the results of Yang and Kwang (1997)).

Cron–Sherman model coincides with the directivity pattern of the source, modulo the cosine of the vertical angle. The appearance of the cosine is due to the angle at which a surface element of radiation is visible from the point of observation. This simple relationship between the directivity patterns of the received and the radiated signal is due to the fact that in our calculations, we have not taken into account either absorption or scattering in the medium, which, naturally, change the characteristics of the directivity on the propagation of sound. In a general formulation of the problem, one has to explain how the vertical angular distribution of the radiation from the sources is transformed as a result of oceanic factors (absorption, scattering, interaction with the bottom and with the surface, etc.) to the observed dependence of the intensity field on the vertical angle.

For problems in a waveguide, when the sound field is a sum of propagating modes, the angular distribution of intensity is replaced by the modal distribution, and here one can speak about the angular distribution, having in mind the corresponding Brillouin angles for the propagating modes. The vertical directivity of the noise is a very important and informative quantity, and therefore great attention is paid to its study, both experimental (Buckingham (1980), Buckingham and Jones (1987), Carbone *et al.* (1998), Cavanagh and Renner (1980), Aredov (1988)) and theoretical (Yang and Kwang (1997), Hamson (1985)). In particular, Buckingham and Jones (1987) presented experimental data from which it was clear that in the vertical plane the intensity distribution may have a maximum at angle zero from the horizontal. Yang and Kwang (1997) present results from measurement of the angular dependence of the intensity, some of which are shown in Figure 7.8. For the case of deep water, Cavanagh and Renner (1980) considered the connection between the vertical directivity and the vertical intensity distribution of the noise field in a ray approximation.

Certain properties of the angular distribution of the intensity can be easily understood from simple considerations; for example, the value of the angle χ_0 from the horizontal, at which the intensity is zero (becomes small), corresponds to the total reflection angle from the bottom, so that a knowledge of that dependence provides additional information about the properties of the bottom. In deep water, a technique for determining properties of the reflection coefficient from the bottom from the directivity characteristics of the noise field was proposed and realised in Aredov and Furduev (1994). Carbone *et al.* (1998) undertook a detailed study of the effect of the waveguide parameters on the vertical directivity for shallow water. They showed, in particular, that the source directivity characteristics and properties of the bottom have the greatest effect on the noise directivity, where one intrinsic special point in the angular dependence of the intensity is given by the value of the total reflection angle. Absorption in the bottom in the presence of shear waves also has a significant effect. These factors change the angular dependence of the coefficient of reflection from the bottom and thus also the angular characteristics of the received noise. The sound speed profile affects the noise directivity to a lesser extent. However, its effect is highly typical. One can qualitatively analyse (in the language of rays) the properties of the angular distribution for different types of sound speed profiles: for a negative gradient (near-bottom channel) and for a positive gradient (near-surface). In the first case, rays emanating from sources on the surface tend towards the bottom, and at the point or receipt, there are no rays propagating horizontally (with zero grazing angle). This corresponds to the presence of a trough in the angular dependence (noise notch effect). In the case of a surface channel, rays emanating from sources tend towards the surface, and horizontally directed rays occur at the point of receipt. For certain values of the waveguide parameters and characteristics of the sources, this may even lead to the appearance of a maximum at zero. Corresponding numerical examples, indicating the possibilities for determining parameters of the bottom from the angular characteristics of the noise are given in Carbone *et al.* (1988). Hamson (1985) presents numerical calculations of the characteristics of the array directivity for various environmental conditions in the case of a pressure-release surface and a solid bottom, for summer and winter sound speed profiles at a fixed frequency $f = 480$ Hz.

In addition to the sound speed profiles, the parameters of the bottom and the characteristics of the sources, the various mechanisms for scattering of sound on propagation (random inhomogeneities in the ocean depth, and also roughnesses of the bottom and the surface) are important. These factors are incorporated in a general formulation of the problem with scattering in Kuryanov (1984) within the framework of the radiation transfer equation (RTE). Based on numerical analysis, Kuznetsova (1989) gives an interpretation of experimental data obtained in Buckingham and Jones (1987), taking into account both attenuation in the bottom and absorption in the water, and also scattering on a disturbed ocean surface.

In this section, in a shallow-water model, we have obtained a number of equations, determining how the redistribution of the angular intensity takes place as a consequence of angle-dependent absorption in the bottom and scattering on random inhomogeneities.

We shall consider the following waveguide model. Suppose we have a water layer of thickness H, with sound speed $c(z)$ and density ρ, lying on an homogeneous fluid half-space (c_1, ρ_1, α). The sound speed in the water layer is a random function of coordinates and time. $c = \langle c \rangle + \delta c$, which produces equation (4.8) for the refractive index. In the (r, z) coordinate system the received is located at the point $(0, z)$. We are interested in the angular distribution of the intensity of sound, arriving from the whole of the waveguide surface. When there are no inhomogeneities, this distribution can be found from equation (7.54), and it is clear that the modal intensity is proportional to the expression $q_\omega^2 N_0 \psi_l^2(z_s)/2\rho c q_l \gamma_l$, which, in the WKB approximation and assuming a dipole nature of the noise sources, gives $2N_0 M_\omega^2 q_l \sigma_l(0)/\rho c D_l \gamma_l$. Let us further consider the formation of the noise field by a distributed source, taking into account random inhomogeneities. In these discussions, we have posited the diffusion approach. Our analysis will be somewhat simplified, with the assumption of azimuthal symmetry, taking into account only the discrete spectrum (in other words, the far field is assumed to be dominant). In this case, for the spectral component of the sound field at the point $(0; z)$, we shall use expression (7.19), where as the Green function we take a random function given as a decomposition into modes:

$$S(0, z, z_s; \omega) = \int Q_\omega(\vec{r}) \sum_l c_l(r) \frac{\psi_l(z) \exp[i(q_l r + \pi/4)]}{\sqrt{8\pi q_l r}} d\vec{r} \tag{7.68}$$

where the mode amplitudes $c_l(r)$ are random functions of the coordinates. Corresponding stochastic equations can be obtained for these, from which, after averaging procedures, one obtains equations for the quantities $W_l(r) = \langle |c_l(r)|^2 \rangle$, proportional to the intensity transported by the mode with index l.

On the assumption (7.51) that the sources in the plane are δ-correlated, we obtain for the vertical correlation function

$$B_\omega(z, z') = \frac{q_\omega^2 N_0}{4} \sum_l \int_0^\infty dr\, W_l(r) \frac{\psi_l(z)\psi_l(z')}{q_l} \tag{7.69}$$

When there are no random inhomogeneities, $W_l(r) = \psi_l^2(z_s)\exp(-\gamma_l r)$, and from equation (7.69) we obtain equation (7.53). The spectral energy density (mean square pressure) or the average intensity is also expressed in terms of $W_l(r)$:

$$I_\omega(z) = \frac{q_\omega^2 N_0}{2\rho c} \sum_l \int_0^\infty dr\, W_l(r) \frac{\psi_l^2(z)}{q_l} \tag{7.70}$$

A system of equations for determining $W_l(r)$ was also given in Chapter 4, where the value of $W_l(0)$ for a monopole was given for the 'initial' conditions. If we consider the dipole as source, then $W_l(0) = 2q_l\sigma_l(0)/D_l$. Assuming a sufficiently large number of propagating modes, we can move from a discrete variable (the mode index) to a continuous variable and obtain the diffusion equation (4.46) for $W_p(r)$. The diffusion coefficient a_p is determined by the waveguide parameters and by random inhomogeneities. For example, when we have a waveguide of constant depth, with a constant average sound speed, instead of the mode index, we consider the continuous variable p: $0 \leq p \leq p_{\max} \sim kH/\pi$. The value of $p_{\max}$ is determined by the angle of total internal reflection from the bottom χ_c. The connection between p and the physical natural quantity, the grazing angle of a sound ray with the bottom, is determined in the simplest case of a rigid bottom by the relations:

$$p = \frac{kH}{\pi} \sin \chi \quad p_{\max} = \frac{kH}{\pi} \sin \chi_c \tag{7.71}$$

In the case of a continuous variable, the summation formulae of (7.52)–(7.54) are replaced by integrals, where the value of $W_p(r)$ is determined from equations (4.46) with the corresponding initial conditions. Assuming dipole sources in a waveguide with constant sound speed, $W_p(0) = (2\pi^2 z_s^2/H^3)p^2$. For the boundary conditions, we must take

$$\left.\frac{\partial W_p(r)}{\partial_p}\right|_{p=0} = 0 \quad W_p(r)\big|_{p=p_{\max}} = 0 \tag{7.72}$$

In the calculation we also perform averaging over the position of the receiver, which gives a smooth dependence on depth for the intensity.

Thus, equation (7.70) gives the total intensity of the noise field at depth z. The integral analogue of this formula has the form

$$I_\omega(z) = \frac{q_\omega^2 N_0}{\rho c} \int_0^{p_{\max}} dp \int_0^\infty dr \frac{W_p(r)}{\sigma_p(z) D_p} \tag{7.73}$$

Based on this formula, the angular distribution of the intensity, in which we are interested, has the form

$$I_\omega(p, z) = \frac{q_\omega^2 N_0}{\rho c} \int_0^\infty \frac{W_p(r)}{\sigma_p(z) D_p} dr = \frac{q_\omega^2 N_0}{\rho c} \frac{Y_p(z)}{\sigma_p(z) D_p(z)} \tag{7.74}$$

where

$$Y_p = \int_0^\infty W_p(r) dr \tag{7.75}$$

We shall consider the value of Y_p as a function of the parameter p (or the angle in the vertical plane). We note that Y_p satisfies an ordinary differential equation, which can be obtained by integrating equation (4.47) over r from 0 to ∞, assuming $W_p(\infty) = 0$:

$$\frac{d}{dp}\left(a_p \frac{dY_p}{dp}\right) - \gamma_p Y_p + W_p(0) = 0 \tag{7.76}$$

The 'initial' conditions for Y_p have the form

$$\left.\frac{dY_p}{dp}\right|_{p=0} = Y_p\big|_{p=p_{\max}} = 0 \tag{7.77}$$

We then obtain concrete results within the model, when $\langle c \rangle$ = constant and correspondences for the unperturbed functions and the eigenvalues (i.e. without taking random inhomogeneities into account) are obtained in the Pekeris model. We shall further suppose that the noise field is due to uniformly distributed point dipoles (or monopoles located close to the surface). We shall also assume, as in Section 4.5, that the propagation of the radiation in the waveguide takes place in a small sector of grazing angles, so that the diffusion coefficient for these values can be assumed to be independent of p: $a_p = a =$ constant. The loss coefficient corresponding to the Pekeris model was given in (2.88); thus, the equation for the function Y_p has the form

$$a\frac{d^2 Y_p}{dp^2} - \frac{s\pi^2}{k^2 H^3} p^2 Y_p = -\frac{2\pi^2 z_s^2}{H^3} p^2 \tag{7.78}$$

with the initial conditions (7.77). The solution of this equation is expressed in terms of a Bessel function of an imaginary argument. According to (7.74) for $I_\omega(p)$, one can obtain the expression (for the Pekeris model, independent of depth)

$$I_\omega(p) = \frac{\rho\omega^3 N_0 M_\omega^2}{4Hsc^2}\left[1 - \sqrt{\frac{p}{p_{\max}}}\frac{I_{-1/4}(\hat{\varepsilon}\pi^2 p^2/k^2 H^2)}{I_{-1/4}(\hat{\varepsilon}\pi^2 p_{\max}^2/k^2 H^2)}\right] \tag{7.79}$$

where $\hat{\varepsilon} = (k/2\pi)\sqrt{sH/a}$.

Since, in what follows, we shall be interested only in the form of the vertical directivity of the noise, we represent equation (7.78) in the form:

$$I_\omega(p) = I_\omega(0)F(p) \tag{7.80}$$

where the function $F(p)$ is normalised so that $F(0) = 1$ defines the vertical directivity of the noise field. If in equation (7.78) we move to the argument of the angle from the horizontal χ, according to equation (7.71), then for the directionality function we obtain the expression

$$F(\chi) = \frac{C}{\cos\chi}\left[1 - \sqrt{\frac{\sin\chi}{\sin\chi_0}}\frac{I_{-1/4}(\hat{\varepsilon}\sin^2\chi)}{I_{-1/4}(\hat{\varepsilon}\sin^2\chi_0)}\right] \tag{7.81}$$

where

$$C = \left\{1 - 2^{1/4}\Gamma(3/4)\left[(\hat{\varepsilon}\sin^2\chi_0)^{1/4} I_{-1/4}(\hat{\varepsilon}\sin^2\chi_0)\right]^{-1}\right\}^{-1}$$

where $\Gamma(\ldots)$ is Euler gamma-function.

It is clear that the nature of the dependence on the angle χ is determined by the parameter $\hat{\varepsilon}$, including both the geometric parameters of the channel and the absorption coefficient γ and the diffusion coefficient, which takes effective account of random inhomogeneities. Depending on the values of the parameters of the problem, the coefficient ε ranges from zero to infinity.

Let us consider the limiting cases. If $\hat{\varepsilon} \to 0$, which corresponds to the situation when absorption is small in comparison with scattering (diffusion of modes), then

$$F(\chi) = \begin{cases} 1/\cos\chi(1 - \sin^4\chi/\sin^4\chi_0) & \chi < \chi_0 \\ 0 & \chi > \chi_0 \end{cases} \tag{7.82}$$

The other limit $\hat{\varepsilon} \to \infty$ corresponds to large absorption or sometimes a small effect of random inhomogeneities. Here the limiting case is the dependence

$$F(\chi) = \begin{cases} 1/\cos\chi & \chi < \chi_0 \\ 0 & \chi > \chi_0 \end{cases} \tag{7.83}$$

We note that the intensity (7.83) is non-zero at $\chi = 0$, although the intensity of an individual dipole source tends to zero as $\chi \to 0$. This ultimate value is obtained as a result of integration over the whole surface, taking into account the nature of the attenuation. In other model situations, the results will be different.

In the general case, an arbitrary value of the parameter $\hat{\varepsilon}$ is found between the above limiting cases. The figure shows a number of dependences $F(\chi)$. The dependences corresponding to the parameter values $\hat{\varepsilon} = 0.1$ and $\hat{\varepsilon} = 100$ practically coincide with the limiting cases.

The qualitative nature of the dependence obtained in (7.81) corresponds to the numerical results obtained in Kuznetsova (1989): in the absence of inhomogeneities there is a minimum in the region of small angles χ. This is due to the minimum in the angular distribution of the intensity of the radiation of an individual source. In the presence of random inhomogeneities there is a jump in the energy in the region of small grazing angles. As the concentration of random inhomogeneities increases (growth of the diffusion coefficient), the minimum becomes less pronounced and for sufficiently large values of a becomes a maximum.

7.6 EFFECT OF THE SOUND SPEED PROFILE ON THE DEPTH DEPENDENCE OF THE NOISE FIELD INTENSITY

Experimental data from measurement of the vertical distribution of the intensity of low-frequency noise (Morris (1978), Lyubchenko and Petnikov (1991)), together with theoretical studies of this question (Kuperman and Ingenito (1980), Weston (1980), Klyachin and Kuryanov (1981)) attest to the dependence of average values of the noise level on depth, associated with changes in the sound speed profile.

The best known measurements of the depth dependence of the noise intensity in deep water are given in Morris (1978). The data presented there have served as material for a theoretical analysis in various works (Arase (1967), Weston (1980)). Based on an averaged approach in the framework of ray theory, the depth distribution of the intensity was considered in Weston (1980), where it was shown that in the presence of a deep-water sound channel the 'law of conjugate depths' is satisfied; that is, the intensities are identical at depths with the same sound speed up to the so-called critical depth – the depth at which the sound speed is equal to the sound speed on the surface. The maximum value of the noise intensity occurs somewhat below the critical depth, and the minimum value of the intensity occurs on the channel axis. These general laws are also described by theory (Arase (1967)) and agree with calculated data (Tonoyan (1991)). The spatial distribution of the noise intensity was considered in Klyachin and Kuryanov (1981) in the framework of the radiation transfer equation, where, in particular, agreement with Morris's experiment was also obtained.

The question of the connection between the depth dependence of the noise intensity and the sound speed profile in shallow water has been studied theoretically by Kuperman and Ingenito (1980), who considered three types of sound speed profile: constant, linear with a negative gradient, corresponding to a near-bottom sound channel, and near-surface sound channel. The curves of intensity against depth, in their nature, approximately restate the sound speed profile in the waveguide, with a constant and linear depth dependence. As the frequency increases, the slope of the curves increases. For a near-surface waveguide, the nature of the dependence is different, and the maximum intensity value occurs on the surface, in the region with minimum sound speed. Qualitatively, this behaviour can be understood from the following arguments. Naturally, in a waveguide with constant sound speed, all the modes excited by sources on the surface have on average a constant amplitude with depth and thus the intensity distribution with depth will be approximately constant, apart from in areas near the bottom and the surface, where boundary conditions play a role. If the sound speed profile has a negative gradient, only some of the modes (rays) are excited by the surface sources; namely, only those which graze the surface. Each of these modes has a smooth envelope (for example, in the WKB method, this envelope is proportional to $[\sigma(z)]^{-1/2}$, where $\sigma(z)$ is the transverse wave number). This envelope decreases towards the bottom for all modes, leading to a decrease in the intensity on summation. Near-bottom modes, which might contribute near to the bottom, are not excited by surface sources. (A relative increase in the noise intensity near the bottom, and even the occurrence of a maximum there, may arise in the presence of a mechanism for the delivery of energy from excited near-surface waves to near-bottom waves; for example, scattering on random inhomogeneities.) For a near-surface channel, the maximum intensity value on the surface is again due to the presence of near-surface modes (rays) excited by surface sources.

Let us now consider experimental results from measurement of the vertical profile of the noise field intensity in shallow-water conditions in the Barents Sea. A detailed description of the conditions of the experiment was given in Lyubchenko and Petnikov (1991).

The depth dependence of the intensity was obtained for measurement of the noise field on a vertical drifting array of length 40 m, comprising 11 equidistantly positioned receivers (hydrophones). The upper end of the array was positioned at a depth of 40 m. As one moved to the left or the right of the point of observation, the sound speed profile changed somewhat, exhibiting a close to linear dependence. The slope of the bottom in the given region was insignificant and amounted to less than $0^\circ.007$.

The wind speed at the time of measurement was 7.6 m/s. The rate of drift of the array was approximately 2 knots, which led to a deviation of the array from the vertical by approximately 30°. Measurement of the average slope angle was performed using two depth sensors located at the ends of the receiving array. Figure 7.10 shows graphs of the vertical distributions of the average values of the

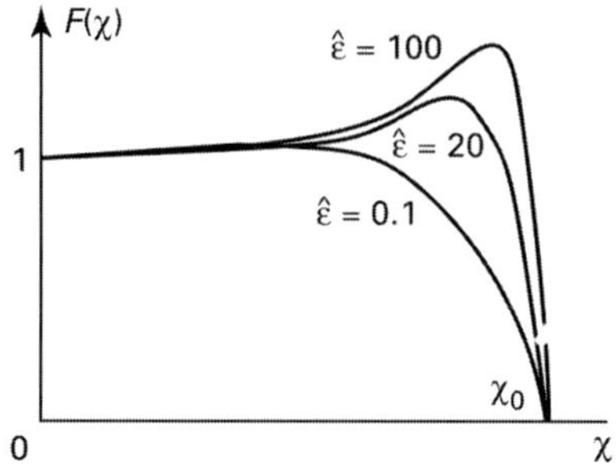

Figure 7.9. Dependence of the noise intensity I on the angle from the horizontal (grazing angle). Values of the parameter ε are shown in the figure, χ_0 corresponds to the total reflection angle.

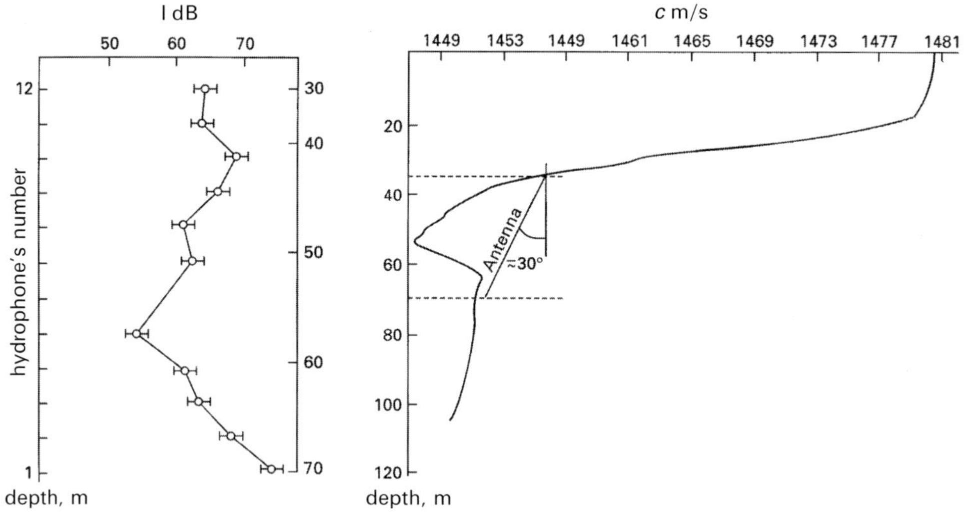

Figure 7.10. Sound speed profile in the region of the experiment of Lyubchenko and Petnikov (1991), and corresponding measured distribution of the noise intensity, averaged in the band 200–600 Hz.

noise level and the corresponding sound speed distribution with drift for drift of the vertical array in the region of the sea banks in the area where the underwater sound channel is formed. The noise field was averaged in the 200–600 Hz band. The averaging time for acquisition of the average spectral noise levels was 20 min, and the number of signals averaged was 60.

Taking into account the inclination of the array, comparison of the vertical sound speed profiles and profiles of the average noise intensity shows that minimum values of the sound speed profile correspond to minimum values in the vertical distribution of the noise. Similar measurements of the vertical distribution of the average intensity values of the noise field were performed in shallow-water regions, where there is no underwater sound channel and the sound speed profile

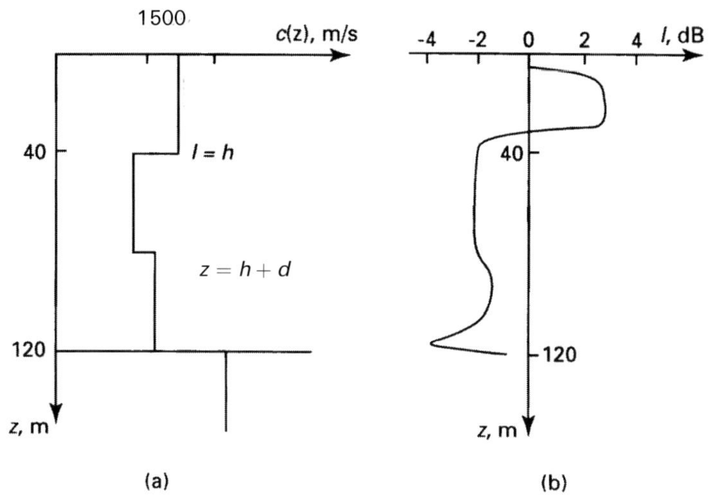

Figure 7.11. Model sound speed profile (*a*) and corresponding intensity distribution, averaged in the band 200–600 Hz.

is characterised by a small negative gradient. In this case, with experimental error, the vertical distribution of the intensity of the noise was monotonic and practically constant.

For the interpretation of the experimental dependence of the noise intensity let us consider a model of a regular (constant depth) waveguide with the model sound speed profile shown in Figure 7.11:

$$c(z) = \begin{cases} c = 1480\,\text{m/s} & 0 < z < h \\ c' = 1445\,\text{m/s} & h < z < h + d \\ c'' = 1450\,\text{m/s} & h + d < z < H \end{cases} \tag{7.84}$$

We further suppose that the layer with this sound speed profile lies on a fluid homogeneous absorbing half-space with corresponding parameters c_1, ρ_1, α. To calculate the intensity we use equation (7.60) with summation of the modal intensities. In this case the eigenfunctions have the form:

$$\psi(z) \begin{cases} \sin \sigma z & 0 < z < h \\ B \sin \sigma'(z - h) + C \cos \sigma'(z - h) & h < z < h + d \\ G \sin \sigma''(z - h - d) + E \cos \sigma''(z - h - d) & h + d < z < H \\ V \exp[-\kappa(z - H)] & z > H \end{cases} \tag{7.85}$$

where

$$B = \frac{\sigma}{\sigma'} \cos \sigma h \quad C = \sin \sigma h \quad G = \frac{\sigma'}{\sigma''}(B \cos \sigma' d - C \sin \sigma' d)$$

$$E = B \sin \sigma' d + C \cos \sigma' d \quad V = G \sin \sigma''(H - h - d) + E \cos \sigma''(H - h - d)$$

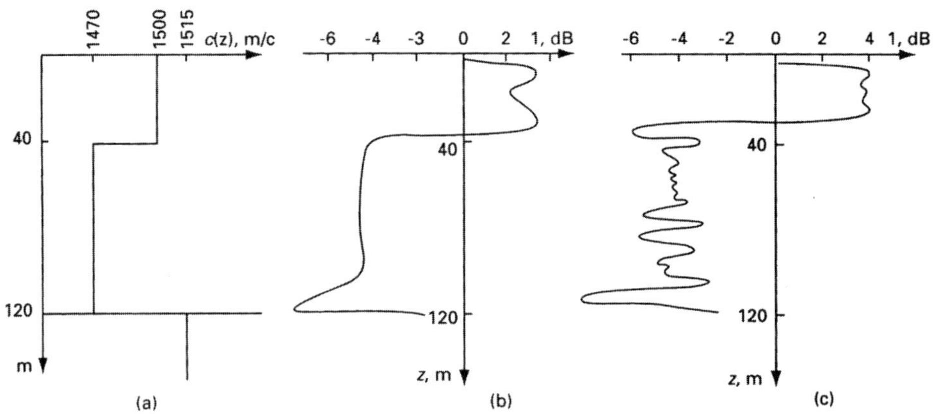

Figure 7.12. Model sound speed profile (*a*) and intensity distribution at frequency 300 Hz (*c*) and averaged in the band 200–600 Hz (*b*).

$\sigma, \sigma', \sigma''$ are the transverse wavenumbers in the corresponding regions of depth of the waveguide: $\xi = \sqrt{k^2 - \sigma^2} = \sqrt{k'^2 - \sigma'^2} = \sqrt{k''^2 - \sigma''^2}$, and the normalisation coefficient is found from the condition (2.33).

The eigenvalues are found from the transcendental equation obtained from the continuity condition

$$\tan \sigma''(H - h - d) + \frac{(m\sigma''/\kappa)R + (\sigma''/\sigma')T}{R - (\sigma''/\sigma')T} = 0 \tag{7.86}$$

where $R = (\sigma/\sigma') - \tan \sigma h \cdot \tan \sigma' d$, $T = (\sigma/\sigma') \tan \sigma' d + \tan \sigma h$.

Based on this model, numerical calculations of the depth dependence of the noise field intensity, averaged over frequency in the band 200–600 Hz, were performed. The dependence $I(z)$, averaged in the band 200–600 Hz, for a chosen sound speed profile and bottom parameters $c_1 = 1{,}525\,\text{m/s}$, $\rho_1 = 1.7\,\text{g/cm}^3$, $\alpha = 0.015$ is shown in Figure 7.11. Clearly, the intensity curve approximately restates the sound speed profile. A qualitative explanation of the nature of this behaviour was given earlier. It is interesting to note that scattering on random inhomogeneities, which may lead to redistribution of the intensity across the modes, and, correspondingly, by depth, did not occur in this case. In particular, this relates to scattering on surface waves. This is explained by the fact the near-bottom modes (not grazing the surface), which must play a significant role in the redistribution of the field, are not excited on the surface and, hence, do not interact with near-surface modes (which are excited).

Let us further consider how the change in sound speed profile with depth affects the averaged noise intensity for a negative (overall) gradient. We shall compute the intensity of the noise in a model example of a waveguide with sound speed profile represented by a two-step function (Figure 7.12(*a*)). This figure shows the vertically averaged distribution of the intensity (Figure 7.12(*b*)) and also the non-averaged

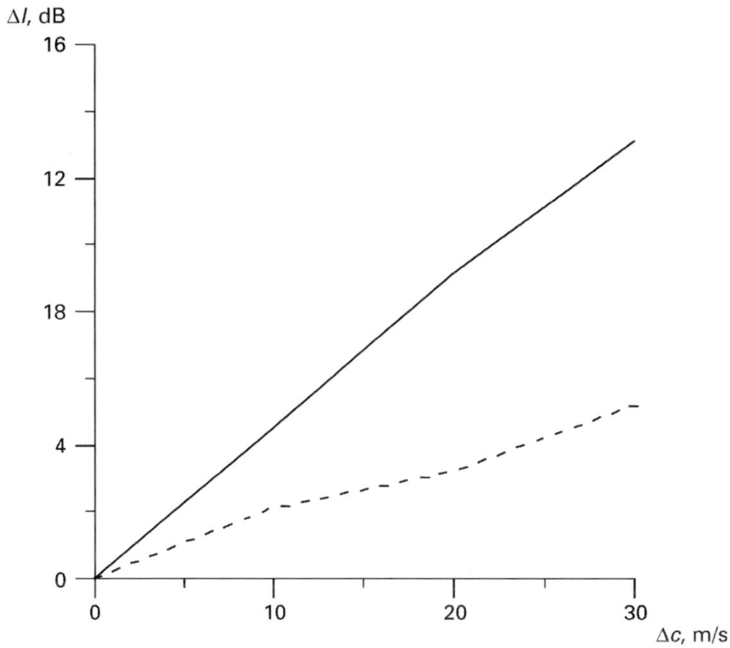

Figure 7.13. Dependence of the jump in the averaged noise intensity on the fall in sound speed. Curve 1 corresponds to $c_1 = 1{,}515\,\text{m/s}$, curve 2 to $c_1 = 1{,}530\,\text{m/s}$.

dependence of the intensity at frequency $f = 300\,\text{Hz}$ (Figure 7.12(*c*)). As we see in Figure 7.12, the jump in the sound speed $\Delta c = c - c'$ at depth $z = h$ corresponds to a change in the averaged intensity ΔI. (In our case $\Delta c \sim 30\,\text{m/s}$ at depth 40 m, and the corresponding jump is $\Delta I \sim 6\,\text{dB}$). Let us determine the value of this change as the difference between the maximum value I_{max} above the level $z = h$ and the minimum value below that level, The quantities I_{max} and I_{min} are determined far from the waveguide boundary $\Delta I = I_{\text{max}} - I_{\text{min}}$. The dependence of ΔI on the jump in the sound speed Δc is shown in Figure 7.13 for two different values of the sound speed in the bottom. Clearly, the change in the noise field intensity at the level corresponding to jump in the sound speed (intensity) will be greater the less the sound speed in the bottom differs from the sound speed on the surface of the waveguide. The parameters in the calculation are: $c = 1,500\,\text{m/s}$, the continuous curve corresponds to $c_1 = 1,505\,\text{m/s}$, the dashed curve to $c_1 = 1,515\,\text{m/s}$.

Thus, results from the study of noise given in this chapter demonstrate the possibility of using measurements of noise characteristics to determine geoacoustic parameters of shallow water (generally speaking averaged over sufficiently large water areas).

Thus, in this chapter, we have formulated models and methods for describing the properties of the noise of surface sources (wind-generated waves) in shallow water, which properties are determined by the wave-guide nature of the propagation and are largely unrelated to the generation mechanism. The main factors determining the shallow-water properties of the acoustic noise field itself are the sound-speed profile, the given near-bottom or near-surface waveguide, and also the absorption properties of the bottom. According to the general approach, the theoretical description and interpretation of the results are performed in the language of modes, in a development of the analysis of Chapter 4, the main quantity determining the space–time variability of the noise field is the distribution function of the intensity with respect to modes. In conclusion we note that certain resultts from the study of waves presented in this chapter show that the measured characteristics of the noise can be used to determined the geoacoustic parameters of shallow water (generally speaking averaged over sufficiently large water areas). The results given are particularly important for the study of the scattering of sound on a localized object (Chapter 8), since the scattered signal must always be recorded and identified against the background of noise, mainly due to surface waves.

8

Diffraction of acoustic waves on localised inhomogeneities in shallow-water sound channels

8.1 FORMULATION OF THE PROBLEM AND GENERAL EQUATIONS

In analysing the propagation of sound in shallow water it is impossible not to turn attention to the properties of the diffraction of sound waves on localised inhomogeneities in sound channels. The recent topicality of this problem is associated with the development of systems for acoustic monitoring of cetaceous mammals in shallow-water gulfs and bays, and also of acoustic tomography systems for monitoring drifting large-scale ice floes and icebergs, submerged fragments from shipwrecks, etc. It is well known that this internal boundary problem of diffraction in waveguides is extremely complex from the mathematical point of view, and its solution involves the use of particular approximation methods. (See, for example Shenderov (1989), Ingenito (1987), Uzunoglu *et al.* (1982), Kravtsov *et al.* (1984), Sarkissian (1997), Gorskaya *et al.* (1988), Bostrom (1984), Hackman *et al.* (1988), Kvyatkovskii (1988), Dawson *et al.* (1990), Tuzhilkin *et al.* (1995)). The detailed study of these methods goes far beyond the scope of this book, and here we shall consider only one such (Kravtsov *et al.* (1984), Sarkissian (1997)) which, in our view, is among the most transparent from the physical point of view, in which the problem is divided into two stages: the first stage is the solution of the diffraction problem in an homogeneous space on bodies of a complex shape, and the second is the solution of the problem of the propagation of sound in waveguides, including smoothly irregular waveguides. We note that in this method, one can graphically demonstrate the diffraction of sound waves in shallow-water sound channels.

Suppose a waveguide comprises a water layer of depth H, characterised by a density and a sound speed profile, $\rho(z)$ and $\rho(z)$, respectively, lying on an absorbing bottom, which can be characterised by various parameters (density, sound speed, attenuation coefficients of acoustic waves). In other words, we suppose that we are

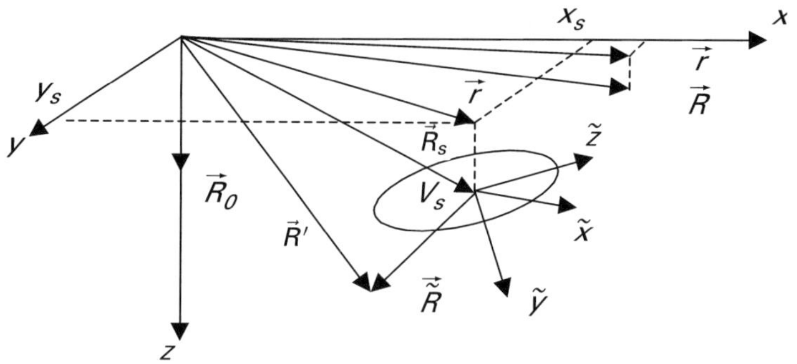

Figure 8.1. Coordinate system used to describe the diffraction of acoustic waves in waveguides.

dealing with a regular waveguide with a constant stratification.[1] As in Chapter 2, we introduce the coordinate system associated with the waveguide (see Figure 8.1), and suppose that a point source is located at the point $(0, 0, z_0)$, and a receiver at the point of observation (x, y, z). Let us also suppose that in this waveguide, there is a scatterer, occupying a spatial volume V_s and characterised by a set of acoustic parameters; for example, density ρ_s and sound speed[2] $c_s(\vec{R})$. We introduce coordinate system $(\tilde{x}, \tilde{y}, \tilde{z})$ associated with the scatterer, whose orientation is due to the symmetries and position of the scatterer. The origin of this coordinate system is placed at a point symbolically called the centre of the scatterer (x_s, y_s, z_s). In what follows, for brevity, this point will be referred to as the coordinates of the scatterer, and its radius vector will be denoted by $\vec{R}_s$. Correspondingly, the radius vector of an arbitrary point in the coordinate system, associated with the scatter, will be denoted by $\vec{\tilde{R}}$. In addition, let us define the radius vectors in the horizontal plane, respectively $\vec{r}_s$ and $\vec{r}$ for the positions of the scatterer and the receiver ($|\vec{r}| \equiv r$, $|\vec{r}_s| \equiv r_s$).

Let us now consider the field of a pure-tone point source of frequency (see p. 11), in connection with which, in what follows, we shall omit the time part of the dependence $e^{-i\omega t}$. For the complex amplitude function $P(\vec{R})$ of the sound field we have the equation

$$[\Delta + K^2(\vec{R})]P(\vec{R}) = -A\delta(\vec{R} - \vec{R}_0) \tag{8.1}$$

where

$$K(\vec{R}) = \begin{cases} k(z) & \vec{R} \notin V_s \\ k_s(\vec{R}) & \vec{R} \in V_s \end{cases} \qquad k_s(\vec{R}) = \frac{\omega}{c_s(\vec{R})}$$

A is a coefficient depending on the power of the source, and is given by equation

[1] Here, we have assumed a constant stratification of the waveguide, solely to simplify the formulae. The algorithm given below can be easily transferred to the case of a smoothly irregular waveguide, as in Section 2.6.

[2] This theory admits the introduction of a more complex model of the scatterer.

(2.40). Here, on the boundary (surface $\tilde{S}$) of the scatterer, the sound field must satisfy the conditions:

$$\left[P_+(\vec{R}) = P_-(\vec{R}), \frac{1}{\rho}\frac{\partial P_+(\vec{R})}{\partial n} = \frac{1}{\rho_s}\frac{\partial P_-(\vec{R})}{\partial n}\right]_{\vec{R}\in\tilde{S}} \tag{8.2}$$

where $P_\pm$ is the amplitude of the field, inside and outside the surface of the scatterer, respectively.

One common approach to the scattering problem is to subdivide the complex amplitude of the sound field at the point of receipt $P(\vec{R})$ into the 'incident' field $P_0(\vec{R})$ and the 'scattered' field $P_s(\vec{R})$

$$P(\vec{R}) = P_0(\vec{R}) + P_s(\vec{R}) \tag{8.3}$$

where, by the incident field, we mean the field formed by the source in the absence of the scatterer. In our formulation of the problem, when the source is a pure-tone point source, this field will be $P_0(\vec{R}) = A\Psi(\vec{R}, \vec{R}_0)$, where $\Psi(\vec{R}, \vec{R}_0)$ is the Green function, being the solution of equation (1.13) with corresponding boundary conditions on the bottom and at the surface. The main methods for calculating the scattered sound field are associated with various integral representations (or equivalent integral equations) (see Hönl (1961), Taylor (1972)). We shall derive one of these below. We introduce the operators

$$\hat{L} = \Delta + K^2(\vec{R}), \quad \hat{L}_0 = \Delta + k^2(z), \quad \hat{L}_s = \Delta + k_s^2(\vec{R}) \tag{8.4}$$

For the operator $\hat{L}$ and arbitrary functions u, v, we have Green's formula for some region of space V:

$$\int_V (u\hat{L}v - v\hat{L}u)dV = \int_S \left(u\frac{\partial v}{\partial n} - v\frac{\partial u}{\partial n}\right)dS \tag{8.5}$$

where S is the surrounding surface. If we perform the integration over the whole infinite space, then because of the decrease in the field at infinity, the surface integral and, hence, the right-hand side of equation (8.5), must be zero. Further, let us take for the function v the Green function $\Psi(\vec{R}', \vec{R})$, and as the function u the desired amplitude of the sound field P, which is also written in the form of a function of two coordinates $P = P(\vec{R}', \vec{R}_0)$. After subtracting the region V_S occupied by scatterer from the volume integral, and taking into account (2.2), (8.1) and also the relations $\hat{L}_s P|_{\vec{R}'\in V_s} = 0$ and $\hat{L}_0 \Psi|_{\vec{R}'\in V_s} = 0$, which are valid in view of the absence of scatterers in the volume V_s, we obtain:[3]

$$P(\vec{R}, \vec{R}_0) = A\Psi(\vec{R}, \vec{R}_0) + \int_{V_s} P(\vec{R}', \vec{R}_0)\hat{L}_s\Psi(\vec{R}', \vec{R})d\vec{R}' \tag{8.6}$$

[3] In deriving equation (8.6) we have also used the invertibility of problem (2.2) in relation to the position of the sound source and receiver; i.e. the relation $\Psi(\vec{R}_0, \vec{R}) = \Psi(\vec{R}, \vec{R}_0)$.

or, in a rather different form, without giving the dependence of the sound field amplitude on the source coordinates explicitly:

$$P(\vec{R}) = A\Psi(\vec{R}, \vec{R}_0) + \int_{V_s} P(\vec{R}')\delta k^2(\vec{R}')\Psi(\vec{R}', \vec{R})d\vec{R}' \tag{8.7}$$

where $\delta k^2(\vec{R}) = k_s(\vec{R}) - k^2(z)$. Thus, the scattered field according to (8.3) is equal to

$$P_s(\vec{R}) = \int_{V_s} P(\vec{R}')\delta k^2(\vec{R}')\Psi(\vec{R}', \vec{R})d\vec{R}' \tag{8.8}$$

We note that the field with complex amplitude $P(\vec{R})$, satisfying the equation (8.1) is proportional to the Green function of the problem in the presence of a scatterer, and in this sense, equations (8.6) and (8.7) are integral equations of the Lippman–Schwinger type in scattering theory (Taylor (1972)). If we perform the integration solely over the region not occupied by the scatterer, then from equation (8.5) we obtain an integral equation of Helmholtz type (see, for example, Hönl (1961)), including an integral over the surface of the scatterer. This last form of the integral equation was used in Sarkissian (1994, 1997).

Equations (8.6) and (8.7) can be solved in perturbation theory assuming smallness of the scattered field in comparison with the incident field, in which case one can obtain a Born series, as often used in scattering theory. If we introduce into the analysis the so-called T-operator, then we can write down the corresponding equation for the T-matrices, which is also used in certain works (Bostrom (1984), Hackman *et al.* (1988), Kvyatkovskii (1988)).

We proceed as follows. Without making any assumptions about the structure of the scatterer, we introduce the function $\nu(\vec{R}) = P(\vec{R})\delta k^2(\vec{R})$, which we shall assume is to be defined. For the scattered field, from equation (8.8) we obtain the relation

$$P_s(\vec{R}) = \int_{V_s} \nu(\vec{R}')\Psi(\vec{R}, \vec{R}')d\vec{R}' \tag{8.9}$$

which can be interpreted as an integral equation assigning the function $\nu(\vec{R}')$, which has the sense of the density of secondary fictitious sources, distributed in the volume of the scatterer[4]. To compute the unknown function $\nu(\vec{R}')$, the boundary conditions (8.2) for the field on the surface $\hat{S}$ are used.

For the diffraction problem in a waveguide, the system of equations (8.2) is often solved numerically (see, for example, Sarkissian (1994)) in an approximation of the unknown function $\nu(\vec{R}')$ as being piecewise constant. This approach is based on the assumption that in a neighbourhood of the scatterer the medium varies smoothly, and that in this neighbourhood, we can assume that the Green function $\Psi(\vec{R}, \vec{R}')$ coincides with the Green function $\Psi(\vec{R}, \vec{R}')$ for free space (2.3). In this case, the function $\nu(\vec{R}')$ is defined based on the solution of the problem of the scattering of a plane wave of unit amplitude on the object itself in an homogeneous space. The

[4] Starting from the Helmholtz integral equation, we can introduce an analogous surface density of the fictitious sources.

density of the sources in the scattering of the plane wave (we shall denote it by $\nu_0(\vec{k}_i, \vec{R}')$, where $\vec{k}_i$ is the wave vector of the incident wave) depends on the orientation of the scatterer with respect to the incident wave. According to Kravtsov (1984), to determine $\nu_0(\vec{k}_i, \vec{R}')$ we use equations (8.7) and (8.8), where for the incident wave we take a plane wave of unit amplitude. The boundary conditions (8.2) in this case give the following system of equations:

$$\left[\exp(i\vec{k}_i\vec{R}) + \int_{V_s} \nu_0(\vec{k}_i, \vec{R}')\Psi_0(\vec{R}, \vec{R}')d\vec{R}'\right]_{\vec{R}\in\tilde{S}} = P_-(\vec{k}_i, \vec{R})\Big|_{\vec{R}\in\tilde{S}}$$

$$\frac{1}{\rho}\frac{\partial}{\partial n}\left[\exp(i\vec{k}_i\vec{R}) + \int_{V_s} \nu_0(\vec{k}_i, \vec{R}')\Psi_0(\vec{R}, \vec{R}')d\vec{R}'\right]_{\vec{R}\in\tilde{S}} = \frac{1}{\rho_s}\frac{\partial}{\partial n}P_-(\vec{k}_i, \vec{R})\Big|_{\vec{R}\in\tilde{S}} \tag{8.10}$$

Here we note that for scattering of a plane wave in an homogeneous infinite space, the scattered sound wave in the far zone for large $|\vec{R}_s|$ and $\vec{R} - \vec{R}_s|$ ($|\vec{R}_s| \gg L^2/\lambda$, $|\vec{R} - \vec{R}_s| \gg L^2/\lambda$ (L is the characteristic size of a scatterer), satisfies the Sommerfeld radiation condition. According to (8.9) for a wave of unit amplitude, it can be written in the form

$$P_s(\vec{R}) = \int_{\tilde{\vec{R}}'\in V_s} \nu_0(\vec{k}_i, \vec{R}')\frac{\exp(ik|\vec{R} - \vec{R}'|)}{4\pi|\vec{R} - \vec{R}'|}d\vec{R}'$$

$$\cong \left[\frac{1}{4\pi}\int_{\tilde{\vec{R}}\in\tilde{V}_s} \nu_0(\vec{k}_i, \tilde{\vec{R}})\exp(-i\vec{k}_s\tilde{\vec{R}})d\tilde{\vec{R}}\right]\frac{\exp(ik|\vec{R} - \vec{R}_s|)}{|\vec{R} - \vec{R}_s|} \tag{8.11}$$

where $\vec{k}_s$ is the wave vector of the scattered wave, $\vec{R}' = \vec{R}_s - \tilde{\vec{R}}$, $|\tilde{\vec{R}}| \ll |\vec{R}_s|$, $|\tilde{\vec{R}}| \ll |\vec{R} - \vec{R}_s|$. The first factor in equation (8.11), enclosed in square brackets, is the scattering amplitude $F(\vec{k}_i, \vec{k}_s)$:

$$F(\vec{k}_i, \vec{k}_s) = \frac{1}{4\pi}\int_{\tilde{\vec{R}}\in V_s} \nu_0(\vec{k}_i, \tilde{\vec{R}})\exp(-i\vec{k}_s\tilde{\vec{R}})d\tilde{\vec{R}} \tag{8.12}$$

which is one of the main diffraction characteristics of the scattering body. For solids of revolution with an homogeneous structure and a relatively simple shape – for a example, a sphere, or a spheroid – solutions of the diffraction problem in an homogeneous infinite medium are known, and analytic expressions have been obtained for the scattering amplitude. For bodies with a more complex structure and shape with given wave dimensions, this characteristic is measured experimentally for free space. We note that the scattering amplitude can be expressed in terms of the angles θ and φ in the spherical coordinate system associated with the scatterer, which determine the direction of arrival of the incident (index i) and the scattered (index s) plane wave $F(\vec{k}_i, \vec{k}_s) \equiv F(\theta_i, \varphi_i, \theta_s, \varphi_s)$.

We shall further use the modal description of the sound field in the waveguide (see Section 2.1), and to determine the function $\nu(\vec{R})$ and thereby the scattered field, we find the incident field near the scatterer, where we assume that the medium is homogeneous and has a low absorption. For the functions appearing

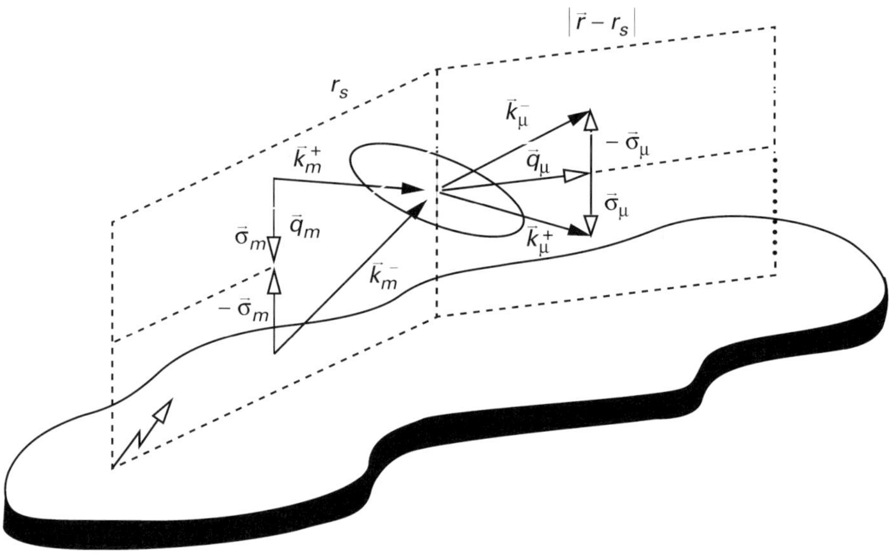

Figure 8.2. Wave vectors in the problem of scattering in a waveguide.

in the expression for the incident wave $(\vec{R}_s + \vec{R}'')$ in the form (2.41), for small changes in the arguments, denoted by a double prime, we can obtain:

$$\psi_m(z_s + z'') \cong a_m^+ \exp(i\vec{\sigma}_m \vec{R}'') + a_m^- \exp(-i\vec{\sigma}_m \vec{R}'')$$

$$\frac{\exp[i\xi_m|\vec{r}_s - \vec{r}''|]}{\sqrt{\xi_m|\vec{r}_s - \vec{r}''|}} \cong \frac{\exp[iq_m|\vec{r}_s - \vec{r}''|]}{\sqrt{q_m|\vec{r}_s - \vec{r}''|}} \exp(-\gamma_m|\vec{r}_s - \vec{r}''|/2)$$

$$\cong \frac{\exp[iq_m r_s]}{\sqrt{q_m r_s}} \exp[i\vec{q}_m \vec{R}''] \exp(-\gamma_m r_s/2) \tag{8.13}$$

Then

$$P_i(\vec{R}_s + \vec{R}'') = \sum_m a_m \psi_m(z_s + z'') \frac{\exp[iq_m|\vec{r}_s - \vec{r}''|]}{\sqrt{q_m|\vec{r}_s - \vec{r}''|}} \exp[-\gamma_m|\vec{r}_s - \vec{r}''|/2]$$

$$\cong \sum_m b_m(r_s) \left[a_m^+ \exp(i\vec{k}_m^+ \vec{R}'') + a_m^- \exp(i\vec{k}_m^- \vec{R}'')\right] \tag{8.14}$$

where $a_m = \sqrt{\rho_0 c_0 W_0} e^{i\pi/4} \psi_m(z_0)$, $\vec{\sigma}_m$ is a vector directed along the z-axis with value $\sigma_m = \sqrt{k^2(z_s) - q_m^2}$, $a_m^{\pm} = \frac{1}{2}[\psi_m(z) \pm (d\psi_m/dz)/i\sigma_m]_{z=z_s}$, $\vec{k}_m^{\pm}$ are the wave vectors with horizontal and vertical components, respectively, $(\vec{q}_m, \pm\vec{\sigma}_m)$ (see Figure 8.2), $b_m(r_s) = a_m[\exp(iq_m r_s)/\sqrt{q_m r_s}] \exp(-\gamma_m r_s/2)$.

Substituting the incident field (8.14) in condition (8.2) for determining the unknown function $\nu(\vec{R})$, and taking into account that $P_+ = P_i + P_s$, we obtain

$$\nu(\vec{R}') = \sum_m b_m(r_s)[a_m^+ \nu_0(\vec{k}_m^+, \vec{R}') + a_m^- \nu_0(\vec{k}_m^-, \vec{R}')] \tag{8.15}$$

Then computing the scattered field (8.9) with a Green function for the waveguide, in which we made an approximation for the eigenvalues in the region of the scatterer analogous to (8.13), we obtain for the scattered field at the point of receipt:

$$P_s(\vec{r}, z) = \sqrt{\frac{i}{8\pi}} \sum_{\mu} b_\mu(r_s)\psi_\mu(z) \frac{\exp[iq_\mu|\vec{r} - \vec{r}_s|]}{\sqrt{q_\mu|\vec{r} - \vec{r}_s|}} \exp[-\gamma_\mu|\vec{r} - \vec{r}_s|/2] \tag{8.16}$$

where

$$b_\mu(r_s) = \sum_m S_{\mu m} b_m(r_s) \tag{8.17}$$

where $S_{\mu m}$ is the scattering matrix, equal to:

$$S_{\mu m} = 4\pi\left[a_m^+ a_\mu^+ F(\vec{k}_m^+, \vec{k}_\mu^-) + a_m^+ a_\mu^- F(\vec{k}_m^+, \vec{k}_\mu^+) + a_m^- a_\mu^+ F(\vec{k}_m^-, \vec{k}_\mu^-) + a_m^- a_\mu^- F(\vec{k}_m^-, \vec{k}_\mu^+)\right] \tag{8.18}$$

where μ is the index of a mode of the scattered field.

This approach is valid under the conditions that the multiple scattering effects are small and that the medium is homogeneous in a neighbourhood of the scatterer. In other words, the scattering body must be far from the waveguide boundary at distance h, greater than its horizontal dimensions L, ($h \geq L$), and the changes in the vertical wave vector $\vec{\sigma}_m$ along the z-axis at a distance equal to the vertical dimension d must be small; i.e. $d < L_{\sigma_m} \equiv \sigma_m/[d\sigma_m/dz]$. If we take into account the fact that $\sigma_m \cong m\pi/H$ (this expression is satisfied for the Pekeris model with a pressure-release bottom, see (2.50)), and $L_{\sigma_m} \cong \{\sigma_m^2/[k|dk/dz|]\}_{z=z_s}$, then the last condition can be rewritten in the form:

$$d < \frac{c\lambda^2}{4H^2|dc/dz|} \tag{8.19}$$

Thus, equations (8.16)–(8.18) give a method of determining the scattered field in the waveguide based on a knowledge of the characteristics of the scatterer (scattering amplitude) in the free space and the characteristics of the waveguide (eigenvalues and eigenfunctions). Of course, even after one has successfully separated the problem of the scattering on the object from that of the propagation of sound in the waveguide, in a real waveguide this problem can only be solved numerically, for both components, in both the determination of the scattering amplitude and the construction of the Green function for the waveguide. In addition, in the framework of the problem studied below, we can also compute in analytic form the characteristics of the field scattered on some object in the waveguide. We have in mind the important subcase (for monitoring) of the formulation of the diffraction problem, when a scatterer (Sc) crosses a stationary acoustic track, moving in the (x, y)-plane along the normal ($\hat{\gamma} = 90°$) to the baseline joining the sound source (So) and the receiver (R) (see Figure 8.3). We shall further suppose that the scatterer is located in the far zone with respect to the sound source and receiver and is a thin planar rectangular screen of width L and height d (horizontal size 'in the (x, y)-plane' and vertical size 'along the z-axis', respectively). We shall also assume that a

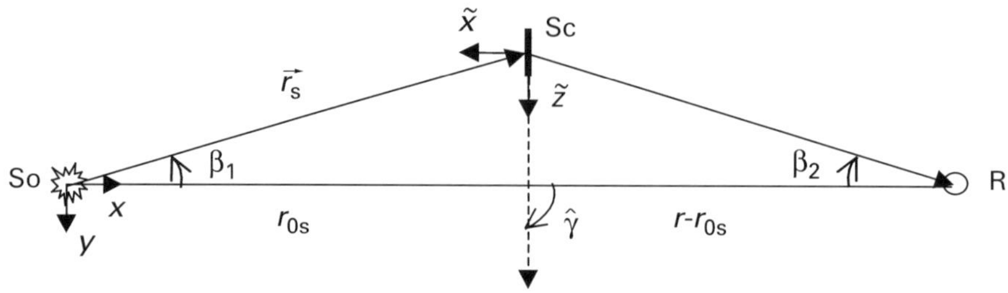

Figure 8.3. Relative position of source (So), receiver (R) and scatterer (Sc) in a model of the diffraction problem.

pure-tone signal is radiated and the wavelength of the sound is related to the dimensions of the scatterer as follows:

$$L \gg \lambda \cong d \tag{8.20}$$

In this case, we may suppose that the scattering body, as it were, contracts to a straight line section L, and that the density of the secondary fictitious sources, distributed on the screen, $\nu(\vec{k}_i, \vec{R}')$ is proportional to the incident quasi-plane wave[5]

$$\nu(\vec{k}_i, \vec{R}') = A'B\frac{\exp(i\vec{k}_i\vec{R}')}{k_iR'} \cong A'\frac{\exp(i\vec{k}_i\vec{R}_s)}{k_iR_s}\nu_0(\vec{k}_i, \vec{\tilde{R}}) \tag{8.21}$$

Here $\nu_0(\vec{k}_i, \vec{\tilde{R}}) = B\exp(i\vec{k}_i\vec{\tilde{R}})$, B and A' are coefficients of proportionality, depending on the acoustic properties of the body and the power of the source, respectively. Substituting (8.21) in (8.12) and using the evident relations for the infinitely thin screen (see Figure 8.3), $\vec{\tilde{R}} = (0, \tilde{y}, \tilde{z})$, $|\vec{R}_s| \equiv R_s \cong r_{0s}$, $|\vec{R} - \vec{R}_s| \cong r - r_{0s}$, valid for $y_s \ll r_{0s}$, $z_0 = z_s$, we obtain:

$$\begin{aligned} F(\vec{k}_i, \vec{k}_s) &\cong \frac{1}{4\pi}\int_{-L/2}^{L/2} \exp\left(ik\tilde{z}\frac{y_s}{r_{0s}}\right)\exp\left(ik\tilde{z}\frac{y_s}{r - r_{0s}}\right)d\tilde{z}\int_{-d/2}^{d/2} d\tilde{y} \\ &= \frac{1}{4\pi}BLd\frac{\sin(\pi y_s L/F^2)}{\pi y_s L/F^2} \end{aligned} \tag{8.22}$$

where r_{0s} is the distance between the sound source and the point at which the scatterer crosses the stationary track, and F is the radius of the first Fresnel zone, $F = \sqrt{\lambda r_{0s}(r - r_0)/r}$. As can be seen from (8.22), the modulus $|F(\vec{k}_i, \vec{k}_s)|$ in this case is described by a dependence of type $\sin\Omega/\Omega$ (this fact was first mentioned in Gorskii *et al.* (1995)) and one can employ an analogy with the study of long linear arrays; i.e. compare the scattering amplitude to the directivity characteristic of an array in the form of a line segment. The density of the fictitious sources in this case can be treated as the complex amplitude distribution of the sound field with respect to the array aperture with phase compensation in the direction of the wave incidence. In this connection, we note that long scattering solids of revolution, differing in shape from

[5] This statement stems from the known Huygens–Fresnel principle.

a rectilinear screen and corresponding to the condition (8.20), can also be put in correspondence with a linear array, albeit with an even more complex amplitude distribution (Kuzkin (1997)). For example, for a spheroid the corresponding modulus of the amplitude distribution has the form $[1-(2\tilde{z}/L)^2]$. Of course, here the scattering amplitude will be described by an equation more complex than (8.22). However, as shown in the theory of linear arrays, the angular width of the directivity pattern (the scattering amplitude in our case) increases insignificantly.

Turning to scattering in shallow-water waveguides, we note that in the above approach the diffraction of acoustic waves can be treated as a rescattering of waveguide modes of the incident sound field to waveguide modes of the scattered field. As mentioned earlier in the book (see, for example, Section 5.1) for shallow water for large distances between receiver and source in the sound channel there actually exist only a few low modes with approximately equal eigenvalues. This corresponds to satisfaction of the inequalities $|\vec{k}_1^{\pm} - \vec{k}_M^{\pm}| \ll k$, $|\sigma_M^{\pm}/\vec{q}_M| \ll 1$, where M is the number of such modes. Here, for the subcase considered earlier in conditions of waveguide propagation, the scattering amplitude appearing in expression (8.18) in a first approximation is independent of the indices of incident and scattered modes and is described to within a constant by the expression[6] $\sin(\pi y_s, LF^{-2})/\pi y_s LF^{-2}$.

We further note that in acoustic monitoring it is usual to recorded perturbations of the sound field ΔP due to the presence of a scatterer $\Delta P = |P_0 + P_s| - |P_-|$. For $|P_0| \gg |P_s|$, where the value of ΔP is equal to:

$$\Delta P \cong \frac{P_0 P_s^* + P_0^* P_s}{2|P_0|} \tag{8.23}$$

Substituting P_0 in the form (2.41) and P_s (8.16) in (8.23) we obtain for the given subcase of scattering in a waveguide when $q_{l,m,\mu} \gg \gamma_{l,m,\mu}$:

$$\begin{aligned} \Delta \cong \sqrt{\rho_0 c_0 W_0} BLdC(r, z_0, z) \frac{\sin(\pi y_s L/F^2)}{\pi y_s L/F^2} \\ \times \sum_{l,m,\mu} E_{l,m,\mu}(r, r_s, z_0, z_s, z) \cos(q_m |\vec{r}_s| + q_\mu |\vec{r} - \vec{r}_s| - q_l r) \end{aligned} \tag{8.24}$$

Here

$$\begin{aligned} C(r, z_0, z) = \Bigg(& \sum_l \frac{\psi_l^2(z_0)\psi_l^2(z)}{q_l r} \exp(-\gamma_l r) \\ & + \sum_l \sum_{m=l+1} \frac{\psi_l(z_0)\psi_l(z)\psi_m(z_0)\psi_m(z) \exp(-(\gamma_m + \gamma_l) r/2)}{(q_m q_q)^{1/2} r} \cos((q_m - q_l) r) \Bigg)^{-1/2} \end{aligned} \tag{8.25}$$

[6] We note that here we assume isotropic scattering in the plane $\tilde{x}\tilde{y}$ in the band of angle defined by small Brillouin grazing angles of propagating low modes, and also associated with a small vertical screen size d.

$$E_{l,m,\mu}(r, r_s, z_0, z_s, z) = \sqrt{8\pi}\psi_l(z_0)\psi_l(z)\psi_m(z_0)\psi_m(z_s)\psi_\mu(z_s)\psi_\mu(z) \times \exp(-(\gamma_l r + \gamma_m r_{0s} + \gamma_\mu(r - r_{0s}))/2) \times (q_l q_m q_\mu r r_{0s}(r - r_{0s}))^{-1/2} \tag{8.26}$$

It is significant that the amplitude of the perturbations of the sound field (8.24) is described by the expression $\sin(\pi y_s LF^{-2})/\pi y_s LF^{-2}$. This means that for a given rate of movement of the screen ν_s the greatest values of ΔP occur in the immediate neighbourhood of the baseline and their duration T is equal to:

$$T = \frac{2F^2}{L\nu_s} \tag{8.27}$$

We also note here that when equation (8.24) is normalised and averaged with respect to the depth of the source, the receiver and the scatterer and, additionally, averaging over distances r and r_s in ranges greater than the period of the interference beats, for $y_s = 0$ and a rigid bottom, we arrive at the following equation[7]:

$$\left\langle \frac{\Delta P}{|P_0|} \right\rangle \sim \frac{2S_1}{S_F} = \frac{Ld}{HF} = \frac{Ld\sqrt{r}}{H\sqrt{\lambda r_{0s}(r - r_{0s})}} \tag{8.28}$$

where S_1 is the area of the shaded outline of the screen, and S_F is the area of the first Fresnel zone. According to (8.28), the perturbations of the sound field grow as the scatterer nears the sound source or receiver, as usually happens with Fresnel diffraction.

It is also important to identify the properties of the phase behaviour of the perturbation ΔP. In conditions of propagation of few modes, which are typical for shallow water, the phases of all harmonic components in (8.24) are approximately equal and equal to the phase of the total perturbation φ_s:

$$q_m|\vec{r}_s| + q_\mu|\vec{r} - \vec{r}_s| - q_l r \cong k\frac{y_s^2}{2r_{0s}} + k\frac{y_s^2}{2(r - r_{0s})} = \frac{\pi y_s^2}{F^2} = \varphi_s \tag{8.29}$$

In this case, on movement of the screen the recorded perturbation of the sound field is a signal with a frequency which varies according to the law

$$\omega = \left|\frac{d\varphi_s}{dt}\right| = \frac{2\pi\nu_s^2}{F^2}|t - t_0|$$

(Gorskii *et al.* (1995)). (Here, t_0 is the time at which the baseline is crossed). Thus, the recorded perturbation $\Delta(t)$ is a signal with symmetric linear frequency modulation and an envelope of type $\sin\Omega/\Omega$. According to (8.24), this signal can be written in

[7] Equation (8.28) can be obtained in a simpler way (Gorskii *et al.* (1995)), based on the Huygens–Fresnel principle.

the following form:

$$\Delta P(t) = \hat{A}\frac{\sin(\pi\nu_s(t-t_0)L/F^2)}{\pi\nu_s(t-t_0)L/F^2}\cos(\pi\nu_s^2(t-t_0)^2/F^2+\varphi_0) \tag{8.30}$$

Here, $\hat{A}$ and φ_0 are constant values of the signal amplitude and phase.

8.2 SCATTERING OF THE SOUND FIELD BY A SOFT SPHEROID IN THE WAVEGUIDE

Based on the equations obtained in the previous section, we shall consider further the more general case of solution of the problem of diffraction on a body located in a shallow-water waveguide. Here, we shall not make special approximations, associated with the propagation of small number of modes, we shall not choose a planar screen as the scatterer, and we shall impose only quite soft constraints on the dimensions of the body (8.19). As far as the movement of the body is concerned, we shall not restrict ourselves to the case in which it crosses a stationary track at right angles. In this case it is not possible to obtain simple analytic expressions to describe the scattered field, and the problem can only be solved numerically. As the scattering object we choose an acoustically soft spheroid, drawn out in the horizontal plane, with minor axis d and major axis L. The choice of the model of the scatterer is due to the fact that, on the one hand, the problem of diffraction on a spheroid admits a rigorous analytic solution (Babilov *et al.* (1998)), while on the other hand, this model best describes the scattering of sound on the bodies of sea mammals and, in particular, on the bodies of blue whales. In other words, we shall attempt to model diffraction perturbations of the sound field caused by these mammals, when they cross the stationary track between the sound source and receiver in a shallow-water waveguide. The sound source is assumed to be pure tone.

According to Babilou *et al.* (1998) the scattering amplitude of a spheroid in free space is equal to:

$$F(k_m^{\pm}, k_\mu^{\pm}) = \frac{2i}{k}\sum_{n=0}^{\infty}\sum_{l=n}^{\infty}\varepsilon_n S_{nl}(\chi, \cos\theta_m^{\pm})S_{nl}(\chi, \cos\theta_\mu^{\pm})$$
$$\times\frac{m_s^{-1}R_{nl}^{(1)}(\chi,\vartheta)R_{nl}^{(1)\prime}(\chi_s,\vartheta)-R_{nl}^{(1)\prime}(\chi,\vartheta)R_{nl}^{(1)}(\chi_s,\vartheta)}{m_s^{-1}R_{nl}^{(1)\prime}(\chi_s,\vartheta)R_{nl}^{(3)}(\chi,\vartheta)-R_{nl}^{(3)\prime}(\chi,\vartheta)R_{nl}^{(1)}(\chi_s,\vartheta)}\cos[n(\varphi_m^{\pm}-\varphi_\mu^{\pm})] \tag{8.31}$$

where $\varepsilon_0 = 1$ for $n = 0$ and $\varepsilon_n = 2$ for $n \neq 0$ is the Neumann symbol; S_{nl}, $R_{nl}^{(1)}$, $R_{nl}^{(3)}$ are the prolate angular and the spheroidal radial functions of the first and third kind (the prime on the symbol for the radial function denotes the derivative with respect to ϑ); $\chi_s = (k_s/2)\sqrt{L^2-d^2}$ ($k_s = 2\pi f/c_s$, c_s is the sound speed in the spheroid), $m_s = \rho_s/\rho$ (ρ_s is the density of the spheroid), $\chi = (k/2)\sqrt{L^2-d^2}$, $\vartheta = L/\sqrt{L^2-d^2}$ are parameters of the spheroid characterising its acoustic properties, the wave sizes and the prolateness:

$$\theta_m^+ = \theta_m^- = \arccos\left(\frac{q_m \cos(\hat{\gamma} + \beta_1)}{k}\right)$$

$$\theta_\mu^+ = \theta_\mu^- = \arccos\left(\frac{q_\mu \cos(\hat{\gamma} - \beta_2)}{k}\right)$$

$$\vartheta_m^\pm = \pi \mp \arctan\left(\frac{\sigma_m}{q_m \sin(\hat{\gamma} + \beta_1)}\right)$$

$$\vartheta_\mu^\pm = \pi \mp \arctan\left(\frac{\sigma_\mu}{q_\mu \sin(\hat{\gamma} - \beta_2)}\right) \tag{8.32}$$

where $\beta_{1,2}$ are azimuthal angles determining the position of the scatterer relative to the source and the receiver (Figure 8.3).

For $\chi \gg 1$, we can used the following asymptotic approximation for the prolate spheroidal radial functions (Komarov *et al.* (1976)):

$$R_{nl}^{(1)}(\chi, \vartheta) \approx \sqrt{\frac{\pi}{2\chi\vartheta}} J_n\left(\chi\zeta - \frac{2e+1}{2}\arctan\zeta\right) \tag{8.33}$$

$$R_{nl}^{(3)}(\chi, \vartheta) \approx \sqrt{\frac{\pi}{2\chi\vartheta}} H_n^{(1)}\left(\chi\zeta - \frac{2e+1}{2}\arctan\zeta\right) \tag{8.34}$$

where J_n and $H_n^{(1)}$ are, respectively, the Bessel and Hankel functions of nth order; $\zeta = \sqrt{\vartheta^2 - 1}$, $e = l - n$. Unlike $R_{nl}^{(1),(3)}$, the present asymptotics for the prolate angular spheroidal functions S_{nl} do not give a uniform convergence as the parameter $\eta = \cos\theta_{m,\mu}^\pm$ is varied. Thus, the values of the function S_{nl} are determined by numerical solution of the singular Sturm–Liouville problem (Komarov *et al.* (1976)):

$$\left[\frac{d}{d\eta}(1-\eta^2)\frac{d}{d\eta} + \chi^2(1-\eta^2) - \frac{n^2}{1-\eta^2} + \lambda_{nl}\right] S_{nl}(\chi, \eta) = 0 \quad |\eta| \le 1 \tag{8.35}$$

where the function $S_{nl}(\chi, \eta)$ is bounded as $\eta \to \pm 1$ and normalised: $\int_{-1}^{1} S_{nl}^2(\chi, \eta) d\eta = 1$, λ_{nl} are eigenvalues.

The sum of the infinite double series (8.31) is in practice limited by harmonics with indices: $n = 0, \ldots, N_\infty$, $l = n, \ldots, l_\infty$, where $L_\infty = [-0.5n^2 + \chi]$, and N_∞ is the greatest number for which the inequality $N_\infty \le [-0.5N_\infty^2 + \chi]$ is satisfied (here, the square brackets denote the integer part of the argument). As the number of harmonics increases, the value of the scattering amplitude is refined by at most 1%.

For numerical modelling, we choose a plane-layered waveguide of depth $H = 50$ m, bounded above by a free surface $z = 0$ and below by an absorbing fluid bottom. We shall assume that the sound speed in the waveguide and the density do not depend on depth and are equal to $c_1 = 1{,}480$ m/s and $\rho = 1.00$ g/cm^3, respectively. The sound speed in the bottom is taken to be equal to $c_1 = 1{,}780$ m/s, the density $\rho_1 = 1.8$ g/cm^3, and the attenuation parameter $\alpha = 0.015$. We shall also assume that a point source with emission power

$W_0 = 500$ W is statically located on the bottom of the waveguide and a receiver at distance $r = 10$ km. The emission frequency is taken to be $f = 300$ Hz. The parameters of the spheroid are chosen as follows: major axis (length) $L = 25$ m, minor axis (maximum diameter) $d = 3.3$ m (typical size of blue whales (*balaenoptera musculus*) (Belkovich *et al.* (1972))), the sound speed in the body of the spheroid is $c_s = 1,540$ m/s, the density of the spheroid is $\rho_s = 1.05$ g/cm^3 (average value for the bodies of mammals (Bergmann (1954))), the depth of submersion is $z_s = 25$ m. The remaining parameters of the problem (receiver depth z, distance r_{0s}, horizontal distance from the baseline y_s, the angle γ at which the stationary track is crossed) are varied during the calculations.

The results of the modelling are shown in Figures 8.4–8.7. Figure 8.4 shows the dependence of $\Delta P = |P| - |P_0|$ on time for various depths of receipt and for various distances r_{0s} with the scatterer moving with speed $\nu = 1$ m/s perpendicularly to the baseline joining the sound source and receiver. As one can see from the figure, when the model scatterer crosses the stationary acoustic track, it induces perturbations of the sound field of several hundredths of a Pascal, with a duration of several hundred seconds. Analysis of the figure shows that the model of the scattered signal, leading to equation (8.30), is only partially valid. In particular, the estimate (8.27) of the effective signal duration T and the quadratic dependence of the signal phase on time is valid. To make the last statement graphically clear, we turn to Figure 8.5(*a*), where the dots show the time dependences for the three signals given in Figure 8.4(*b*). The continuous curve in Figure 8.5(*a*) shows the analogous dependence, determined by equation (8.30). As far as the signal envelope is concerned, it is satisfactorily described by functions of the form $\sin(\Omega)/\Omega$ only in the region of the main maximum. We also stress here that equation (8.28) is only valid for averaging over vertical and horizontal coordinates. The interference of normal modes in a shallow-water waveguide practically masks the weak dependence of the perturbations of the wave field on the distances to the sound source and receiver, observed for Fresnel diffraction. Figure 8.6 shows the value of ΔP at the time the spheroid crosses the stationary track ($y_s = 0$) for various distances r_{0s} at an angle of 90°.

Perturbations of the sound field when the scatterer moves at an acute angle to the baseline are shown in Figure 8.7. As one can see from the figure, the amplitude of the perturbations and their duration vary insignificantly, although the nature of the variations is markedly more complicated. Equation (8.30) no longer describes the observed dependence $\Delta P(t)$, which in the general case becomes asymmetric relative to the time t_0. This dependence can no longer be referred to as a signal with linear frequency modulation. (The dependence of the phase on time for this case is shown in Figure 8.5(*b*).)

Thus, when a soft spheroid with acoustic characteristics similar to the characteristics of cetaceans crosses a stationary acoustic track, characteristic diffraction perturbations of the sound field, whose amplitude is somewhat greater than or equal to (see Chapter 7) the level of natural acoustic noise, are induced at the point of receipt. Evidently, in view of their long duration and the nonlinear dependence of their phase upon time, in order to record these perturbations in full-scale conditions against a noise background, it is expedient to use an optimal correlation receiver. Such a

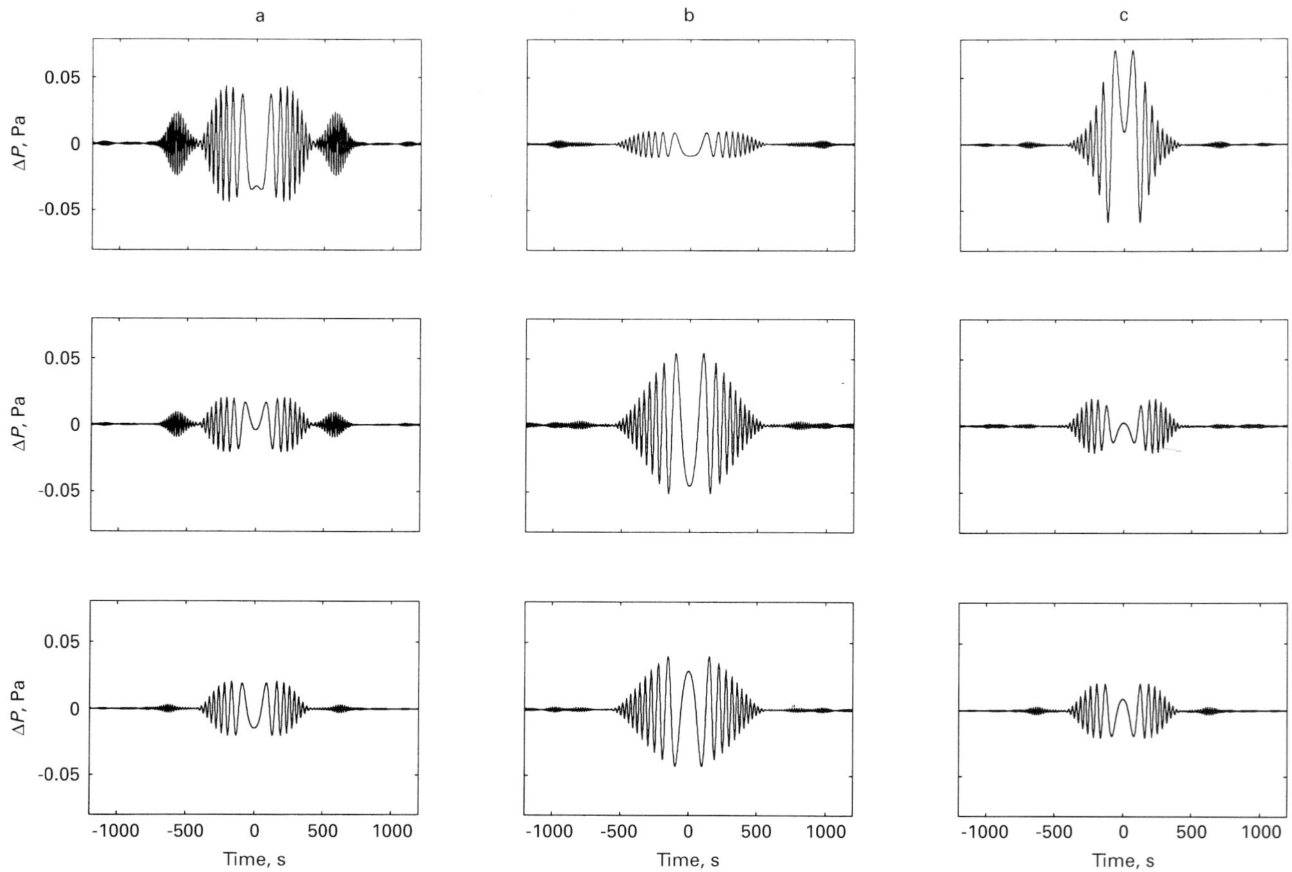

Figure 8.4. Dependence of the sound field perturbations on time $t - t_0$ for a scatterer moving perpendicularly to the baseline. *a*, *b*, *c* are vertical columns of the figures corresponding to sources r_{0_s} at 2.5 km, 5 km and 7.5 km, respectively. The top line of figures corresponds to a depth of receipt of 10 m, the middle line to 30 m, and the bottom line to 50 m.

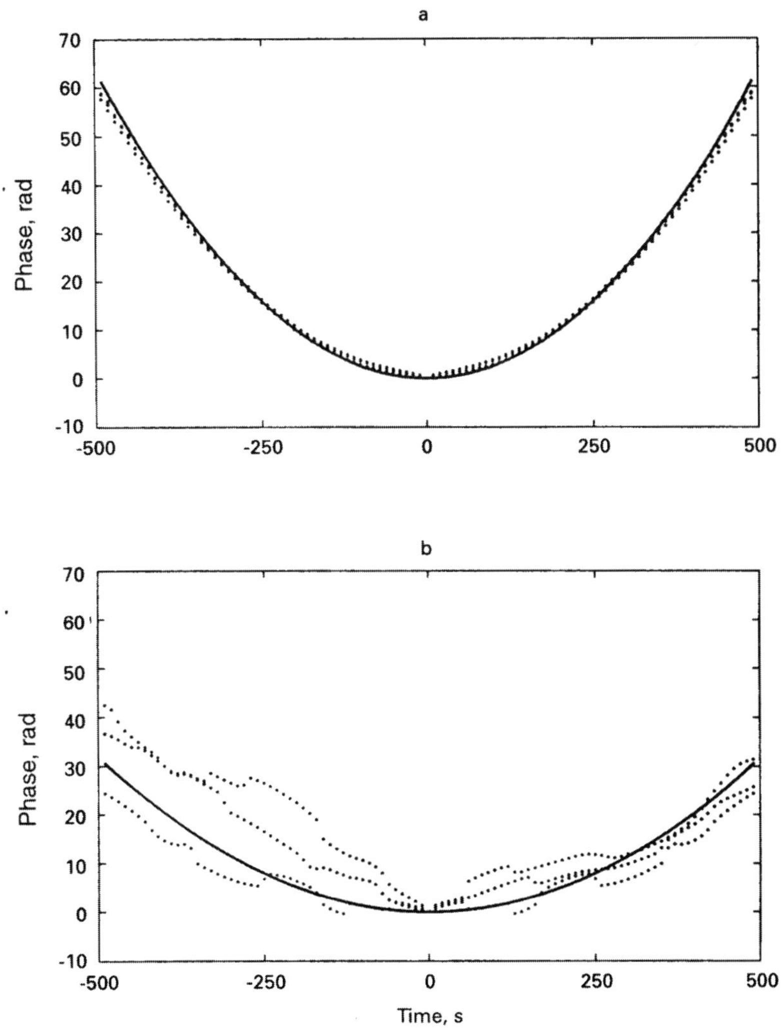

Figure 8.5. Time dependence of the phase of the sound field perturbations when a scatterer crosses a stationary track at right angles (*a*) and at an angle of 45° (*b*). The dots relate to perturbations calculated for different depths by numerical modelling. The continuous curve corresponds to model perturbations determined by equation (8.24).

receiver is also effective when the initial phase of the received signal is not known exactly (see, for example, Gonorovskii *et al.* (1977)). For the reference signal in an optimal receiver, we need to use the calculated signal (8.23), analogous to that shown in Figures 8.4 and 8.7. (Here, we note that when the scatterer moves at right angles, a simplified version of (8.30) can be used, as demonstrated in Kuzkin (1997) and Zverev *et al.* (1995). Of course, in this case, one has to choose the *a priori* unknown parameters of the movement of the scatterer: the speed v_s, depth z_s,

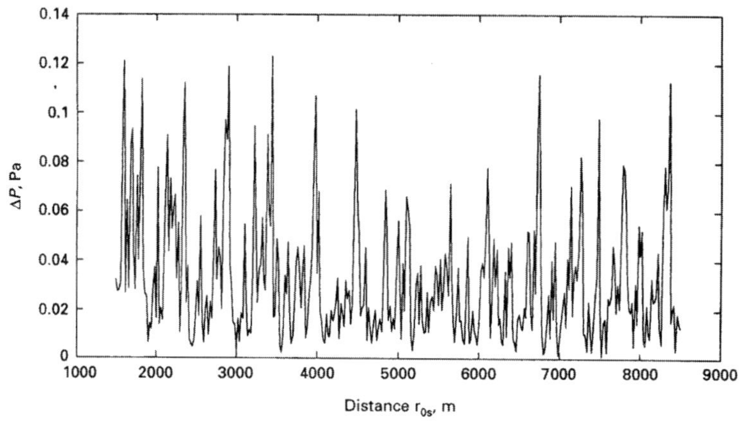

Figure 8.6. Dependence of the amplitude of the sound field perturbations at the time the stationary track is crossed on the distance r_{0_s} between scatterer and sound source. Depth of receipt, 30 m.

distance r_{0s}, angle $\hat{\gamma}$, which are inherently of most interest for studies of the migration of cetaceans in individual shallow-water regions. The maximum response of the optimal receiver will correspond to the actual set of parameters, provided, of course, that the characteristics of the medium are precisely known. Thus, not only diffraction perturbations of the sound field recorded, but also unknown parameters, are determined. In other words, we are talking about acoustic tomography, as described in Chapter 6, based on matched field signal processing. The difference here is that the quantities to be estimated are the parameters of the scatterer.

In conclusion, we stress that in this chapter we have only described a method for calculating diffraction perturbations of sound fields in shallow-water waveguides and demonstrated certain general patterns in their behaviour. Estimation of all the parameter of a scatterer (for example, the distance to it) by the technique of matching observed and theoretical data is a separate problem, in which, for a given set of sound sources and receivers, one has to prove that each parameter is uniquely determined and to give the corresponding accuracy estimation. However, solution of that problem is beyond the scope of this book.

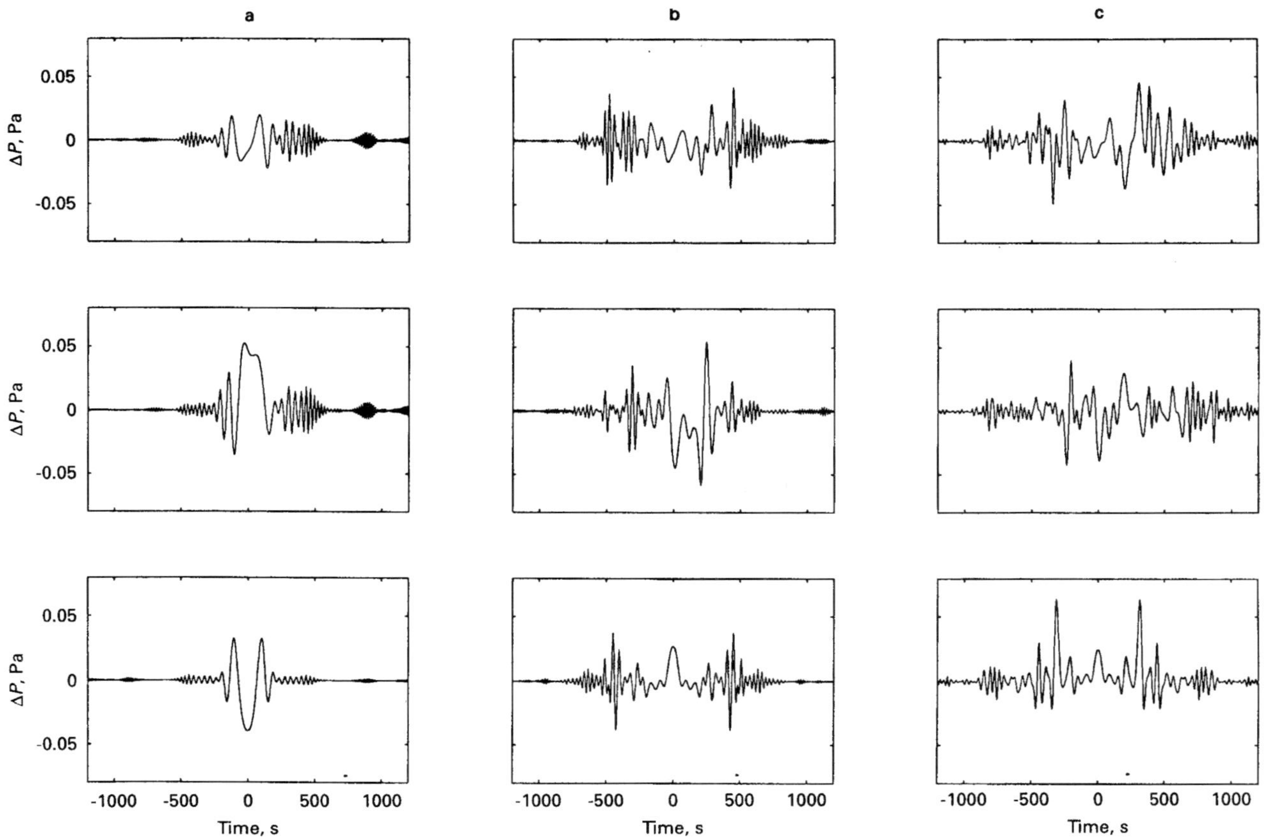

Figure 8.7. Dependence of the sound field perturbations on the time $t - t_0$ when the scatterer crossed the baseline at an angle. $r_{0_s} = 5\,\text{km}$. *a*, *b*, *c* are vertical columns of figures corresponding to angle $\hat{\gamma} = 80°$, $60°$ and $45°$, respectively. The top line of figures corresponds to a depth of receipt of 10 m, the middle line to 30 m, and the bottom line to 50 m.

9

Conclusion

We shall present a brief summary of this book. We have considered theoretical models and methods, as well as the characteristics of the formulation and the results of a number of experiments on oceanic waveguides at a low depth, which correspond mainly to the conditions of the sea shelf. This division of the acoustics of the ocean has recently acquired the name of shallow-water acoustics and has become, in its own, right, an individual division of Earth sciences. This has happened, first, by virtue of its importance for man's industrial activity, as well as for the ecology of the coastal zone of the World Ocean. One other reason for the emergence of this division is associated with the specifice characteristics of the propagation of sound on the oceanic shelf, which, in turn, have led to the development of of the theory of the propagation of sound in shallow-water waveguides and of the technique of full-scale physical experiment, and to the development of computational algorithms and signal processing methods. In conclusion, we again stress these characteristics and the consequences arising from them.

The geoacoustic properties of the ocean shelf zone have certain specific traits in comparison with the deep-water area:

- the sound-speed profile is largely characterised by monotonic behaviour (the function $c(z)$ has a mainly positive or negative gradient as a function of the time of year)., which is determined by either a near-bottom or a near-surface sound channel.
- a spatially variable bathymetry (depth of the ocean), where the change in depth (more or less smooth) may be comparable with the depth itself;
- a complex (possibly layered) bottom structure, usually comprising sedimentary rocks, absorption in the bottom is significant and depends on the sound frequency;
- there is a significant spatial anisotropy of the mesoscale hydrodynamic processes (internal waves, tides, frontal zones), associated with the bottom relief. In

particular, this is due to the conisderable effect of the shelf-break and the coastline on the formation and duration of these processes;
- clearly expressed specific geographical characteristics of all the geoacoustic properties or other properties of one particular region of the shelf, being sharply distinguished from others in terms of its geoacoustic characteristics.

As far as models and theory, describing the propagation of sound in shallow water (in comparison with the 'older' acoustics of the deep ocean) are concerned, we stress two aspects.

- The use of a principally modal approach, where the sound pressure field is represented in the form of the sum of a small enough (usually, no more than a few tens) number of waveguide modes. Here, the main characteristics (including measureable ones) of the sound field are given in the langiage of modes: mode propagation constants, mode attenuation coefficients, mode amplitudes, the spatial scales of beating, etc.
- Account is taken, in a number of cases, of mode coupling, associated with a considerable spatial inhomogeneity of the sound channel. This relates both to the deterministic case and to randomly caused occurrences of mode coupling, where a special role is played by inhomogeneities of the boundaries with the sea surface and the bottom.
- The use of special methods for calculating the sound field, for example, the method of vertical modes and horizontal rays to calculate refraction. It is important to note that the listed properties of the channel have led to the development of methods for calculating the sound field using a parabolic equation, written down for the three-dimensional case.

One result of the listed properties of the shallow-water medium, the model and the methods, is the existence of a number of specific acoustic effects, determining the main properties of the sound field. These are, in particular

- the so-called 'dying mode' effect, due to a difference in mode attenuation coefficients, which leads to the fact that at cmparatively large distances only one mode (or group of modes) remains, with the least attenuation coefficients. The index of this mode (or the index of the group of modes) is determined by the sound-speed profile;
- the presence of 'waveguide' dispersion, giving a dependence of the eigenvalues (mode propagation constants) and, consequently, of the phase and group velocities on frequency, which leads to differen characteristics of the interference pattern in the frequency domain.
- a higher level of noise (in comparison with deep water) of the surface disturbance and its dependence, determined by the sound speed profile.

As far as particular elements in the methods and techniques of full-scale experiment are concerned, these are mainly the following.

- The possibility (when vertical emitting and receiving arrays are used) of covering practically the whole depth of the waveguide, which makes it possible to carry out selective excitation and receipt of individual modes.

- The performance of specific experiments in stationary acoustics with static sound sources and receivers, which is relatively easy to do in shallow water.
- The possibility, due to two factors, of observing a comparatively stable interference patterm of the sound field both in space (in the vertical plane) and in the frequency domain.

The development of computational algorithms and signal processing techniques in shallow-water acoustics is mainly associated with problems in the monitoring of the environment. This has to do with the following.

- Global acoustic monitoring, when the average temperature or the salinity of sea water is measured over a large area of the sea. This monitoring is essentially possible because of the dependence of the sound speed on temperature and salinity, as indicated in Chapter 1. Presently, acoustic methods permit measurement of the temperature with accuracy up to 0.01° K and of salinity with accuracy 0/05‰. For example, such monitoring is carried out in the framework of the ATOC (Acoustic Thermometry of Ocean Climate) (Krolik and Narasimhan (1996)), and in addition, a method for acosutic monitoring if the flow of refreshed water through the Bering Strait (Avilov *et al.* (1999)) is being developed. Here we note that even when such monitoring is carried out on the scales of individual areas of the World Ocean, the sound receivers and sources are often located on the sea shelf and
- sound propagates over several tens of kilometres through shallow water, where, all the laws of shallow-water acoustics are satisfied.
- On the monitoring of large-scale hydrodynamic peturbations in shallow water (tidal and internal waves, frontal zones, currents, etc.), see Lynch *et al.* (1966) and Parsons *et al.* (1996). We note that this monitoring is always very useful for marine biology, since it provides the possibility of acoustic diagnosis of the dynamics of biogenic material and accumulations of plankton, closely associated with the movement of water masses.
- On estimates of the parameters of the sea bed from the characteristics of the acoustic field in a near-bed sound channel.
- On observation of the movement of large marine mammals (whales, dolphins, etc.) on the sea shelf.

In these problems, in addition to classical acoustic tomography, in order to determine the characteristics of the sound channel, the tomography has been developed, based on matched field signal processing, the essence of which is studied in this book.

Let us also mention that matched field processing in shallow water can be successfully used to control sound fields (focussing if a field in a given region of the waveguide), as demonstrated in Song *et al.* (1998), where a method based on an interference invariant of the sound field was developed for this control. We again stress that for this control, source arrays are used which cover the whole (or almost the whole) depth of the waveguide.

Furthermore, as can be seen from the characteristics of the shelf zone listed above, in order to solve applied and fundamental problems associated with the propagation of sound in shallow water, we have to take into account (sometimes simultaneously) a variety of phenomena: absorption of sound in the bottom and in the water layer, scattering on random and deterministic homogeneities, the presence of noise for various reasons, etc. In this book, we have attempted to present the methods and techniques for solving these problems, in such a way that they were considered, (somewhat coarsely) independently from one another: the division of the book into chapters corresponds to this. In the sometimes quite coarse presentation of the mathematcs, we have paid particular attention to an attempt to distinguish simple analytic formulae and estimates, permitting a qualitative understanding of the processes which occur, and also of the main aspects of experiment design.

In conclusion, the authors hope that this book will be useful both for specialists working in the domain of shallow-water acoustics and for students wishing to become acquainted with the rapidly developing and practically very promising area of Earth sciences.

Notation

a	diffusion coefficient
b_s	soliton half width
$B(\vec{R}, \vec{R}')$	sound field spatial correlation function
$B(\tau)$	envelope of the joint correlation function of the received and emitted signals ($B_e(\tau)$ – experimental, $B_{\text{th}}(\tau)$ – calculated)
$B_c(\vec{R}, \vec{R}')$	correlation function for the sound speed
$B_d(z, z')$	correlation coefficient in the vertical direction
$B_s(\vec{R}, \vec{R}')$	correlation function for surface disturbance
$B_q(\vec{r}, \omega; \vec{r}', \omega')$	joint spectral correlation function for surface noise sources
$c(z)$	sound speed in water
$c(z) = c_s(1 - \varepsilon z)$	vertical linear profile of the sound speed
c_0	sound speed at the source
c_s	sound speed at the surface
c_H	sound speed on the bottom
c_n	sound speed in a multilayered bottom
f	frequency
f_l	cut-off frequency of lth mode
$\Delta = \dfrac{\partial^2}{\partial x^2} + \dfrac{\partial^2}{\partial y^2} + \dfrac{\partial^2}{\partial z^2}$	operator
Δf_m	Doppler frequency shift for mth mode
δf	width of spectral line
f_0	position of first maximum in the spectrum of an explosive signal
$F(\vec{k}_i, \vec{k}_s)$	scattering amplitude
$g(\xi) = (i/\omega\rho) Z_1(\xi)$, $Z_1(\xi)$	is the impedance of the bottom
$g(\vartheta)$	directivity function for sources in the plane

G_{ml}	mode interaction coefficient
$h(t)$	depth variations
h_n	layer thickness
H	depth of sea
I	sound intensity
$\bar{I}$	sound intensity averaged over space
i	imaginary unit
j	index of hydrophone
J	ray invariant
J_h	number of receiving hydrophones
k	value of wave vector
$\vec{k}_i$	incident wave wave vector
$\vec{k}_s$	scattered wave wave vector
k_s	wave vector on sea surface
k_n	wave vector in multilayered bottom
K	value of the numerical criterion for comparing proximity
K_0	coefficient of proportionality, depending on the type of the explosive material
l, m	mode indices
l_j	distance from jth hydrophone to anchor on bottom
L	length of acoustic antenna
L_0	distance from depth sensor to sea bed
$\tilde{L}$	characteristic scale for variations of waveguide parameters
$\tilde{L}_H = H/H_y$	characteristic scale for variations of depth ($H_y = \partial H/\partial y$)
$\tilde{L}_c = c/c_y$	characteristic scale for variations of sound speed ($c_y = \partial c/\partial y$)
M	maximum number of propagating modes
$\vec{M}(t)$	vector of time-varying dipole moment
m_1	ρ_1/ρ
$\nabla = \left(\dfrac{\partial}{\partial x}, \dfrac{\partial}{\partial y}, \dfrac{\partial}{\partial z}\right)$	operator
$n(z) = c_s/c(z)$	refractive index in waveguide
$n_1 = c_H/c_1$	refractive index in bottom
n	index of layer in multilayered bottom
N	number of layers in bottom
N_l	normalization coefficient
$\hat{N}_m = \sum_j^J \psi_m^2(z_j)$	gain of vertical chain of hydrophones
$\tilde{N}$	buoyancy frequency
p	continuous parameter corresponding to the mode index

$p(r,z,t)$	real (measured) sound pressure
P_m	pressure amplitude for *m*th mode
$P(r,z,t)$	complex sound pressure
$\lvert P(r,z,t)\rvert$	sound pressure amplitude
P_i	sound pressure in incident wave
P_s	sound pressure in scattered wave
q	exponent determining the frequency dependence of sound absorption in the bottom
$Q_m(r, r', z_j, z', \omega)$	transfer function of an optimal filter designed to receive a signal of the *m*th mode
$Q(\vec{R}, t)$	volume density of emitted signal
$Q(\vec{r}, t)$	surface density of emitted signal
$\tilde{Q}$	coefficient, depending on the physical properties of the water
r, ϑ, z	cylindrical coordinate system
R, ϑ, φ	spherical coordinate system
$\hat{r}$	distance, for which the comparison criterion has a maximum
r_b	distance beyond which dispersion distortions become marked
r_{bm}	distance, beyond which sound signals corresponding to two adjacent modes m and $m+1$ are distinguishable according to their propagation time in the channel
r_{im}	distance, beyond which intramode signal distortions are marked
r_0	1 m
r_0^*	reference distance
r_1	transitional distance, beyond which spherical divergence of a wavefront ends and the 'three by two' law for attenuation is valid
r^*	distance beyond which the 'three by two' law ceases to be satisfied
Δr_d	discreteness of measurements of the sound field in terms of distance
Δr_j	displacement of the *j*th hydrophone
$\vec{r}$	radius vector in the plane
$\vec{R}$	radius vector in space
$R_m = r/\cos\chi_m$	'length of the *m*th mode'
$R_\parallel$	longitudinal radius of correlation of inhomogeneities
$R_\perp$	transverse radius of correlation of inhomogeneities
s	sound loss coefficient in the bottom
$s(\vec{r}, t)$	deviation of surface due to surface waves
$S(\omega)$	emitted signal spectrum
$S(\vec{k})$	spatial spectrum

$S(\vec{k}, \omega)$	space–frequency spectrum
$S(\vec{R}, \omega)$	signal spectrum in waveguide
$S_{\mu m}$	scattering matrix
t_m	signal propagation time corresponding to *m*th mode
T	duration of emitted signal
$\hat{T}$	accumulation (measurement) time
T_1	period of first pulsation of gas bubble
$\tilde{T}(r, z)$	temperature field in waveguide
u	phase velocity of an internal wave
ν	speed of ship
ν_m^{ph}	phase velocity of *m*th mode
ν_l^{gr}	group velocity of *l*th mode
V	reflection coefficient from bottom
$V(t)$	volume speed of pulsed exchange
V_ω	spectral function of volume speed
w	charge mass
W	Wronskian of a system of equations
W_0	source power
W_l	coefficient determining the proportion of sound field energy associated with the *l*th mode
x, y, z	cartesian coordinate system
$Y(r, \omega)$	spectrum of the complex signal envelope at output from an optimal filter
z_0	depth of sound source
z_j	depth of receiving hydrophone
z_s	depth of plane of noise sources
α	parameter determining the absorption coefficient in a model of a fluid bottom
α_0	angle between the directions of propagation of internal and sound waves
β	attenuation coefficient for sound in a waveguide, in the presence of a number of energy-bearing modes with the same attenuation coefficient
β_1	attenuation coefficient for sound in a waveguide, where the 'three by two' law applies
δ	relative sound field intensity in a waveguide
δ_r	horizontal angle of refraction
Δ	horizontal displacement of a ray on reflection from the bottom
q_l	horizontal propagation constant of the *l*th adiabatic mode
κ	propagation constant in the bottom for modes of the continuous spectrum
κ_l	coefficient of attenuation in the bottom for the *l*th mode

λ	length of a sound wave
$\Lambda_{ml} = 2\pi/\lvert q_m - q_l \rvert$	period of intermode pulsations ($\Lambda_l = 2\pi/\lvert q_l - q_{l+1} \rvert$)
$\mu(r, z, t)$	random component of the square of the refractive index in a waveguide
$\nu(\vec{k}, \vec{R})$	density of fictitious sources
$\xi_m = q_m + (i\gamma_m/2)$	eigenvalue of the mth mode
$\vec{\xi}$	wave vector in the horizontal plane
ρ	water density
ρ_1	density in the bottom
$\tilde{\sigma}$	value of the numerical criterion for comparing proximity
σ_l	vertical wavenumber for the lth mode
$\vec{\sigma}$	wave vector in the vertical plane
$\Delta\tau = 2\pi/\Delta\omega$	pulse duration at the output of an optimal receiver
τ_ν	correlation time of random changes in ship speed
$\varphi(r, z, t)$	signal phase
$\Delta\varphi_0$	angle of inclination of hydrophone chain
φ_m	phase of mth mode
$\delta\varphi(t)$	variations of signal phase
Φ_l	function of lth adiabatic mode
$\tilde{\Phi}_{m\vec{\kappa}}(z)$	internal wave mode with index m and wave vector $\vec{\kappa}$
χ_m	grazing angle of a Brillouin ray corresponding to the mth mode
$\psi_m(z)$	eigenfunction of the waveguide
$\Psi(r, z; \omega)$	field of an acoustic point source of frequency ω, Green function, waveguide transfer function
$\Psi_c(r, z)$	field of a point source, corresponding to the continuous spectrum
ω	cycle frequency
$\Delta\omega$	bandwidth for signal frequency
Ω_l	cycle frequency of an internal wave
$\zeta(\vec{r}, z, t)$	vertical displacement of a surface of constant density

Acronyms

CCF	Cross-correlation function
CTD probe	Conductivity, Temperature, Depth probe
EJP branch cut	Ewing–Jardetsky–Press branch cut method for doing branch cut line in complex plane
GM	Generalized Modes
GPI RAS	General Physics Institute of Russian Academy of Sciences
IW	Internal Waves
JUSREX	Joint US–Russian Experiment
KdV	Korteveg de Vries
LFM	Linear Frequency Modulation
MFP(T)	Matched Field Processing (Tomography)
RTE	Radiation Transfer Equation
SACLATCEN	NATO SACLANT (Supreme Allied Commander Europe) Reserach Centre, La Spezia, Italy
SA method	Simulated Annealing method
SNAP	SACKLANT Normal mode Acoustic Programme
SWARM	Shallow Water Acoustics Random Media experiment, conducted in 1995
TL	Transmission Loss
WHOI	Woods Hole Oceanographic Institution
WKB approximation	Wentzel–Kramers–Brillouin approximation for solving differential equations

Bibliography

Abdullaev S.S., Niyazov B.A. and Khabibullaev P.K. (1986) Spatial correlation of surface noise in an acoustic waveguide. *DAN SSSR* **288**, 223–225 (in Russian).

Ageeva N.S. (1974) Noise field of a lumpy source in the ocean. In: L.M. Brekhovskikh (ed.) *Ocean Acoustics* (pp. 163–229). Moscow: Nauka (in Russian)

Ageeva N.S. and Krupin V.D. (1979) Structure of an infrasonic field in a shallow sea. *Sov. Phys.-Acoust.* **25**(3), 192–195.

Ageeva N.S. and Krupin V.D. (1980) Effect of the bottom on the formation of the sound field in shallow water. *Akusticheskii Zhurnal* **26**(2), 161–166 (in Russian)

Ageeva N.S. and Krupin V.D. (1981) Frequency response curves for normal modes in a shallow sea with a layered absorbing bottom. *Sov. Phys. Acoust.* **27**(5), 372–376.

Ageeva N.S., Krupin V.D., Perelygin V.P. and Studenichnik N.V. (1994) Development of a geoacoustic model of shallow sea floor. *Acoust. Phys.* **40**(2), 159–165.

Andreev M.Yu. Katsnelson B.G., Kulapin L.G. and Petnikov V.G. (1996) Influence of hydrodynamic variations in a shallow sea on sound field phase. *Acoust. Phys.* **42**(4), 402–406.

Arase E.M. and Arase T. (1967) Ambient sea noise in the deep and shallow water. *J. Acoust. Soc. Am.* **42**(2), 73–77.

Aredov A.A. (1988) Anisotropy characteristics and spatial correlation of the noise field in the ocean. *Sov. Phys. Acoust.* **34**(1), 56–59.

Aredov A.A. and Furduev A.V. (1994) Angular and frequency dependencies of the bottom reflection coefficient from the anisotropic characteristics of a noise field. *Acoust. Phys.* **40**(2), 176–180.

Arons A.B. (1948), Secondary pressure pulses due to gas globe oscillation in underwater explosions: II. Selection of adiabatic parameters in the theory of oscillation. *J. Acoust. Soc. Am.* **20**, 277.

Artelnyi V.V. and Raevskii M.A. (1987) Multiple scattering of low-frequency sound waves by ocean turbulence. *Sov. Phys. Acoust.* **33**(1), 6–12.

Artelnyi V.V. and Raevskii M.A. (1984) Statistical characteristics of normal modes in a waveguide with volume random inhomogeneities. *Izv. Vuzov. Radiofizika* **37**(9), 1142–1150 (in Russian).

Artelnyi V.V. and Raevskii M.A. (1987) Multiple scattering of waves on anisotropic fluctuations of the refractive index in an infinite medium and refraction waveguides. *Izv. Vuzov. Radiofizika* **30**(7), 866–874 (in Russian)

Ashley C.E., Siegman W.L. and Jacobson M.J. (1987) Random-bottom structured and topographical effects on sound transmission in shallow channel. *J. Acoust. Soc. Am.* **81**(3), 650–659.

Avilov K.V. (1995) Pseudodifferential parabolic equations of sound propagation in the slowly range-dependent ocean and their numerical solutions. *Acoust. Phys.* **41**(1), 1–7.

Avilov K.V., Krupin V.D., Petnikov V.G. and Sabinin K.D. (1999) Estimation of the possibility of acoustic monitoring of the Bering Strait. *Acoustical Physics V* **45**(5), 530–534.

Babilov E.P. and Kanevskii V.A. (1988) Sound scattering by a gas-filled spheroidal swim. *Sov. Phys. Acoust.* **34**(1), 11–13.

Babii V.V. (1983) *Fine-scale Structure of the Sound Speed Field in the Ocean* (200 pp.). Leningrad: Gidrometeoizdat (in Russian)

Baggeroer A.B. and Kuperman W.A. (1992) Matched field processing in ocean acoustics. In: Josa M.F. Moura and Isabel M.G. Lourtie (eds) *Acoustic Signal Processing for Ocean Exploration* (pp. 79–114). Dordrecht: Kluwer Academic.

Badiey M., Bongiovanni K.P. and Siegmann W.L. (1997) Analysis and model/data comparison of broadband acoustic propagation at the Atlantic Generating Station (AGS) site. *J. Acoust. Soc. Am.* **101**(4), 1921–1935.

Badiey M. and Jaya I. (1994) Shallow-water acoustic/geoacoustic/ experiments at the New Jersey Atlantic Generating Station site. *J. Acoust. Soc. Am.* **96**(6), 3593–3604.

Barridge R. and Weinberg G. (1977) Horizontal rays and vertical modes. In: J.B. Keller and J.S. Papadakis (eds) *Wave Propagation and Underwater Acoustics* (pp. 86–150). Springer-Verlag.

Beardsley R.C. *et al.* (1997) Shelfbreak PRIMER – an integrated acoustic and oceanographic field study in the Middle Atlantic Bight. *Proc. Int. Conf. on Shallow-water Acoustics, Beijing*.

Beebe J.N. and McDanial S.T. (1980) Geoacoustic models of the seabed to support range-Scotian Shelf. In: W.A. Kuperman and F.B. Jensen (eds) *Bottom-interacting Ocean Acoustics* (pp. 507–523). New York: Plenum Press.

Belov A.I. and Komarov A.G. (1996) Mode analysis of the sound field in weakly irregular waveguides. *Acoust. Phys.* **42**(20), 232–233.

Belov A.I., Serebryany A.N. and Zhuravlev V.A. (1997) Observation of internal wave effects on acoustic signals in a shallow sea with a weak thermocline. *Proc. Int. Conf. on Shallow-water Acoustics, Beijin*.

Belkovich V.M. and Yablokov A.V. (1972) *Whales and Dolphins* (472 pp.). Moscow: Nauka (in Russian).

Beran M.J. and Whitman A.M. (1978) Scattering in the depth direction from a quadratic sound channel. *J. Acoust. Soc. Am.* **64**(2), 658–664.

Bergmann L. (1954) *Ultrasound and Its Application in Science and Technology*. Zurich (in German).

Biot M. (1956) Theory of propagation of elastic waves in a fluid-saturated porous solid. *J. Acoust. Soc. Am.* **28**(1), 168–191.

Box G. and Jenkins G. (1970) *Time Series Analysis. Forecasting and Control*. Holden-Day.

Bondar L.F., Borisov S.V., Gritsenko A.V., Zakharov V.A., Ilichev V.I., Kovzel D.G., Morgunov Yu.N. and Rutenko A.N. (1994) Intensity and phase fluctuations of low-frequency acoustic signals along stationary tracks in shelf regions of the Sea of Japan. *Acoust. Phys.* **40**(4), 497–504.

Borodin V.V., Zhuravlev V.A., Kobozev I.K. and Kravtsov Yu.A. (1993) Average characteristics of sound fields in ocean waveguides. *Sov. Phys.-Acoust.* **38**(4), 332–336.

Bostrom A. (1984) Transmission and reflection of acoustic waves by an obstacle in a waveguide. *Wave Motion V* **12**(2), 167–184.

Boutin P., Kemp J., Liberatore S., Lynch J., Witzell N., Metzger K. and Webb D. (1989) *Results of the Lake Seneca Directivity Source Level and Pulse Response Tests on the MIT 400 Hz Webb Tomography Sources* (Technical Memorandum No. WHOI-1-89). Woods Hole Oceanographic Institution, MA.

Boyles C.A. (1983) Coupled mode solution for a cylindrically symmetric oceanic waveguide with a range and depth dependent refractive index and a time varying rough sea surface. *J. Acoust. Soc. Amer.* **73**, 800–805.

Brekhovskikh L.M. (1965) The average field in an underwater sound channel. *Sov. Phys. Acoust.* **11**(2), 126–134.

Brekhovskikh L.M. (1980) *Waves in Layered Media* (2nd edn). New York: Academic Press.

Brekhovskikh L.M. and Lysanov Yu.P. (1991) *Fundamentals of Ocean Acoustics* (270 pp.) Springer-Verlag.

Brekhovskikh L.M. and Godin O.A. (1992) *Acoustics of Layered Medium*. Springer-Verlag.

Brekhovskikh L. and Goncharov V. (1994) *Mechanics of Continued and Wave Dynamics* (342 pp.). Springer-Verlag.

Buckingham M.J. and Jones S.A. (1987) A new shallow-ocean technique for determining the critical angle of the seabed from the vertical directionality of the ambient noise in the water column. *J. Acoust. Soc. Am.* **81**(4), 938–946.

Buckingham M.J. (1980) A theoretical model of ambient noise in a low loss, shallow water channel. *J. Acoust. Soc. Am.* **67**(4), 1186–1192.

Bunkin F.V., Katsnelson B.G., Kravtsov Yu.A., Kulapin L.G., Petnikov V.G., Rivelis E.A., Reznikov V.M, Sabirov O.I. and Sidenko A.V. (1989) Average estimates of sound absorption in shallow ocean waveguides. *Sov. Phys.-Acoust.* **35**(1), 1–4.

Bunkin F.V., Gindler I.V., Kozelskii A.R., Kravtsov Yu.A. and Petnikov V.G. (1989) Dispersion distortion of complex acoustic signals in shallow water ocean waveguides. *Sov. Phys. Acoust.* **35**(5), 461–464.

Bunkin F.V., Vavilin A.V., Zhuravlev V.A., Kozelskii A.R., Kravtsov Yu.A., Kuzkin V.M., Kulapin L.G., Lyubchenko A.Yu., Omelchenko N.N., Petnikov V.G., Sabirov O.I., Slavinskii M.M. and Khudiev L.T. (1984) Investigation of bottom propagation of acoustic signals with highly stable. *Izvestiya, Atmospheric and Ocean Physics* **20**(2), 159–161.

Burlakova I.B., Saichev A.I. and Slavinskii M.M. (1987) Spectrum of a signal with the elimination of frequency wandering. *Sov. Phys, Acoust* **33**(3), 236–241.

Burov V.A. and Sergeev S.N. (1991) Processing of acoustic fields in oceanic waveguides with compensation of an unknown distortion of the profile of the vertical receiving antenna. In: *Formation of Acoustic Fields in Waveguides* (pp. 214–218). Nizhnii Novgorod: Izd-vo IPF AN SSSR (in Russian).

Burov V.A. and Sergeev S.N. (1992) Modern methods of perturbation theory in calculation of hydroacoustic fields. *Vestn. Mosk. un-ta Seriya 3* **33**(2), 49–56 (in Russian).

Buryakovskii L.A., Dzhafarov I.S. and Kerimov V.B. (1991) *Search for and Exploration of Sea Sites for Petroleum and Gas*. Moscow: Nedra (in Russian).

Bjorno L. (1981) Inhomogeneities and instabilities in underwater sound propagation. In: L. Bjorno (ed.) *Underwater Acoustic Signal Processing* (pp. 29–40). Holland: D. Reidel.

Caille G.W. *et al.* (1997) Overview of the joint China–U.S. 'Yellow Sea '96' experiment. *Proc. Int. Conf. on Shallow-water Acoustics, Beijing.*

Candy J.V. and Sallivan E.J. (1993) Sound velocity profile estimation: A system theoretic approach. *IEEE J. Oceanic Engineering* **18**(3), 240–252.

Carbone N.M., Deane G.B. and Buckingham M.J. (1998) Estimating the compressional and shear wave speeds from the vertical coherence of ambient noise in the water column. *J. Acoust. Soc. Am.* **103**(2), 801–813.

Cavanagh R.G. and Renner W.W. (1980) Vertical directionality and depth dependence of averaged acoustic signals and noise. *J. Acoust. Soc. Am.* **68**(5), 1467–1474.

Chen C.-T. and Millero F.J. (1977) Sound speed of seawater at high pressure. *J. Acoust. Soc. Am.* **62**(3), 1129–1135.

Chernov L.A. (1975) *Wave Propagation in a Medium with Random Inhomogeneities* (172 pp.) Moscow: Nauka (in Russian).

Chuprov S.S. (1982) Interference structure of the sound field in a stratified ocean In: L.M. Brekhovskikh (ed.) *Ocean Acoustics. Current Status* (pp. 71–91). Moscow: Nauka (in Russian).

Clay C. and Medwin H. (1977) *Acoustical Oceanography*. Wiley.

Collins M.D. (1993) The adiabatic mode parabolic equation. *J. Acoust. Soc. Amer.* **94**(4), 2269–2278.

Cron B. and Sherman C H. (1962) Spatial-correlation functions for various noise models. *J. Acoust. Soc. Am.* **34**(11), 1732–1736.

Crouch W.W. and Burt P.J. (1972) The logarithmic dependence of surface generated ambient shelf noise spectrum level on wind-speed. *J. Acoust. Soc. Am.* **51**(3), 1066–1072.

Davydov A.S. (1976) *Quantum Mechanics* (translated, edited and with additions by D. ter Haar, 2nd edn). Oxford: Pergamon Press.

Dacol D.K. (1997) Coupled-amplitude approach to solving the Helmholtz equation. *J. Acoust. Soc. Am.* **101**(5), 2566–2570.

Daugherty J.R. and Lynch J.F. (1990) Surface wave, internal wave, and source motion effects on matched field processing in a shallow water waveguide. *J. Acoust. Soc. Am.* **87**(6), 203–252.

Dawson T.W. and Fawcett J. (1990) A boundary integral equation method for acoustic scattering in waveguide with nonplanar surfaces. *J. Acoust. Soc. Am.* **87**(3), 1110–1125.

Denham R.N. (1986) Intensity decay laws for near-surface sound sources in the ocean. *J. Acoust. Soc. Am.* **79**(1), 60–63.

Derevnin V.A. *et al.* (1986) Acoustic bottom stations for studying signals and noise in a deep ocean. *Deep-water Systems and Complexes, Cherkassy, Proc. Seminar 'Deep-water systems and complexes'* (pp. 52–55) (in Russian).

Derevyagina E.I., Katsnelson B.G. and Lyubchenko A.Yu. (1994) Vertical structure of the intensity of low-frequency noise in shallow water. *Acoust. Phys.* **40**(3), 343–346.

Derevyagina E.I. and Katsnelson B.G. (1995) The effect of random inhomogeneities on the vertical directionality of surface noise in shallow water. *Acoust. Phys.* **41**(2), 205–209.

Derevyagina E.I. and Katsnelson B.G. (1988) Influence of an absorbing bottom on the frequency spectrum of the noise field in a shallow sea. *Sov. Phys.-Acoust.* **34**(1), 95–97.

Derevyagina E. and Katsnelson B.G. (1988) Effect of the shape of the region of distribution of surface noise sources on the spectrum of the noise field in shallow water. *Sudostroitelnaya promyshlennoct'. Seriya 'Akustika'* **3**, 50–53 (in Russian).

DeSanto J.A. (ed) (1979) *Ocean Acoustics*. Springer-Verlag.

Dickey T.D. (1991) The emergence of concurrent high-resolution physical and biological measurements in the upper ocean and their application. *Rev. Geophys. Res.* **29**(3), 383–414.

Doolittle R., Tolstoy A. and Buckingham M. (1988) Experimental confirmation of horizontal refraction of CW acoustic radiation from a point source in wedge-shaped ocean environment. *J. Acoust. Soc. Am.* **83**(6), 2117–2125.

Dosso E.D., Yeremy M.L., Ozard J.M. and Chapman N.R. (1993) Estimation of ocean-bottom properties by matched field inversion of acoustic field data. *[IEEE J. Oceanic Engineering* **18**(3), 232–239.

Dozier L.B. and Tappert F.D. (1978) Statistics of normal mode amplitudes in a random ocean. 1. Theory. *J. Acoust. Soc. Am.* **63**(2), 353–365.

Eliseevnin V.A. (1981) Operation of a vertical line array in a water layer. *Sov. Phys. Acoust.* **27**(2), 125–127.

Essen H.H., Schirmer F. and Sirkes S. (1987) Acoustic remote sensing of internal waves in shallow water. *Int. J. Remote Sensing* **4**(4), 33–47.

Essen H.H., Schirmer F. and Schmalfeldt B. (1983) Influence of tides on acoustic continuous wave (CW) propagation in shallow water. *Acustica* **48**(5), 285–292.

Ewing W.M., Jardetzky W.S. and Press F.E. (1957) *Elastic Waves in Layered Media.* New York: McGraw Hill.

Fedorov K.N. (1976) *Fine Thermohaline Structure of Oceanic Water* (184 pp.). Leningrad: Gidrometeoizdat (in Russian).

Fedorov K.N. (1978) *Fine Structure of Hydrophysical Field in the Ocean* (Oceanic Physics, pp. 113–147). Moscow: Nauka 1 Gidrofizika (in Russian)

Field R., Mire C., Chandler H. and Broadhead M. (1997) The SEASAME Experiments – the effects of internal solitons on the Marine shelf. *Proc. Int. Conf. on Shallow-water Acoustics, Beijing.*

Flatte S.M. (ed.) (1979) *Sound Transmission through a Fluctuating Ocean.* Cambridge University Press.

Forsithe G., Malcolm M. and Moler C. (1977) *Computer Methods for Mathematical Computations.* Prentice-Hall.

Focke K. and Weston D. (1978) Problem of the caustic in range averaged ocean sound channels. *J. Acoust. Soc. Am. Suppl.* **23**(1), 64.

Frisk G.V. (1988) *Inverse Methods in Ocean Bottom Acoustics* (WHOI, 32 pp.). Woods Hole Oceanographic Institution, MA.

Furduev A.V. (1966) Undersurface cavitation as source of noise in the ocean. *Izvestiya, Atmospheric and Oceanic Physics* **2**(5), 314–320.

Furduev A.V. (1974) Oceanic noise. In: L.M. Brekhovskikh (ed.) *Oceanic Acoustics* (pp. 615–691). Moscow: Nauka (in Russian).

Garrett C. and Munk W.H. (1975) Space–time scales of internal waves: A progress report. *J. Geophys. Res.* **80**, 291.

Gasparovic R.V. and Etkin V.S. (1994) An overview of the Joint U.S. Russia Internal Wave Remote Sensing Experiment. *Proc. of IGARSS '94, Pasadena* (pp. 744–746).

Gazarian Yu.L. (1958) The sound field generated by a point source in a lying on a half-space. *Sov. Phys. Acoust.* **4**(3), 237–242.

Gershfeld D.A. and Eller A.L. (1985) Geometric considerations in determining the optimum frequency of acoustic propagation in a shallow water waveguide. *J. Acoust. Soc. Am.* **78**(2), 632–641.

Gindler I.V. (1990) Perturbation theory for plane-layered open acoustic waveguides. *Sov. Phys. Acoust.* **36**(4), 349–352.

Godin O.A.S. (1998) A note on differential equations of coupled-mode propagation in fluids. *J. Acoust. Soc. Am.* **103**(1), 159–169.

Goff J.A. (1988) Stochastic modeling of seafloor morphology: Inversion of sea beam data for second order statistics. *J. Geophys. Res.* **93**, 13589–13608.

Goff J.A. (1995) The relationship between local- and global-scale scattering functions for fractal surfaces under a separation of scales hypothesis. *J. Acoust. Soc. Am.* **97**(3), 1586–1595.

Gonorovskii I.S. (1977) *Radio-engineering Chains and Signals*. Moscow: Sov. Radio (in Russian).

Gorskaya N.V., Nikolaev G.N., Rychova T.A. and Salin B.M. (1981) Spectral analysis in the investigation of fields of harmonic sources in acoustic waveguides. *Sov. Phys. Acoust.* **27**(2), 110–112.

Gorskaya N.S. and Raevskii M.A. (1984) Influence of a random internal-wave field on sound propagation in the ocean. *Sov. Phys. Acoust.* **30**(2), 108–113.

Gorskaya N.V, Gorskii S.M., Zverev V.A., Nikolaev G.N., Kurin V.V. and Khilko A.I. (1988) Short-wave diffraction in a multimode layered waveguide. *Sov. Phys. Acoust.* **34**(4), 29–32.

Gorskaya N.S. and Raevskii M.A. (1990) Low-frequency sound field in an ocean waveguide with an irregular bottom. *Sov. Phys. Acoust.* **36**(3), 235–237.

Grachev G.A. (1983) Specific characteristics of signal attenuation in a shallow sea. *Sov. Phys. Acoust.* **29**(2), 160–161.

Graves R.D., Nagl A. and Uberall H. (1975) Range dependent normal modes in underwater sound propagation. *J. Acoust. Soc. Am.* **58**(4), 1171–1177.

Grigorev V.A., Katsnelson B.G. and Petnikov V.G. (1996) Frequency dependence of the effective bottom sound absorption in the Barents Sea. *Acoust. Phys.* **42**(5), 627–629.

Gurevich A.V. and Tsedilina E.E. (1973) *Ultra Long Range Propagation of Short Waves in the Ionosphere* (243 pp.). Moscow: Nauka (in Russian).

Gorskii S.M. Zverev V.A., Matveev A.L. and Mityugov V.V. (1995) Incoherent accumulation of acoustic diffraction signals. *Acoust. Phys.* **41**(2), 190–197.

Hackman R.H. and Sammelmann G.S. (1988) Multiple-scattering analysis for a target in an oceanic waveguide. *J. Acoust. Soc. Am.* **84**(5), 1813–1825.

Hall M.V. (1995) Depth variation of acoustic horizontal refraction theory a cold-core ocean eddy. In: O. Diachok (ed.) *Full Field Method in Ocean and Seismo Acoustics* (pp. 267–272). Kluwer Academic.

Hall J. and Watt J. (ed) (1976) *Modern Numerical Methods for Ordinary Differential Equations*. Oxford: Clarendon Press.

Hamilton E.L. (1985) Sound velocity as a function of depth in marine sediment. *J. Acoust. Soc. Am.* **78**(4), 1348–1355.

Hampton L.D. (ed.) (1973) *Physics of Sound in Marine Sediments*. New York: Plenum Press.

Hamson R.M. (1985) The theoretical response of vertical and horizontal line arrays to wind induced noise in shallow water. *J. Acoust. Soc. Am.* **78**(5), 1702–1709.

Headrick R.H., Lynch J.F., Kemp J.K., Newhall A.E., v.d. Heydt K., Apel J., Badiey M., Ching-sang Chiu, Finette S., Orr M., Pasewark B., Turgot A., Wolf S. and Tielbuerger D. (2000) Acoustic normal mode fluctuation statistics in the 1995 SWARM internal wave scattering experiment. *J. Acoust. Soc. Am.* **107**(1), 201–220.

Honl H., Maue A.V. and Westfal K. (1961) *Theorie der Beugung*. Berlin: Springer-Verlag (in German).

Ingenito F. and Wolf S.N. (1976) Acoustic propagation in shallow water overlying a consolidated bottom. *J. Acoust. Soc. Am.* **60**(3), 611–617.

Ingenito F. (1987) Scattering from an object in a stratified medium. *J. Acoust. Soc. Am.* **82**(6), 2051–2059.

Ishimaru A. (1978) *Wave Propagation and Scattering in Random Media* New York: Academic Press.

Jensen F.B. (1981) Sound propagation in shallow water. *J. Acoust. Soc. Am.* **70**(5), 1397–1406.

Jensen F.B. and Ferla M.C. (1979) *SNAP: The SACLANTSEN Normal-Mode Acoustic Propagation Model* (SACLANTSEN Memorandum SM-121). La Spezia, Italy: Saclant ASW Research Center.

Jensen F.B. and Kuperman W.A. (1983) Optimum frequency by sound propagation in shallow water. *J. Acoust. Soc. Am.* **73**(3), 813–819.

Karnovskii A.M., Loginov K.A., Rivelis E.A. and Freiman M.E. (1981) A model of the noise field in a waveguide. *Sov. Phys. Acoust.* **27**(5), 417–420.

Katsenelenbaum B.Z. (1961) On the theory of irregular acoustic waveguides. *Sov. Phys. Acoust.* **7**, 159–164.

Katsenelenbaum B.Z. (1960) *Theory of Irregular Waveguides with Slowly Varying Parameters* (216 pp.). Moscow: Izd. AN SSSR (in Russian).

Katsnelson B.G. and Pereselkov S.A. (1996) Shallow water internal wave field model for underwater acoustics. *Proc. Int. Con. OCEANS '96, MTS/IEEE, Florida* (pp. 27–31).

Katsnelson B.G. and Kulapin L.G. (1984) Average sound-intensity decay law in an irregular underwater acoustic waveguide. *Sov. Phys. Acoust.* **30**(5), 379–382.

Katsnelson B.G., Kravtsov, Yu.A., Kulapin, L.G., Petnikov, V.G. and Sabirov O.I. (1985) Average decay law of sound in an irregular near-bottom sound channel. *Sov. Phys. Acoust.* **4**, 321–322.

Katsnelson B.G., Kulapin L.G., Migulin A.A. and Petnikov V.G. (1992) Influence of hydrodynamic fluctuations on the vertical interference structure of a sound field in a waveguide. *Sov. Phys. Acoust.* **38**(2), 164–168.

Katsnelson B.G., Kuzkin V.M., Pereselkov S.A. and Petnikov V.G. (1998) Sound wave attenuation in shallow water with rough boundaries. *ICA/ASA Proc., Seattle* (pp. 2713–2714).

Katsnelson B.G., Kravtsov, Yu. A., Kuzkin V M, Kulapin L.G. and Petnikov V.G. (1984) Simplified theory of sound propagation in shallow water. *Trudy FIAN* **156**, Issledovaniya no gidrofizike, 41–55 (in Russian).

Katsnelson B.G. and Sidenko A.V. (1988) Decay in radiation intensity in a multimode waveguide with random inhomogeneities and absorbing boundaries. *Radiofizika* **31**(4) 433–438 (in Russian).

Katsnelson B.G., Kravtsov Yu.A. and Petnikov V.G. (1987) The primary methods of acoustic propagation theory in stratified horizontally inhomogeneous absorbing boundary waveguides. In: *Oceanic Remote Sensing* (Proceedings of the Institute of General Physics Academy of the Sciences of the USSR, Vol. 1, pp. 176–217). New York: Nova Science.

Katsnelson B.G. and Pereselkov (2000) Horizontal refraction of low frequency sound field due to soliton packets in shallow water. *Acoustical Physics* **46**(6), 779–788.

Keller J.B. and Papadakis J.S. (1977) *Wave Propagation and Underwater Acoustics* (287 pp.). Springer-Verlag.

Klyachin B.I. and Kuryanov B.F. (1981) Effect of sound scattering on the depth dependence of low-frequency oceanic noise. *DAN SSSR* **260**(4), 1009–1012 (in Russian).

Knobles D.P. and Mitchell S.K. (1994) Broadband localization by matched fields in range and bearing in shallow water. *J. Acoust. Soc. Am.* **95**(3), 1813–1820.

Knudsen V.O., Alford R.S. and Emling J.W. (1948) Underwater ambient noise. *J. Marine Res.* **7**, 410–429.

Koch R., Vidmar P. and Lindfray J. (1983) Normal mode identification for impedance boundary condition. *J. Acoust. Soc. Am.* **73**(5), 1567–1570.

Kobozev I.K., Kravtsov Yu.A. and Ogurtsov A.V. (1988) Perturbation theory for calculating the characteristics of normal modes in near-bottom waveguides. *Sov. Phys. Acoust.* **34**(1), 58–61.

Kohler W. and G. Papanicolaou (1977) Wave propagation in a randomly inhomogeneous ocean. In: J.B. Keller and J.S. Papadakis (ed.) *Wave Propagation and Underwater Acoustics* (pp. 153–223). Springer-Verlag.

Konyaev K.V. and Sabinin K.D. (1992) *Waves within the Ocean* (p. 271). St Petersburg: Gidrometeoizdat (in Russian)

Krolik J.L. and Narasimhan S. (1996) Performance bounds on acoustic thermometry of ocean climate in the presence of mesoscale sound speed variability. *J. Acoust. Soc. Am.* **99**(1), 254–265.

Konyukhova N.B. and Pak T.V. (1985) *Diffraction of a Plane Sound Wave on a Rigid Prolate Spheroid* (61 pp.). Moscow: Izd-vo VTs AN SSSR (in Russian).

Komarov N.V., Ponomarev L.I. and Slavyanov S.Yu. (1976) *Spheroidal and Coulomb Spheroidal Functions* (319 pp.). Moscow: Nauka (in Russian).

Kravtsov Yu.A., Kuzkin V.M. and Petnikov V.G. (1984) Perturbation calculation of the horizontal refraction of sound waves in a shallow sea. *Sov. Phys. Acoust.* **30**(1), 45–47.

Kravtsov Yu.A., Kuzkin V.M. and Petnikov V.G. (1988) Resolvability of rays and modes in an ideal waveguide. *Sov. Phys. Acoust.* **34**(4), 387–390.

Kravtsov Yu.A., Kuzkin C.M. and Petnikov V.G. (1984) Wave diffraction by regular scatterers in multimode waveguides. *Sov. Phys. Acoust.* **30**(3), 199–202.

Kravtsov Yu.A. and Orlov Yu.I. (1990) *Geometrical Optics of Inhomogeneous Media*. Springer-Verlag.

Krupin V.D. (1972) A special effect in a waveguide having a negative sound velocity gradient. *Sov. Phys. Acoust.* **18**(4), 556–564.

Kryazhev F.I. and Kudryashov V.M. (1984) Effect of scattering on the boundary on the sound field in a waveguide. In: *Problems of Oceanic Acoustics* (pp. 57–69). Moscow: Nauka (in Russian).

Kvyatovskii S.O. (1988) Diffraction of sound waves by a scatterer in a waveguide. *Sov. Phys. Acoust.* **34**(4), 430–431.

Kuperman W.A. and Jensen F.B. (eds) (1980) *Bottom-interacting Ocean Acoustics* (717 pp.). New York: Plenum Press.

Kuperman W.A. and Ferla M.C. (1985) A shallow water experiment to determine the source spectrum level of wind generated noise. *J. Acoust. Soc. Am.* **77**(6), 2067–2073.

Kuperman W.A. and Ingenito F. (1980) Spatial correlation of surface generated noise in a stratified ocean. *J. Acoust. Soc. Am.* **67**(6), 1988–1996.

Kuryanov B.F. (1963) Spatial correlation of fields emitted by random sources in a plane. *Akust. Zhurn.* **9**(4), 441–448 (in Russian).

Kuryanov B.M. and Klyachin B.I. (1984) Application of the theory of radiation transport to problems of oceanic noise propagation. In: *Problems of Oceanic Acoustics* (pp. 16–36). Moscow: Nauka (in Russian).

Kuzkin V.M., Ogurtsov A.V. and Petnikov V.G. (1998) The effect of hydrodynamic variability on frequency shifts of the interference pattern of a sound field in a shallow sea. *Acoust. Phys. Vol.* **44**(1), 77–82.

Kuzkin V.M. (1996) Horizontal refraction of acoustic waves at internal waves in shallow sea. *Acoust. Phys.* **42**(3), 390–392.

Kuzkin V.M. (1997) Characteristics of diffracted acoustic signals in an oceanic waveguide. *Acoust, Phys.* **43**(4), 440–445.

Kuznetsova E.P. (1988) Sound field of surface noise sources in a randomly inhomogeneous waveguide. *Sov. Phys. Acoust.* **34**(5), 511–514.

Kuznetsova E.P. (1989) Calculation of the directivity of noise in a shallow sea. *Sov. Phys. Acoust.* **35**(6), 627–630.

Laville F., Abbott G.D. and Miller M.J. (1991) Underwater sound generation by rainfall. *J. Acoust. Soc. Am.* **89**(2), 715–721.

Lebedev O.V., Pronchatov-Rubtsov N.V. and Simdyankin S.I. (1996) Calculation of the sound field in a waveguide with layered absorbing bottom. *Acoust. Phys.* **42**(1), 65–70.

Leslie C.B. and Sorensen N.R. (1961) Integral solution of the shallow water sound field. *J. Acoust. Soc. Am.* **33**(3), 323–329.

Lindsay C.E. and Chapman N.R. (1993) Matched field inversion method for geoacoustic model parameters using adaptive simulated annealing. *IEEE J. Oceanic Engineering* **189**(3), 224–231.

Lukin A.N., Migulin A.A., Moiseev S.N., Petnikov V.G. and Suslikov O.B. (1993) Model of fluctuations of the envelope of a harmonic signal in a shallow sea. *Acoust. Phys.* **39**(4), 363–366.

Lynch J.L., Jin G., Pawlowich R., Ray D., Plueddemann A.J., Chlu C., Miller J.H., Bourke R.H., Parsons A.R. and Muench R. (1996) Acoustic travel-time perturbations due to shallow water internal waves and tides in the Barents Sea polar front: Theory and experiment. *J. Acoust. Soc. Am.* **99**(2), 803–821.

Lynch J., Frye D., Peal K., Liberatore S., Kery S., Hobart E., Newhal A. and Smith S. (1992) *Real-time Tomography Mooring* (Technical report WHOI092-29). Woods Hole Oceanographic Institution, MA.

Lyubchenko A.Yu. and Petnikov V.G. (1991) Experimental investigations of the vertical distribution of the intensity of low-frequency noise in a shallow sea. *Sov. Phys. Acoust.* **37**(3), 303–304.

Macferson I.D. and Daintith M.J. (1967) Practical model of shallow water acoustic propagation. *J. Acoust. Soc. Am.* **41**(4), 850–854.

Maltsev N.E. (1970) Modifications of the method of cross sections. *Sov. Phys. Acoust.* **1**, 81–85.

Marcuse D. (1972) *Light Transmission Optics*. Princeton, NJ: Van Nostrand Reinhold.

McDaniel S.T. (1978) Calculation of mode conversion rates. *J. Acoust. Soc. Am.* **63**(3), 739–749.

McDaniel S.T. and McCammon D.F. (1987) Mode-coupling and the environmental sensitivity of shallow water propagation loss prediction. *J. Acoust. Soc. Am.* **82**(1), 217–223.

Mendus V.I. and Postnov G.A. (1993) Angular distribution of the intensity of high-frequency dynamic noise in the ocean. *Acoust. Phys.* **39**(6), 582–586.

Mendus V.I. and Postnov G.A. (1993) Intensity of sound radiated by an air bubble. *Acoust. Phys.* **39**(5), 460–463.

Miller J.H., Ching-Sang Chiu and Lynch J.F. (1993) Signal processing for coastal acoustic tomography. *Proc. Int. Conf. on Theoretical and Computational Acoustics, Mustique.*

Mitchell S. and Focke K. (1983) The role of seabottom attenuation profile in shallow water acoustic propagation. *J. Acoust. Soc. Am.* **73**(1), 153–162.

Modeling mode arrivals in the 1995 SWARM experiment acoustic transmissions. *J. Acoust. Soc. Am.* **107**(1), 221–236.

Morris G.B. (1978) Depth dependence of ambient noise in the northeastern Pacific Ocean. *J. Acoust. Soc. Am.* **64**(2), 581–590.

Munk W.H. (1979) Ocean acoustic tomography: A scheme for large-scale monitoring. *Deep-Sea Res.* **26**(2A), 123–161.

Munk W.H., Spindel R.C., Baggeroer A. and Birdsall T.G. (1994). The Heard Island feasibility test. *J. Acoust. Soc. Am.* **96**(4), 2330–2342.

Murphy E.L., Wasilieff A. and Jensen F.B. (1976) Frequency-dependent influence of the sea bottom on the near-surface sound field in shallow sea. *J. Acoust. Soc. Am.* **59**(4), 839–845.

Nechaev A.G. and Khilko A.I. (1987) *Reconstruction of Oceanic Inhomogeneities along an Acoustic Track by the Differential Diagnostics Method* (Preprint No. 178). Institute of Applied Physics of the USSR Academy of Sciences (in Russian).

Nagl A., Uberall H., Hang A. and Zarur G. (1978) Adiabatic mode theory of underwater sound propagation in range-dependent environment. *J. Acoust. Soc. Am.* **63**(3), 739–749.

Nozdrin Yu.V. and Plakhin Yu.A. (1983) On the fine thermohaline structure of the waters of the central portion of the Red Sea. *Izvestiya, Atmospheric and Oceanic Physics* **19**(3), 219–224.

Oguz H.N. (1994) A theoretical study of low-frequency oceanic ambient noise. *J. Acoust. Soc. Am.* **95**(4), 1895–1912.

Parker S.P. (ed.) (1980) *Ocean and Atmosphere* (Encyclopedia). McGraw-Hill.

Parsons A.R. *et al.* (1996) The Barents Sea polar front in Summer. *J. Geophys. Res.* **10**(C6), 14201–14221.

Pekeris C.L. (1948) Theory of propagation of explosive sound in shallow water. *Geol. Soc. Am. Mem.* **27**, 1–117.

Perrone A.J. (1969) Infrasonic and low frequency ambient noise measurements on the Grand Banks. *J. Acoust. Soc. Am.* **47**(3), 762–770.

Petnikov V.G., Katsnelson B.G., Pereselkov S.A. and Sabinin K.D. (1997) Numerical modeling of sound field perturbation caused by internal waves in shallow water. *Proc. Int. Conf. on Shallow-water Acoustics, Beijing.*

Phillips O.M. (1977) *The Dynamics of Upper Ocean* (2nd edn). Cambridge University Press.

Pierce A.D. (1965) Extension of the method of normal modes to sound propagation in almost-stratified medium. *J. Acoust. Soc. Am.* **37**, 19–27.

Pierce A.D. (1982) Guided mode disappearance during upslope propagation in variable depth shallow water overlying a fluid bottom. *J. Acoust. Soc. Am.* **72**(2), 523–531.

Pierson W.J. and Moskovitz L.A. (1964) A proposed spectral form for fully developed wind seas based on the similarity theory of S.A. Kitaigorodsky. *J. Geophys. Res.* **69**(24), 5181–5190.

Piggott G.L. (1965) Ambient sea noise at low frequencies in shallow water of Scotia Shelf. *J. Acoust. Soc. Am.* **36**, 2152–2163.

Porter M. and Reiss E.L. (1984) A numerical method for ocean-acoustic normal modes. *J. Acoust. Soc. Am.* **76**(1), 244–252.

Porter M. and Reiss E.L. (1985) A numeral method for bottom interaction in ocean acoustic normal modes. *J. Acoust. Soc. Am.* **77**(5), 1760–1767.

Porter M.B. (1992) *The KRAKEN Normal Mode Program* (Report NRL/MR/5120-92-6920, 194 pp.). Washington, DC.

Rajan S.D., Lynch J.F. and Frisk G.V. (1987) Perturbative inversion methods for obtaining bottom geoacoustic parameters in shallow water. *J. Acoust. Soc. Am.* **82**(3), 998–1017.

Robinson I.S. (1994) *Satellite Oceanography* (Wiley–Praxis Series in Remote Sensing, 455 pp.). Chichester, UK: Wiley–Praxis.

Ross D. (1976) *Mechanics of Underwater Noise* (375 pp.). Oxford: Pergamon Press.

Rutherford S.R. (1979) An examination of multipath processes in a range dependent ocean environment within the context of adiabatic mode theory. *J. Acoust. Soc. Am.* **66**(5), 1482–1486.

Rutenko A.N. (1997) Seasonal variability of intensity and phase fluctuations of low-frequency underwater sound signals at the shelf of the Sea of Japan. *Acoust. Phys.* **43**(1), 84–91.

Rytov S.M., Kravtsov Yu.A. and Tatarskii V.I. (1989) *Principles of Statistical Radiophysics* (Vol. 4). Springer-Verlag.

Sarkissian A. (1994) Method of superposition applied to scattering from a target in a bounded medium. *J. Acoust. Soc. Am.* **95**(5), 2340–2345.

Sarkissian A. (1997) Extraction of a target scattering response from measurements made over long ranges in shallow water. *J. Acoust. Soc. Am.* **102**(2), 825–832.

Sarkissian A. (1994) Method of superposition applied to scattering from a target in a bounded medium. *J. Acoust. Soc. Am.* **95**(5), 2340–2345.

Schmidt H. and Kuperman W.A. (1986) Estimation of surface noise source level from low-frequency seismoacoustic noise measurements. *J. Acoust. Soc. Am.* **84**(6), 2153–2161.

Shang E.C. and Anderson V.C. (1986) Surface generated noise under low-wind speed at kilohertz frequencies. *J. Acoust. Soc. Am.* **79**(4), 964–971.

Shvachko R.F. (1982) Fluctuations of sound on inhomogeneities in a deep ocean. In: *Oceanic Acoustics. Current Status* (pp. 132–140). Moscow: Nauka (in Russian).

Shenderov E.L. (1989) *Emission and Scattering of Sound* (p. 304). Leningrad: Sudostroenie (in Russian).

Shmelerv A.Yu., Migylin A.A. and Petnikov V.G. (1992) Horizontal refraction of low frequency acoustic waves in the Barents Sea stationary acoustic track experiment. *J. Acoust. Soc. Am.* **92**(2), 1003–1007.

Shmelerv A.Yu. (1990) (Dissertation, Moscow General Physics Institute of the USSR Academy of Sciences) (in Russian)

Smith G.B., Feullade C., Del Balzo D.R. and Byrne C.L. (1989) A nonlinear matched-field processor for detection and localization of quiet source in a noisy shallow-water environment. *J. Acoust. Soc. Am.* **85**(3), 1158–1166.

Smith P.W. (1980) Averaged sound transmission in range dependent channels. *J. Acoust. Soc. Am.* **68**(1), 269–281.

Smith P.W. (1986) Low-frequency roll off in the response of shallow water channels. *J. Acoust. Soc. Am.* **79**(1), 71–75.

Song, H.C., Kuperman, W.A. and Hodgkiss, W.S. (1998) A time-reversal mirror with variable range focusing. *J. Acoust. Soc.* **103**(6), 3234–3240.

Spindel R.C. (1985) Signal processing in ocean tomography. In: H.G. Urban (ed.) *Adaptive Methods in Underwater Acoustics* (pp. 687–710). Boston: D. Reidel.

Sticler D.C. and Ammicht E. (1980) Uniform asymptotic evaluation of the continuous spectrum contribution for the Pekeris model. *J. Acoust. Soc. Am.* **67**(6), 2018–2024.

Sidenko A.V. (1989) Wave propagation in randomly inhomogeneous dissipative waveguides (Dissertation, Voronezh, p. 120) (in Russian).

Taylor J.R. (1972) *Scattering Theory. The Quantum Theory on Nonrelativistic Collisions.* Wiley.

Thorp W.H. and Browing D.G. (1973) Attenuation of law frequency sound in the ocean. *J. Sound and Vibration* **26**(4), 576–578.

Tindle C.T., Hobaeck H. and Muir T.G. (1987) Downslope propagation of the normal modes in a shallow water wedge. *J. Acoust. Soc. Am.* **81**(2), 275–286.

Tindle C.T. and Dean G.B. (1985) Sound propagation over a sloping bottom using rays with beam displacement. *J. Acoust. Soc. Am.* **78**(4), 1366–1374.

Tindle C.T. and Weston D.E. (1980) Connection of acoustic beam displacement cycle distance and attenuation for rays and normal modes. *J. Acoust. Soc. Am.* **67**(5), 1614–1622.

Titchmarsh E.C. (1958) *Eigenfunction Expansion Associated with Second-order Differential Equations*. Oxford: Clarendon Press.

Tolstoy A., Diachok O. and Frazer L.N., (1991) Acoustic tomography via matched field processing. *J. Acoust. Soc. Am.* **89**(3), 1119–1127.

Tonoyan I.P. (1991) Propagation of acoustic noise in the ocean. *Izvestiya, Atmospheric and Oceanic Physics* **27**(5), 570–577.

Turbiner A.V. (1984) The eigenvalue spectrum in quantum mechanics and the nonlinearisation procedure. *Sov. Phys. Uspekhi* **27**(9), 668–694.

Tuzhilkin Yu.I. and Eliseevnin V.A. (1995) Diffraction of a sound field at a plane rectangular vertical screen in a waveguide. *Acoust. Phys.* **41**(2), 214–218.

Ugincius P. (1969) Intensity equation in ray acoustics. *J. Acoust. Soc. Am.* **45**(1), 193–209.

Urick R.J. (1975) *Principles of Underwater Sound* (384 pp.). New York: McGraw-Hill.

Uzunoglu N.K. and Fikioris J.K. (1982) Scattering from an inhomogeneity inside a dielectric-slab waveguide. *J. Opt. Soc. Am.* **72**(5), 628–637.

Vadov R.A. (1994) Characteristics of underwater explosions sound sources characteristics. *Acoust. Phys.* **40**(4), 600–602.

Vavilin A.V., Kozelskii A.R., Petnikov V.G., Reznikov V.M. and Rivelis E.A. (1987) Characteristics of the dispersion distortions of pulse signals in acoustic waveguides with absorption. *Sov. Phys. Acoust.* **33**(5), 830–833.

Vianna M.L. and Soares-Filho W. (1986) Broadband noise propagation in a Pekeris waveguide. *J. Acoustic Soc. Am.* **79**(1), 76–83.

Voitovich N.N. and Shatrov A.D., (1972) Normal mode expansion of the field in an underwater sound channel. *Sov. Phys. Acoust.* **18**(4), 434–438.

Volovov V.I. and Ivakin A.N. (1980) Reflection of sound from a bottom with sound speed and density gradients. *Akust. Zhurn* **26**(2), 194–199 (in Russian).

Voronovich A.G., Goncharov V.V. and Chepurin Yu.A. (1991) Measurement of the mode spectrum of a low-frequency acoustic field in the deep sea. *Doklady, Earth Science Sections* **317**, 250–254.

Voronovich A.G., Goncharov V.V., Nikoltsev A.Yu. and Chepurin Yu.A. (1992) Comparative analysis of methods for the normal mode decomposition of a sound field in a waveguide: Numerical simulation and full-scale experiment. *Sov. Phys. Acoust* **38**(4), 365–370.

Wenz G.M. (1962) Acoustic ambient noise in the ocean: spectra and sources. *J. Acoust. Soc. Am.* **34**(12), 1936–1956.

Weston D.E. (1963) Underwater explosions as acoustic sources. In: V.M. Albers (ed.) *Underwater Acoustics, Proc. of a NATO-sponsored Institute, 31 July to 11 August 1961.* New York: Plenum Press.

Weston D.E. (1980) Acoustic flux formulas for range dependent ocean ducts. *J. Acoust Soc. Am.* **68**(1), 269–281.

Weston D.E. (1980) Wave-theory peaks in range dependent channels of uniform sound velocity. *J. Acoust. Soc. Am.* **68**(1), 282–286.

Weston D.E. (1980) Acoustic flux method for oceanic guided waves. *J. Acoust. Soc. Am.* **68**(1), 287–296.

Weston D.E. (1980) Ambient noise depth-dependence models and their relation to low-frequency attenuation. *J. Acoust. Soc. Am.* **67**(2), 530–537.

Weston D.E. (1976) Propagation in water with uniform sound velocity but variable depth lossy bottom. *J. Sound and Vibr.* **47**, 473–483.

Weston D.E. (1994) Wave shifts, beam shifts, and their role in modal and adiabatic propagation. *J. Acoust. Soc. Am.* **96**(1), 406–416.

Whitman A.M. and Beran M.J. (1978) Numerical calculation of the intensity distribution in sound channels using coherence theory. *J. Acoust. Soc. Am.* **63**(6), 1727–1732.

Williams A. (1980) Mode interaction in an isovelocity ocean of uniformly varying depth. *J. Acoust. Soc. Am.* **67**(1), 177–185.

Wille P.C. and Geyer D. (1984) Measurements of the origin of wind dependent ambient noise variability in shallow water. *J. Acoust. Soc. Am.* **75**(1), 173–185.

Wilson J.H. (1979) Very low-frequency (VLF) wind generated noise produced by turbulent pressure fluctuation in the atmosphere near the ocean surface. *J. Acoust. Soc. Am.* **66**, 1499–1507.

Wolf S.N. and Ingenito F. (1982) Site dependence of wind-generated ambient noise in shallow water. In: R.A. Wagstaff and O.Z. Bluy (eds) *Underwater Acoustic Noise* (Rep. CP-32). La Spezia, Italy: Saclant ASW Research Center.

Yang T.C. and Kwang Yoo (1997) Modeling the environment influence on the vertical directionality of ambient noise in shallow water. *J. Acoust. Soc. Am.* **101**(5), 2541–2554.

Zavadskii V.Yu. (1972) *Computation of Wavefields in Open Regions and Waveguides* (p. 558). Moscow: Nauka (in Russian).

Zhang Z.Y. and Tindle C.T. (1995) Improved equivalent fluid approximation for a low shear speed ocean bottom. *J. Acoust. Soc. Am.* **98**(6), 3391–3396.

Zhou J., Zhang X. and Rogers P.H. (1987) Effect of frequency dependence of sea-bottom attenuation on the optimum frequency for acoustic propagation in shallow water. *J. Acoust. Soc. Am.* **82**(1), 287–292.

Zhou J., Zhang X. and Rogers P.H. (1991) Resonant interaction of sound wave with internal solitons on coastal zone. *J. Acoust. Soc. Am.* **90**(4), 2042–2054.

Zhou J., Zhang X. and Rogers P.H. (1996) Modal characteristics of acoustic signal fluctuations induced by shallow water internal waves. *Proc. Int. Conf. OCEANS '96, MTS/IEEE, Florida* (pp. 1–8).

Zhuravlev V.A., Kobozev I.K. and Kravtsov Yu.A. (1987) Power radiated by an acoustic source in an ocean waveguide. *Sov. Phys. Acoust.* **33**(6), 611–614.

Zhuravlev V.A., Zarubin V.N. and Sochnev O.Ya. (1986) Frequency dependence of the attenuation coefficient of a sound field in a shallow sea. *Sov. Phys. Acoust.* **32**(1), 70–71.

Zhuravlev V.A., Kobozev I.K, Migulin A.A., Petnikov V.G. and Shmelev A.Yu. (1991) Phase fluctuations of low-frequency acoustic waves propagating in an underwater sound channel near the bottom. *Sov. Phys. Acoust.* **37**(6), 635–636.

Zhuravlev V.A., Kobozev I.K. and Kravtsov Yu.A. (1989) Dislocations of a phase front in an ocean waveguide and their manifestation in acoustical measurements. *Sov. Phys. Acoust.* **35**(2), 156–159.

Zverev V.A., Matveev, A.L. and Mitygov, V.V. (1995) Matched filtering of acoustic diffraction signals for incoherent accumulation with vertical antenna. *Acoust. Phys.* **41**(4), 518–521.

Index